Unmanned Aerial Remote Sensing

Unmanned Aerial Remote Sensing

UAS for Environmental Applications

Edited by
David R. Green, Billy J. Gregory, and Alex R. Karachok

CRC Press
Taylor & Francis Group
Boca Raton London New York

CRC Press is an imprint of the
Taylor & Francis Group, an **informa** business

MATLAB® is a trademark of The MathWorks, Inc. and is used with permission. The MathWorks does not warrant the accuracy of the text or exercises in this book. This book's use or discussion of MATLAB® software or related products does not constitute endorsement or sponsorship by The MathWorks of a particular pedagogical approach or particular use of the MATLAB® software.

First edition published 2021
by CRC Press
6000 Broken Sound Parkway NW, Suite 300, Boca Raton, FL 33487-2742

and by CRC Press
4 Park Square, Milton Park, Abingdon, Oxon OX14 4RN

First issued in paperback 2022

ISBN 13: 978-1-03-247434-2 (pbk)
ISBN 13: 978-1-4822-4607-0 (hbk)
ISBN 13: 978-0-429-17241-0 (ebk)

DOI: 10.1201/9780429172410

Typeset in Times
by codeMantra

Contents

Foreword

The extraordinary development of unmanned aerial vehicles (UAVs) has been one of the most striking examples of successful technological change in the last two decades. Driven by the capability to miniaturise on-board sensors, enhanced control mechanisms, improved power sources, and lower cost, a rapidly growing user community has ensured that UAVs are becoming part of the 'taken for granted world'. Nowhere is this truer than in environmental data capture and monitoring of the state of crops. But UAV applications are much wider, notably for military purposes.

For someone who was once a rabid enthusiast for radio-controlled balsa wood model aircraft – now antediluvian – these developments are truly amazing. Yet, despite the rapid progress, the UAV world is one in which major challenges exist. These include the danger of collisions between UAVs and aircraft, effective licensing of the UAV owners, training of their pilots, and minimising the loss of privacy. According to Drone Wars UK (*The Times* newspaper 19 August 2019), there have been over 250 crashes of large military UAVs in the last decade. On a smaller scale, I once found a mangled small UAV under my kitchen window which had wrecked part of my roof.

Other current challenges include building in data ethics and exploiting artificial intelligence. Thus data scientists, earth scientists, lawyers, and many other domain-specific skill sets (e.g. farmers) need to be brought together to maximise UAV value and minimise risk. It is therefore self-evident that there is a real need for education of both current and latent members of the UAV community. I therefore welcome this book by David Green and his colleagues who have long and relevant experience. This book is very timely, and it deserves to become very successful.

Professor David Rhind FRS, Honorary FBA

Preface

The subjects of aerial photography, cartography, photogrammetry, and remote sensing were all introduced to me when I first went to study Geography at the University of Edinburgh in Scotland in the UK. I am not quite sure what attracted me to these subjects except perhaps a long-term interest in travel, graphics, imagery, and art (my mother was an artist) or perhaps the wonderful colours and patterns I saw in satellite imagery, as well as the space-age technology; I have always been excited by technology regularly watching as a teenager a BBC TV programme called Tomorrow's World. Whatever the inspiration was, these subjects all rapidly became key to my undergraduate and postgraduate studies and latterly – as it turned out – my future academic career!

The opportunity to study in North America (USA and Canada) both as an undergraduate and postgraduate furthered my interest in aerial photography and introduced me to aerial photo-interpretation and the extraction of quantitative information from aerial photographs at a time when remote sensing in particular was being widely used in environmental resource management. A period of time spent at the Ontario Centre for Remote Sensing (OCRS) in Toronto introduced me to other airborne aerial sensors such as thermal infrared (TIR) and latterly satellite data and imagery, and digital image processing. By the time I left North America, I had also been exposed to Geographic Information Systems (GIS) software and radio-controlled model aircraft to obtain high-resolution aerial imagery for monitoring, mapping, and modelling. I was also lucky enough to begin to explore the potential of model aircraft for monitoring vineyards in the Niagara escarpment, providing me with an early insight into the potential of Precision Viticulture (PV), an area in which I still have considerable interest. This particular interest first developed when one of my father's friends – Tom Lisher, who owned the CALL Printing Group in Cambridge – established a couple of vineyards in the villages of Little Eversden and Cottenham just outside Cambridge in England with the help of my father's advice and my summer labour – and he just so happened to print some of the UK Remote Sensing Society's publications!

When I returned to the UK, I was lucky enough to find a position in the Geography Department at the University of Aberdeen where it was possible to begin to pull all of these geospatial technologies together and especially at a time when their evolution and development was rapidly taking off. The career path I wanted to pursue soon became a reality, and to this day, all of the geospatial technologies still lie at the centre of my geographical interests and research.

An opportunity to use model aircraft for coastal and estuarine monitoring rekindled my interest in small-scale aerial platforms and sensors, and some years later, with the development of small low-cost multi-rotor drones or unmanned aerial vehicles (UAVs), I soon began to explore their potential for several environmental applications. This began when the first DJI Phantom 1 drone was marketed and softcopy photogrammetric software emerged. Together with a colleague Chris Gibbons and his PhD student Baptiste Marteau, we began to explore the UAV monitoring, mapping, and modelling opportunities for studying an engineered river channel in the Lake District. Other riverine, coastal, and vineyard research applications of UAVs soon followed along with several funded UAV-based MSc and PhD projects.

So began a proposal to produce a small book on UAV applications, largely based on my experience over time with low-cost, off-the-shelf model aircraft, drones, and all the associated technologies.

In part, the driving force behind this book has been not only to document the historical context in the development and evolution of this technology over time from my own experiences but also to take the opportunity to commit to paper an acknowledgement and thanks to all of the people who have contributed to my interest in this fascinating area over the years including the authors of the book chapters and most recently: Johannes Fahrentrapp (ZHAW); Jason Hagon (Geodrone Survey Ltd.); Dave Scott (Bristow UAS); Phil McLachlan (Redwing Aero Ltd./D-CAT Ltd.); Trias Gkikopoulos (University of Dundee); Ruben de Vries (Van Hall Institute Laurenstein); Cristina

Gómez (University of Aberdeen); John Cleave and Harvey Mann (Buzzflyer Ltd.); and Brian Critchley, Dave Harrison, and Francis Harwood (AlphaGeo Ltd. and Geo4D Ltd.). Last but not least of course to also acknowledge the role of my two co-editors in helping to finally get this book into publication.

Firstly, in the summer of 2018, a meeting with Billy Gregory at Findhorn Marina in Scotland to assist me with rigging my MG25 Spring yacht furthered a new chapter in my interests and applications of multi-rotor and fixed-wing drones. Aside from being a qualified Royal Yachting Association (RYA) sailing and powerboat instructor, it turned out that he was also a model aircraft enthusiast and remotely controlled aircraft (RCA) pilot! Naturally, this soon led to him helping out with an MSc module on UAV remote sensing and several PhD research projects utilising drone remote sensing, and his setting up of a small UAV business, DroneLite. Secondly, as mentioned earlier, my interest in model aircraft first originated in Canada. In January 2019, when visiting Canada, I came across the person who would soon become the second co-editor – Alex R. Karachok – who just happened to be flying his DJI Spark drone at the locality we were visiting – and turned out to be a drone enthusiast, UAV pilot, and someone with artistic talent! This chance meeting in 2019 brought me full circle back to Canada where my interest in all this fascinating technology first began! Perfect timing for this book!

With the help and support of these two people, I have finally succeeded in pulling this volume together.

David R. Green
7 October 2019

MATLAB® is a registered trademark of The MathWorks, Inc. For product information, please contact:
The MathWorks, Inc.
3 Apple Hill Drive
Natick, MA 01760-2098 USA
Tel: 508-647-7000
Fax: 508-647-7001
E-mail: info@mathworks.com
Web: www.mathworks.com

Editors

David R. Green is the Director of the Aberdeen Institute for Coastal Science and Management (AICSM), Director of the MSc Degree Programme in Geographical Information Systems and Director of the UAV Centre for Environmental Monitoring and Mapping (UCEMM) (www.abdn.ac.uk/research/ucemm) at the Department of Geography and Environment, University of Aberdeen, Scotland, UK. He is past Chairman of the Association for Geographic Information (AGI), past Editor-in-Chief of *The Cartographic Journal*, past President of the EUCC – The Coastal and Marine Union, past Vice Chair of the European Centre for Nature Conservation (ECNC) Scientific Committee, currently a Director and Chair of EGCP Ltd., and Chair of the CoastGIS 2018 Conference. He is a specialist in the environmental applications of geospatial technologies with interests in geographical information systems, remote sensing (terrestrial and bathymetric), cartography/digital mapping, Internet and mobile GIS, coastal and marine resource management, hydrography, marine spatial planning (MSP), and UAV technology. He has published a number of books on Coastal Zone Management and Geographical Information Systems (GIS). David is also Editor-in-Chief of the *Journal of Coastal Conservation, Planning and Management* (Springer).

Billy J. Gregory is a powerboat instructor, sailing instructor, skier, and Civil Aviation Authority (CAA)-qualified commercial drone pilot. Having flown fixed-wing radio-controlled model aircraft for many years, he began flying small multi-rotor UAVs. As an associate of UCEMM (UAV Centre for Environmental Monitoring and Mapping) at the Department of Geography and Environment, University of Aberdeen, Scotland, UK (www.abdn.ac.uk/research/ucemm), he has been involved with a number of UAV-based research projects and helps to train both MSc and PhD students looking to use drones in their projects. In 2018, he set up DroneLite (www.dronelite.co.uk), a small drone company based in Forres, Scotland, UK. He continues to combine his interests in watersport activities with flying both multi-rotor and fixed-wing drones for a variety of aerial photographic and video projects, and participating in small drone workshops, training events, and conference organisation.

Alex R. Karachok was born in Toronto, Ontario. He has many hobbies, the most significant being geology and photography. Alex acquired his first mini-camera in 2011 and his fist drone in 2014. From there, he has developed a passion for photography. In 2018, he was astounded by one of his scout leaders flying a Mavic Pro. This led him to eventually selling his new game console for a DJI Spark. He now lives with his parents, younger sister, and dog in Oakville, Ontario. He owns a DJI Osmo Pocket and edits his own footage from both the drone and the Osmo hoping to become a filmmaker and to use drones in the field of geology. Alex is also an associate of UCEMM (UAV Centre for Environmental Monitoring and Mapping) (www.abdn.ac.uk/research/ucemm) at the Department of Geography and Environment, University of Aberdeen, Scotland, UK, focusing on raising awareness and educating school-age pupils in safe, legal, and responsible drone flying. He is also an Honorary Pilot for DroneLite.

Contributors

Flor Álvarez-Taboada has a PhD degree in Remote Sensing and Forest Monitoring (University of Vigo, Spain, 2006), Master in Data Mining and Statistics (UNED, Spain, 2009), Master in GIS (University of Girona, Spain, 2005), and degree in Forest Engineering (University of Santiago de Compostela, Spain, 2000). She works as Associate Professor at ULE since 2001. Her teaching and research are related to remote sensing applied to natural resources monitoring. She has participated in several projects focused on multi-scale systems to assess forest disturbances using UAV data.

Hamish J. Biggs is an Ecohydraulics Scientist at the National Institute of Water and Atmospheric Research (NIWA) in Christchurch, New Zealand. His research covers flow around aquatic vegetation, remote sensing, sediment transport, biomechanics, image analysis, and field equipment development. Hamish leads the 'Drone flow' research programme at NIWA, which is developing an aerial surveying system to quantify physical habitat in rivers (i.e. bathymetry, substrate, and surface velocities). His team are also working on aerial surveying of aquatic and riparian vegetation, with class differentiation to species level through the use of hyperspectral aerial imagery.

Paul A. Butcher has been a Senior Research Scientist for 15 years with NSW Fisheries, Australia. His current research includes (i) tagging and tracking white and tiger sharks; (ii) using new technologies like VR4G shark listening stations to provide 'real-time' alerts to the public of dangerous sharks, (iii) optimising the efficiency of SMART drumlines by maximising the catch of target species while minimising stress and bycatch, (iv) using genetics to quantify the size of the east coast Australia White Shark population, and (v) quantifying the use of drones as a bather protection tool and observing how sharks move in the marine environment.

Stefano Campana has been working at University of Siena (Italy), Department of History and Cultural Heritage: campana@unisi.it; he has been working for the past 15 years at the University of Cambridge (UK) and at the University of Siena (Italy) where he is currently engaged in teaching and research as tenured Professor of Landscape Archaeology. His work is focused on the understanding of past landscapes, and he is specializing in remote sensing and archaeological methodology for purposes of research, recording, and conservation. The principal context for his work has been Tuscany, but he has also participated in and led research work in the UK, Spain, Turkey, Palestine, Iraq, Kazakhstan, Uzbekistan, Tajikistan, and China.

Andrew P. Colefax is an Australian Scientist working at Southern Cross University. His research focuses on developing drones and related technologies for shark surveillance to improve beach safety. Andrew is developing efficient shark surveillance procedures and improving the detection rates of animals in the water, in line with conservation values.

Melinda A. Coleman is a Principal Research Scientist and Adjunct Professor working on kelp forest conservation. Her research uses a powerful combination of ecology and cutting-edge genetic techniques to restore and future-proof underwater forests.

Nicholas C. Coops is a Professor and Canada Research Chair in Remote Sensing in the Faculty of Forestry, University of British Columbia (UBC) investing the role of remote sensing for forest production and conservation applications.

Belinda G. Curley is a Manager within the Aquatic Environment team at the NSW Department of Primary Industries, Australia. Belinda has a keen interest in strategic ecological and social research to inform management. She is currently working on the development of a monitoring and evaluation framework for the NSW Marine Estate Management Strategy, with a focus on the social and cultural components of the initiative. In her previous roles, she led the social research programme for the NSW Shark Management Strategy and conducted research on the ecology of temperate reef fishes to inform the design of marine protected areas.

Marco Dubbini received his PhD in Geodetic Sciences and Topographic at University of Bologna where he also received an MSc degree in Civil Engineering. Since 1998, he has been working in the field of geomatics in applications concerning the control and monitoring of the land and infrastructure. He is Technical Researcher and Adjunct Professor at University of Bologna and a founding partner of an innovative start-up company that produces patents and deals with the design and implementation of UAV systems, multispectral sensors, and acquisition and analysis of photogrammetric, multispectral, thermal, and LIDAR data.

Johannes Fahrentrapp received his Diploma in Biology from University of Freiburg, Germany. Following project work on several diseases and pests in grapevine such as the phytoplasma of bois noir and its vectoring grass hopper, Johannes went to ETH Zurich to work on his PhD project on fire blight resistance in wild apple species Malus × robusta 5. Later, he worked as Principal Investigator (PI) on plant–pathogen interaction projects in tomato and pest monitoring methods of soft fruit crops.

Cristina Gómez has a PhD in Forest Systems Conservation and Management (UVA, Spain), an MSc in Applied Geospatial Technologies (UOA, UK), and an MSc in Forestry (UPM, Spain). Her research focuses on integration of remotely sensed data with other spatial data to support forest monitoring activities over a range of temporal and spatial scales. Research interests include forest inventory, structure, and change dynamics from drastic or subtle disturbance to regeneration and stability.

Rocio Ballesteros Gonzalez completed her PhDs in Agricultural Engineering and Forestry Engineering, and MSc studies in Agricultural and Environmental Engineering at Castilla-La Mancha University. Her PhD focused on using high-resolution aerial images for agricultural applications. She has been working with unmanned aerial vehicles (UAVs) in precision agriculture, forestry, and environmental applications since 2009. She performed research internships at Food and Agricultural Organization for the United Nations (FAO, Italy, Rome) and at Cranfield University (UK). Currently, she is Assistant Professor at Castilla-La Mancha University. Dr. Ballesteros Gonzales is the author of more than 20 high-impact papers and many popular science papers aimed at developing new tools for generating useful information using UAVs. She has given several keynotes or plenary talks on UAVs and environmental studies at international conferences. She has ongoing collaborations with international research institutions to promote using new technologies to improve farming management.

Tristan R.H. Goodbody is a Postdoctoral Fellow in the Faculty of Forestry, University of British Columbia (UBC) researching methods for improving forest inventories using remote sensing.

Alice Gorodetsky is a Designer and Artist currently living in San Francisco, CA. While working towards her Bachelor's degree in Architecture at Syracuse University, Alice worked for the Performative Praxis Lab researching the use of heat mapping drones to inform building maintenance and design. After completing her degree in 2017, she has been working for both the DC and California offices of SKB Architecture and Design creating environmentally conscious and biophilic office space for a variety of companies and organisations. Her current goal is to complete Architect Registration Examination (ARE) testing and earn her architecture license.

Jason J. Hagon studied Geography and Geographical Information Systems at the University of Aberdeen between 2011 and 2017. Working alongside the Universities UAS Centre for Environmental Monitoring and Mapping (UCEMM), he is also a Director of GeoDrone Survey Ltd, a company specialising in aerial image data processing. Since 2017, he has worked as a data processor for Andrews Survey, a hydrographic survey company based in Aberdeen.

Brendan P. Kelaher is an Australian Marine Biologist with more than 20 years' experience. He engages in research and teaching that supports evidence-based management of marine systems and fisheries. Professor Kelaher has well-established research programmes focused on marine ecology, impact assessment, and fisheries biology. He also investigates how drones can cost-effectively contribute to shark-bite mitigation, marine management, and wildlife monitoring.

Lammert Kooistra obtained his MSc degree in Soil Science from Wageningen University and his PhD degree from the Radboud University in Nijmegen, the Netherlands. The subject of his dissertation was incorporating spatial variability in ecological risk assessment of contaminated river floodplains. Currently, he is working as Associate Professor at Wageningen University. His main research interest is the application of integrated sensing technology for (agro)environmental conservation and management with special interest in combining unmanned aerial vehicles, sensor networks, and crop modelling. He has contributed to more than 65 peer-reviewed journal papers.

Paul Leinster, CBE, has over 40 years of practical experience in environmental management, science, policy development, and regulation. Paul joined Cranfield in 2015 as Professor of Environmental Assessment. Prior to this, he was Chief Executive of the Environment Agency (EA). Before joining the EA in 1998, Paul worked in the private sector. He has a particular interest in translating research into effective policy, regulatory, operational, and governance measures and in natural capital and ecosystem service approaches to environmental management. He is a member of the government's Natural Capital Committee and chairs the Bedfordshire Local Nature Partnership and the Institute of Environmental Management and Assessment.

Ian Maddock is a Professor of River Science at the University of Worcester and has teaching and research interests in hydromorphology, hydroecology, ecohydraulics, river restoration, and aquatic habitat measurement, mapping, and modelling. The primary focus of his research is on developing methods for measuring, mapping, and monitoring the hydromorphology of streams and how this influences habitat hydraulics. Ian has a research interest in the use of unmanned aerial vehicles and terrestrial laser scanners for ultra-high-resolution remote sensing of river morphology and measurement of river channel change, water velocity, and river discharge.

Francesco Mancini received his MSc degree in Marine Environmental Sciences and a PhD in Geodetic Sciences and Topography from the University of Bologna, Italy. In the past 20 years, he has worked in the field of geomatics, as Researcher (Technical University of Bari, Italy), Associate Professor (University of Modena and Reggio Emilia), and Surveyor in application to ground deformation, natural hazard assessment, and landscape archaeology. He is Professor in Geomatics Technology, GIS, Cartography, and Photogrammetry at schools of Civil and Environmental Engineering. His current interests include precision surveys, spatial data analysis, ground subsidence, surveying in coastal areas, global navigation satellite system (GNSS) positioning, and proximal sensing.

Dmitri Mauquoy has research expertise in the assessment of environmental/climate change using peat-bog deposits. Evidence for change in mire surface wetness, driven by climate change during the Medieval Climatic Optimum and Little Ice Age, from a range of European-raised peat-bog profiles, and a similar profile from Tierra del Fuego, Argentina, has indicated the potential influence of changes in solar activity driving these climatic change events. He has also been involved with the refinement of 14C wiggle-match dating to generate accurate and precise chronologies, in collaboration with international colleagues from the Universities of Amsterdam, Uppsala and Belfast. A 3-year Natural Environment Research Council (NERC) joint standard grant 'Palaeoclimate reconstructions from Tierra del Fuego to detect land-ocean-atmosphere interactions' was awarded in 2011 to Dmitri in order to reconstruct the timing, magnitude, and spatial pattern of long-term climate change, as well as exploring the impact of different causal factors such as changes in atmospheric and ocean circulation, and solar variability on the climate of the study region during the last ~2,000 years. Agence Nationale De La Recherche (ANR)-funded projects have a focus on southern South American palaeoclimate change using peat-bog archive records. He has been involved in a number of Scottish Natural Heritage (SNH) studies into the role of UAV technology for coastal monitoring and mapping.

Euan J. Provost is a Researcher at the National Marine Science Centre, Southern Cross University, Australia. His research investigates the applications of UAVs to improve the sustainability and conservation of coastal ecosystems. In particular, the research focuses on cost-effective generation of quality data to underpin strategic management actions in coastal areas. This research uses UAVs combined with a variety of sensors (e.g. high resolution, thermal) to generate accurate information on human uses of coastal areas, on illegal and legal fishing effort, and for the detection of lost and discarded fishing gear.

Tarek Rakha is an Architect, Building Scientist, and Educator. He is an Assistant Professor of Architecture at Georgia Tech and Faculty at the High Performance Building (HPB) Lab. His research aims to influence architecture, urban design, and planning practices through three areas of expertise: sustainable urban mobility and outdoor thermal comfort, daylighting and energy efficiency in buildings, and building envelope diagnostics using drones. Prior to joining Tech, Dr. Rakha taught at Syracuse University (SU), Rhode Island School of Design (RISD), and MIT. He completed his PhD in building technology at MIT, where he was part of the Sustainable Design Lab as a member of the developing team for UMi, the urban modelling and simulation platform.

Monica Rivas Casado is a Senior Lecturer in Integrated Environmental Monitoring with expertise in the application of statistics to environmental data. Her academic career has been built around the integration of emerging technologies, advanced statistics, and environmental engineering for the design of robust monitoring strategies. Monica has an MSc in Environmental Water Management and a PhD in Applied Geo-Statistics from Cranfield. She is a Chartered Environmentalist, a Chartered Scientist, a Chartered Forestry Engineer, and a Fellow of the Higher Education Academy. She holds a Postgraduate Certificate in Teaching and Learning. Monica is a fully qualified RPQs (drone) pilot.

Peter Roosjen obtained both his MSc and PhD degrees at the Laboratory of Geo-Information Science and Remote Sensing of Wageningen University and Research in the Netherlands, where he worked on multi-angular UAV-based remote sensing for the estimation of biophysical vegetation parameters. Currently, he works as Postdoctoral Researcher at the same laboratory. His research interests include UAV-based remote sensing, object-based image analysis, and object detection. He has (co)authored several articles in peer-reviewed journals.

Enoc Sanz-Ablanedo has a BSc degree in Geology and PhD in Mining Engineering. Since 2006, he has worked as a Professor at the University of León, where he has taught more than 2,500 hours in Geodesy, Topography, and Photogrammetry subjects. As a researcher, he has collaborated in 6 projects of regional or international scope and another 25 projects of local scope. He has published his results in more than 15 scientific papers, most of which are indexed in the Journal Citation Report (JCR). Enoc Sanz Ablanedo is a co-founding researcher of the GI-202 group (Geographical Information Technologies, GEOINCA) of the University of León, at the Ponferrada Campus.

Alastair Skitmore is a recent graduate from Liverpool John Moores University with an MSc in Drone Technologies and Applications. Beforehand, he studied Geology and Physical Geography at Edinburgh University. His recent dissertation project focused on ways to expand the use of fixed-wing UAVs in complex/rugged environments, where previously, only multi-rotor UAVs could be operated. He believes it is of great importance to be able to properly utilise the range and payload benefits which fixed-wing UAVs offer over their multi-rotor counterparts, to assist in large-scale, repetitive environmental monitoring studies.

Andrew Smith: Since a very early age, Andrew has had a keen interest in geography and has combined this passion for all things map related to modern advances in remote sensing and mapping technology. Andrew is a double graduate of the University of Aberdeen, completing a BSc in Geography and MSc in Geographical Information Systems. His research has covered the application of historical maps to chart temporal changes to river geomorphology and utilisation of UAVs to collect, reconstruct, and model fragile dune systems. His current PhD research is concerned with mapping, monitoring, and modelling coastal habitats with UAVs. Andrew currently resides in Aberdeenshire.

Amy Woodget is a Physical Geographer, with broad research interests in the environmental applications of remote sensing and GIS. In particular, her research focuses on exploring novel methods for monitoring and measuring river environments at fine spatial scales. She uses small unmanned aerial systems (UASs) and structure-from-motion (SfM [SFM]) photogrammetry for quantifying a range of physical habitat parameters within river systems. Her aim is to investigate whether these emerging techniques hold value as a tool for robust, reliable, routine assessments of river habitat, especially in light of pressing global challenges such as climate change.

Luca Zanchetta graduated from the University of Padua with a Master's degree in Agricultural Science and also has a Master's in GIS from the University of Aberdeen in Scotland, UK. His work experience has been quite varied, spanning from several years in the rural development sector in Ethiopia leading projects aimed at improving women's conditions, training local farmers and reforestation, to commercial farming. More recently, he had experience as an entrepreneur in the United Arab Emirates (UAE). A passionate traveller, following the Master's in GIS, he has developed a keen interest for UAVs and a focus on how technology can help to reduce the environmental problems in agriculture.

1 Introduction

David R. Green
UCEMM – University of Aberdeen

Billy J. Gregory
DroneLite

Alex R. Karachok
UCEMM – University of Aberdeen

CONTENTS

CONTEXT

As a postgraduate student at the University of Toronto in Canada in the 1980s, I had the chance to work in the Ontario Centre for Remote Sensing (OCRS) – just off Bloor Street – in Toronto. This opportunity was afforded to me courtesy of the then OCRS Director Dr. Simsek Pala (now deceased), Tracey Ellis (now deceased) a research scientist, and my university Supervisor Professor Jap van der Eyk (now deceased) an expert in aerial photography, remote sensing (RS) and photogrammetry, and soil mapping in the Department of Geography at the University of Toronto.

At the time, I was starting to work on a dissertation project for an MSc thesis looking at the use of aerial photography (panchromatic, colour, and colour infrared (CIR)) for land-use/land-cover mapping. Purely by chance, I was allocated to a desk in the office of Ed Wedler, another RS specialist at OCRS, but more importantly for me – and what triggered my interest in drones all these years later – he owned a small company utilising remotely controlled model aircraft (RCMA) for environmental RS.

I had always had an interest in flying as a child, owning a small balsa wood model glider and a tethered model 'fighter' RC aircraft, living close to the Shuttleworth Aircraft Museum in Bedfordshire, occasionally going gliding at Duxford Aerodrome near Cambridge in England, and as a family friend of Professor Professor J.K.S. St. Joseph at the University of Cambridge who flew most of their aerial photography, some of it using the Wallis Autogyro.

A journal paper written by Ed Wedler (Wedler, 1984) provided me with some fascinating insight into the many possibilities of aerial data acquisition using these small platforms which, when coupled with my background in RS, survey and photogrammetry, cartography, digital mapping, and Geographic Information Systems (GIS) – gained from James Young, Roger Kirby, and Tom Waugh – staff at the University of Edinburgh, where I did my BSc degree – took my interest in this technology to a new level!

So, when I found out about the use of all sorts of different radio-controlled (RC) model aircraft for monitoring, mapping, and surveying, I began a lifelong interest in RS with these small-scale aerial (SSA) platforms; it was a technology that just captured my interest and imagination!

A proposal to use model aircraft for a subsequent research thesis, involving vineyard monitoring, at Vineland on the Niagara Peninsula sadly did not materialise! However, when I came back to the UK, and to the University of Aberdeen several years later, I met with a RC model aircraft enthusiast – Norrie Kerr – in Aberdeen who designed, built, and flew his own model aircraft, which were initially used by the then North East River Purification Board (NERPB) in Aberdeen, to monitor some dye tracing experiments in the River Dee tracing the river flow into the sea at Aberdeen Harbour. Norrie also flew some aerial photography of the river channels at Mar Lodge in Deeside for us (David R. Green and Judith Maizels) as part of a river study. Later, working with Professor Dave Raffaelli in Zoology at the University of Aberdeen, we successfully flew several aerial sorties to monitor macro-algal weedmats in the Ythan Estuary using a fixed-wing aircraft and a single-lens reflex (SLR) camera carrying panchromatic, RGB, and filtered films. This was one of the first coastal and estuarine studies using an SSA platform (Green, 1995).

THE BIRTH OF UAVs

Although successful at the time, use of these platforms for RS met with quite a lot of scepticism, and they were not really taken all that seriously, often being treated as 'toys' and regarded as not professional or scientific. However, when the first of the drones or UAVs – the DJI Phantom 1 – became available about 6 years ago, this was the start of the re-emergence and development of small aerial platforms for serious RS.

Comparing the current generation of UAVs or unmanned aerial systems (UASs) to the Phantom 1 now, the Phantom 1 does indeed seem quite 'toy-like' and quite primitive. However, the Phantom 1 and a few other similar low-cost multi-rotor platforms that became available e.g. the 3DR Iris+ soon became recognised as much more than a toy when small cameras e.g. GoPros were attached to them.

With the arrival of custom-made gimbals, this then opened up the beginnings of low-cost airborne RS. Since then, the miniaturisation of the technology – whether it be the platform or the sensor – and the evolution of battery technology, microprocessors, storage, Global Positioning System (GPS), and controllers have revolutionised the whole SSA platform industry. Finally, the potential of these small aerial platforms has just begun to be realised with many applications now emerging.

THIS BOOK

The idea for this book originated with the growing realisation that there was a need to provide a wide range of people from different walks of life and backgrounds with a more informed insight into the rapidly evolving drone or UAV and associated technologies becoming available. Whilst radio-controlled aircraft (RCA) had been around for many years, some of them carrying cameras and other sensors, unmanned airborne vehicles (UAVs) represent a much more up-to-date combination and integration of a number of technologies that have evolved from the miniaturisation of platforms and sensors, together with new battery technology, electronics, and software.

Within a very short period of time, UAVs have reinvented the small-scale airborne platform to a level of public awareness never achieved in the past. In addition, the miniaturisation of sensors, object avoidance, and vision sensor technology have rapidly revolutionised the potential applications of such technology, with a growing number of professional and commercial applications that were not previously possible.

As long-term supporters of this technology and the role that it can clearly play in environmental monitoring, mapping, and modelling, this book is intended to help raise awareness of the many positive aspects that UAVs can bring to society in terms of enhancing our ability to gather high-resolution imagery and to generate high-resolution Digital Surface Models (DSMs) and Digital Terrain Models (DTMs).

In particular, there is a focus on affordable off-the-shelf technology that has empowered a new generation of people – young to old – to be able to fly a small aerial platform and to capture, store, process, and visualise their own aerial photography.

There is now a growing need to help the public to understand the breadth of applications that this drone technology will open up, applications that extend well beyond capturing just video footage and aerial photography, to delivery, search and rescue, agricultural spraying, hydrography, and many more.

Drone technology and artificial intelligence (AI) are very exciting developments, ones that need to be understood by the public and embraced for the tremendous opportunities that they offer.

BOOK CHAPTERS

Whilst there are many more applications out there than the ones it is possible to present here, the following chapters in this book offer a small selection to reveal the practical use of these small platforms and sensors in a range of environmental applications, together with some of the technological developments.

These chapters also provide examples that illustrate the use of off-the-shelf drones, with an emphasis on low-cost RS solutions that capture imagery and provide the basis for a range of products including DTM and DSM.

CHAPTER 1 – INTRODUCTION (DAVID R. GREEN, BILLY J. GREGORY, AND ALEX R. KARACHOK)

This is an introduction to the rationale for this book and its contents.

CHAPTER 2 – FROM RADIO-CONTROLLED MODEL AIRCRAFT TO DRONES (DAVID R. GREEN AND CRISTINA GÓMEZ)

Both fixed-wing and multi-rotor platforms are increasingly being identified as potentially useful airborne platforms for a range of professional environmental RS applications, including both expensive high-end specialist solutions and lower-cost, off-the-shelf commercial options. Small UAVs are ideal for applications where aerial coverage requirements are small, flying experience and expertise is limited, and the operating budget is usually relatively small. With advances in battery technology, navigational controls, and payload capacities, many of the smaller UAVs are now capable of utilising several different sensors to collect photographic data; video footage; multispectral, thermal, and hyperspectral imagery; as well as light detection and ranging (LIDAR). With the aid of specialist digital image processing (DIP) and soft-copy photogrammetry software, aerial data and imagery can be processed into a number of different products including ortho-photos, mosaics, and Digital Elevation Models (DEMs) and analysed to generate useful information for input to a GIS. Aerial photography and video imagery have been acquired using small aerial platforms for over 30 years. Small-scale RC model aircraft and helicopters, using small 35 mm SLR and video cameras, have all been used to acquire panchromatic, colour, CIR, and multispectral aerial photography for a wide range of environmental applications. Chapter 2 begins by examining the history of SSA platforms for aerial image acquisition as the context for the current growth in popularity of UAVs, the associated technologies, and the range of small-scale sensors (e.g. GoPro Hero cameras) currently available for environmental monitoring, mapping, and modelling applications, together with image processing, analysis, and information extraction, and soft-copy photogrammetry software such as Pix4D, AgiSoft, and AirPhotoSE. The past and present advantages and disadvantages of small aerial platforms are also considered, together with some of the emerging technologies and recent developments.

CHAPTER 3 – AQUATIC VEGETATION MONITORING WITH UAS (HAMISH J. BIGGS)

Aerial surveys of aquatic vegetation with UAS provide high-resolution maps at relatively low cost. These maps are used to determine the percentage cover of aquatic vegetation or to estimate total biomass when coupled with ground truth measurements. This data enables effective management of aquatic vegetation that balances practical engineering considerations with ecosystem health. Chapter 3 provides a detailed introduction to aquatic vegetation monitoring with UAS. It covers field deployments, ground truth sampling, hydraulic measurements, processing aerial imagery, and lab work for vegetation biometrics. This data is then pulled together at larger spatial scales to evaluate total cover, total biomass, and hydraulic interactions, which provide inputs for targeted management of aquatic vegetation. Finally, the chapter discusses emerging technologies that have exciting implications for aquatic vegetation monitoring with UAS.

CHAPTER 4 – UNMANNED AERIAL VEHICLES FOR RIVERINE ENVIRONMENTS (MONICA RIVAS CASADO, AMY WOODGET, ROCIO BALLESTEROS GONZALEZ, IAN MADDOCK, AND PAUL LEINSTER)

The key role of rivers in society has been recognised for centuries. However, rivers and freshwater ecosystems in general have always been at the receiving end of anthropogenic activity, from intense pollution to obstruction caused by hard engineering structures. In the last decades, regulatory efforts across the world have focused on the protection and even improvement of the ecological quality of rivers, with some countries granting rivers the same rights as human beings. Perhaps the most relevant regulatory example is the European Union's Water Framework Directive (WFD) that aims to achieve good ecological status or potential of inland and coastal waters. The implementation of such regulatory frameworks requires robust monitoring programmes that provide information concerning the chemical, physico-chemical, biological, and hydromorphological characteristics, amongst others. From a broader perspective, monitoring programmes are also required to better inform planning and management decisions. These monitoring programmes are required from a scientific perspective to improve our understanding of erosion and deposition processes and to develop accurate flood prediction tools. The range of monitoring approaches available is vast and varied, ranging from purely observational in situ methods to RS techniques (e.g. satellite imagery). Regarding methods used for hydromorphological monitoring, a trade-off exists between resolution and wide-area coverage. Typically, RS methods can provide wide-area but low-resolution information, whereas in situ methods provide localised but generally high-resolution data. The starting point for effective environmental management is an appropriate level of understanding of the current state of the environment. This has typically required on-site assessment or investigation. Such activities, including walkover surveys, which provide vital information, can be very resource- and/or labour-intensive. This then reduces the resources available to implement measures to remediate, protect, and improve the environment. In this context, there are several current and potential applications of UAVs that will help in the management of river catchments to better protect and improve these environments for people and wildlife. These could be used routinely by practitioners to increase their effectiveness when carrying out their environmental assessment and management work. UAVs can also be used in combination with other RS technologies, GIS, modelling, and visualisation approaches to further extend their usefulness. Within Chapter 4, we aim to give an insight into the use of UAVs for fluvial applications. We review current data collection and processing considerations, and explore examples of UAV use for riverine feature detection and mapping. Finally, we demonstrate the quantification of physical river parameters from UAV data.

CHAPTER 5 – LOW-COST UAVS FOR ENVIRONMENTAL MONITORING, MAPPING, AND MODELLING OF THE COASTAL ZONE (DAVID R. GREEN, JASON J. HAGON, AND CRISTINA GÓMEZ)

UAVs have been used for many different applications around the world in the last 5 years. As their ease of use improves, the number and range of applications has also increased rapidly. In particular, coastal and marine environments requiring high-resolution local data and information or multi-temporal monitoring data and imagery for change detection are well suited to the use of UAVs carrying a range of different sensors including RGB cameras and thermal and hyperspectral sensors. In recent years, with the widespread availability of small, off-the-shelf, Ready-to-Fly

(RTF) UAVs and accompanying photographic and video sensors, and the rapidly evolving GPS, navigational, and sensor technology, a number of coastal/marine applications have emerged that take advantage and demonstrate the value of this technology for use by coastal researchers and managers. Some of the applications of UAVs to coastal environments have included monitoring coastal and intertidal habitats such as mangroves, saltmarsh, and seagrass to gather ortho-photos, as well as the use of video transects for intertidal bathymetric habitat mapping, to assess inaccessible areas, facilitate repeat surveys, extent and coverage of vegetation, and for monitoring dredging activity and plume extent. Also there are applications that include the generation of three-dimensional (3D) point clouds through stereo-photography or LIDAR, routine marine fauna observation (MFO) including cetaceans and turtle nesting activity using fixed-wing aircraft, minor oil-spill contingency tracking, hyperspectral vegetation classification, detailed engineering inspections of pipelines, offshore structures, thermal imaging, and health and safety intervention. Whilst some of the platforms used for environmental applications have been specialised custom examples, the rapid growth in small, low-cost platforms has led to more of these platforms being standard off-the-shelf UAVs (e.g. the DJI Phantom) has become very popular as an RTF drone. Inevitably the opportunity to monitor changes to the coastal environment has been one of the applications where UAVs have considerable potential given the ease with which it is now possible to acquire multi-temporal photographic imagery not only to see changes to the coastline in terms of erosion and deposition but also to be able to measure the change e.g. cutback and to determine the rate of coastal erosion. In Chapter 5, three different examples are used to illustrate the potential of these small airborne platforms and sensors to gather data on various aspects of the coastal environment. Spanning some 30 years, these provide a good indication of how the aerial platform and sensor technologies have advanced, over a relatively short period of time to the present day, empowering a wide range of people with the potential to acquire and process high-resolution environmental data.

CHAPTER 6 – UNMANNED AERIAL SYSTEM APPLICATIONS TO COASTAL ENVIRONMENTS (FRANCESCO MANCINI AND MARCO DUBBINI)

Since coastal environments became objects of interest for researchers and professionals, satellite-based and aerial RS have been used for a wide range of applications. These include mapping of coastal wetlands and rivers mouths, LIDAR bathymetry, mapping of flooded areas, surveillance, oil-spill tracking, land-use/land-cover delineation of coastal areas, investigations of coastal dynamics, and detection of pollutants. The ability to observe large coastal stretches through remotely sensed information and at spatial and temporal resolutions using satellite-based and airborne imaging systems has been a driving factor here. However, investigations of coastal environments at very high spatial and temporal resolutions are still an issue. To date, satellites have offered the advantage of a synoptic view, with optical and multispectral imaging at increasing spatial and spectral resolution and revisiting times. These factors have suited many applications that have focused on coastal monitoring. However, many applications and studies related to coastal geomorphology and vulnerability analysis of coastal environments continue to require additional spatial resolution. For instance, a hazard that affects a coastal stretch might not be imaged by a satellite sensor due to the orbit design and data acquisition planning. Moreover, airborne surveys over limited areas might come at unsustainable cost. The recent availability of UASs and the joint use of methodologies provided by geomatics engineering, photogrammetry, and computer vision technologies have established new opportunities for environmental monitoring purposes. This is especially the case for traditional survey applications where labour-intensive investigations are required, such as for coastal environments, where logistic difficulties can compromise the quality of the overall product and make the work of the surveyors difficult. The high-resolution and flexible overflights offered by unmanned aerial vehicle (UAV) sensors make these instruments more suitable for a wide range of coastal applications. In particular, all applications that require investigations at desired periods and revisit times of the coastal settings can benefit from the new paradigm offered by UAV surveys. The objective of this chapter is to acquaint the reader with the potential of UAS-based surveys in coastal environments through a discussion of examples available in the very recent literature and through the skills achieved by the authors in this field. After a short introduction to UAV models that can be used to perform aerial surveys and a short description of the available methodologies to process the data acquired by visible and multispectral sensors, the chapter will focus on several UAS applications referred to coastal environments.

CHAPTER 7 – UAV IMAGE ACQUISITION USING STRUCTURE FROM MOTION TO VISUALISE A COASTAL DUNE SYSTEM (ANDREW SMITH)

Availability of accurate high-resolution DSMs within coastal environments has for many years been the realm of expensive airborne survey techniques. This project aims to utilise the UAV approach to surveying a coastal dune system that is rapid, can be flown autonomously, collects a large dataset, is cost-effective, and improves spatial accuracy compared to traditional methods. Structure-from-Motion (SfM [SFM]) software is employed to process the imagery collected from two surveys, creating a 3D dense cloud totalling 68.6 million points, production of a DSM, a high-quality two-dimensional (2D) orthomosaic image, a 3D mesh model, and an average ground sampling distance (GSD) of 2.66 cm/pixel or 7.02 cm². With increasing elevation uncertainties within the coastal margin, comparisons made between the UAV-SfM (SFM) DSM and a 2012 Ordnance Survey DSM, LIDAR DSMs from 2012 and 2014 find on average a 2.36 m discrepancy in fixed elevation points between data sets. Surveying methods include a Real-Time Kinematic Global Navigation Satellite System (RTK GNSS) to validate 35 ground control points (GCPs) used for SfM (SFM) georectification, producing accuracy tolerances of 0.12 m in X, 0.09 m in Y, and 0.06 m in Z.

CHAPTER 8 – MONITORING, MAPPING, AND MODELLING SALTMARSH WITH UAVS (DAVID R. GREEN, DMITRI MAUQUOY, AND JASON J. HAGON)

Coastal and estuarine saltmarsh has long been recognised as having key physical, ecological, and recreational values, acting as sediment and nutrient traps and as natural coastal protection structures functioning as protective buffers between the land and the sea. Recent work suggests that saltmarshes are particularly good for demonstrating how the coast can change in response to environmental influences, including relative sea-level rise. In Scotland, the pace of current relative sea level may lead to inundation of coastal saltmarsh, with the potential for rapid change and loss. It is therefore critical to be able to monitor the response of saltmarsh to sea-level rise, map saltmarsh topography, and model rates of marsh elevation change on a real-time basis. However, changes in saltmarsh vegetation differ according to local circumstances and can be very difficult to map from the land, and even if there is enough survey time to walk vegetation community boundaries using GPS, some areas are so soft and muddy that they cannot be surveyed safely on foot. Aerial imagery can provide an overview of saltmarsh that cannot be obtained from the ground. With geo-rectified photography, it becomes possible to map both area and volumetric changes accurately over time. In contrast to conventional vegetation mapping, aerial imagery is non-selective in its data capture, and multispectral and hyperspectral imagery adds even more information about these changes for scientists who are attempting to interpret coastal change from a variety of perspectives. UAV sensors can: (i) offer comprehensive (i.e. non-selective) high-resolution, spatio-temporal data capture over inaccessible areas of saltmarsh; (ii) create 3D models of saltmarsh micro-topography, soil, and vegetation; and (iii) enhance knowledge and understanding of saltmarsh functioning and the spatial extent of vegetation composition and habitat change over time that affects ecosystem functioning. Such data can: (i) inform coastal policy for improved and sustainable coastal management and (ii) raise awareness and educate the public about the importance of saltmarsh, the benefits of natural coastal protection, and adaptation management. The aims and objectives of Chapter 8 are to demonstrate the practical potential of using UAV-based RS platforms and sensors to monitor, map, and model coastal and estuarine saltmarsh.

CHAPTER 9 – AUTONOMOUS UAV-BASED INSECT MONITORING (JOHANNES FAHRENTRAPP, PETER ROOSJEN, LAMMERT KOOISTRA, DAVID R. GREEN, AND BILLY J. GREGORY)

Drosophila suzukii has become a serious pest in Europe attacking many soft-skinned crops such as several berry species and grapevines since its spread in 2008 to Spain and Italy. An efficient and accurate monitoring system to identify the presence of Drosophila suzukii in crops and their surroundings is essential for the prevention of damage to economically valuable fruit crops. Existing methods for monitoring Drosophila suzukii are costly, time- and labour-intensive, prone to errors, and typically conducted at a low spatial resolution. To overcome current monitoring limitations, we are developing a novel system consisting of sticky traps that are monitored by means of UAVs and an image processing pipeline that automatically identifies and counts the number of Drosophila suzukii per trap location. To this end, we are currently collecting high-resolution RGB imagery of Drosophila suzukii flies in sticky traps taken from both a static position (tripod) and a UAV, which are then used as input to train deep learning models. Preliminary results show that a large part of the Drosophila suzukii flies that are caught in the sticky traps can be correctly identified by the trained deep learning models. In the future, an autonomously flying UAV platform will be programmed to capture imagery of the sticky traps under field conditions. The collected imagery will be transferred directly to cloud-based storage for subsequent processing and analysis to identify the presence and count of Drosophila suzukii in near real time. This data will be used as input to a decision support system (DSS) to provide valuable information for farmers.

CHAPTER 10 – UAV IMAGERY TO MONITOR, MAP, AND MODEL A VINEYARD CANOPY TO AID IN THE APPLICATION OF PRECISION VITICULTURE TO SMALL-AREA VINEYARDS (LUCA ZANCHETTA AND DAVID R. GREEN)

UAVs have become a popular low-cost and very convenient small-scale RS platform for aerial monitoring, image acquisition, and more recently survey applications. In Precision Agriculture (PA), many studies have already proved the potential of small, low-cost, off-the-shelf aerial platforms for small-area studies providing valuable information on crops and crop status for farm managers. With developments in the technology, specifically platforms, sensors, and apps, there is now the capability to capture imagery and convert into information in near real time. Applications of this technology have also been demonstrated to yield useful information in viticulture for vineyard managers and small vineyards in both the UK and Switzerland. As the drone and related technologies continue to evolve, there are many more applications now to exploit the potential of this technology in applied roles. Precision Viticulture (PV) is based on the existence of spatial and temporal variability within vineyards and is made possible by using four important geospatial technologies: (i) GPSs; (ii) RS, DIP, and soft-copy photogrammetry; (iii) GIS; and (iv) digital mapping. In the example reported here, two off-the-shelf RTF UAVs, carrying two types of low-cost commercial miniaturised cameras, were used to acquire imagery of a small vineyard in Italy, at a number of different spectral wavelengths, to demonstrate their potential role in the provision of easy-to-acquire information of use to the vineyard manager. In particular, this study focused on analysing intra-vineyard variability to provide information about differential management practices. The primary intention of this research was to increase awareness about how such low-cost technologies can be of benefit to the vineyard manager and potentially lead to a change in some viticultural practices that may in the longer term reduce costs and/or improve the quality of the wine and increase profits.

CHAPTER 11 – FOREST ECOSYSTEM MONITORING USING UNMANNED AERIAL SYSTEMS (CRISTINA GÓMEZ, TRISTAN R.H. GOODBODY, NICHOLAS C. COOPS, FLOR ÁLVAREZ-TABOADA, AND ENOC SANZ-ABLANEDO)

Monitoring forest ecosystems is a complex task that requires data at multiple scales. UAV systems provide very fine-scale, local information, acquired from both passive and active sensors, which can be used to complement the capacity of satellite and aerial RS and to fill key information gaps. UAV systems constitute a safe, flexible, and relatively low-cost technology for acquisition of frequent and very high spatial resolution data. UAV technologies can bridge the field and RS scales with above canopy perspective, providing data for calibration and validation of RS monitoring systems. UAVs may support forest inventory over small areas when used alone or complement inventory activities at broader scales. Current limitations to the use of UAVs are imposed by battery duration, payload weight, and local regulations, as well as massive data processing capability. In the remaining chapters, we focus on four forestry applications in which UAV technologies contribute to get useful insights for sustainable management. An update of precision inventories of forest stands, assessing growth rates and volume increments near Williams Lake (British Columbia, Canada), characterisation of coniferous regeneration in Quesnel (British Columbia, Canada), monitoring health of Pinus radiata in Fresnedo (León, Spain), and identification of invasive species in Viana do Castelo (Portugal).

CHAPTER 12 – MONITORING OIL AND GAS PIPELINES WITH SMALL UAV SYSTEMS (CRISTINA GÓMEZ AND DAVID R. GREEN)

UAVs are now evolving as highly effective tools for tackling the requirements of oil and gas pipeline monitoring, a specific environmental application of the UAV technology. This chapter describes the current use of UAV platforms and sensors, as well as the foreseen potential of small UAV systems for monitoring oil and gas pipelines. The chapter will include sections covering the following: (i) an overview of the characteristics of UAVs; (ii) the use of UAVs for oil and gas pipeline monitoring to date with particular attention paid to the strengths and successes, as well as the weaknesses; (iii) considerations and developments in the technology of SSA platforms and sensors specifically tailored to oil and gas pipeline monitoring applications, including battery, sensor, navigation, software, and platform; and (iv) future prospects for UAV development and application.

CHAPTER 13 – DRONE-BASED IMAGING IN ARCHAEOLOGY: CURRENT APPLICATIONS AND FUTURE PROSPECTS (STEFANO CAMPANA)

Traditional platforms for close-range RS (balloons and kites etc.) have been joined in the last decade by highly sophisticated automated systems (UAVs or drones), providing fully automatic, highly reliable, ready-to-use, pocket-size, multi-sensor, and – last but not least – highly affordable equipment with a genuine daily-fly capacity. In the last two decades, archaeologists have been testing both platforms and sensors, in particular for 3D documentation of archaeological excavations, the detailed survey of monuments and historic buildings, the 3D imaging of archaeological sites and landscapes, as well as for aerial exploration, airborne geophysics, and LIDAR-based archaeological survey of woodland areas. The scale of applications has ranged from individual sites to landscapes up to 20/30 km^2 in extent. The role of UAV platforms in survey and documentation at a wide range of scales, and in diagnostics more generally, has become a key focus of attention in the inexorably growing field of expertise and practice within archaeological survey and recording.

CHAPTER 14 – UNMANNED AERIAL SYSTEM (UAS) APPLICATIONS IN THE BUILT ENVIRONMENT: TOWARDS AUTOMATED BUILDING INSPECTION PROCEDURES USING DRONES (TAREK RAKHA AND ALICE GORODETSKY)

UASs – a.k.a. drones – have evolved over the past decade as both advanced military technology and off-the-shelf consumer devices. There is a gradual shift towards public use of drones, which presents opportunities for effective remote procedures that can disrupt a variety of built environment disciplines. UAS equipment with RS gear present an opportunity for analysis and inspection of existing building stocks, where architects, engineers, building energy auditors as well as owners can document building performance, visualise heat transfer using infrared (IR) imaging, and create digital models using 3D photogrammetry. Chapter 14 presents a comprehensive review of various literature that addresses this topic, followed by the identification of a standard procedure for operating a UAS for energy audit missions. The presented framework is then tested on the Syracuse University campus site based on the literature review to showcase: (i) pre-flight inspection procedure parameters and methodologies; (ii) during-flight visually identified areas of thermal anomalies using a UAS equipped with IR cameras; and (iii) 3D Computer Aided Design or Drafting (CAD) modelling developed through data gathered using UAS. A discussion of the findings suggests refining procedure accuracy through further empirical experimentation, as well as study replication, as a step towards standardising the automation of building envelope inspection.

CHAPTER 15 – THE APPLICATION OF UAVS TO INFORM COASTAL AREA MANAGEMENT (EUAN J. PROVOST, PAUL A. BUTCHER, ANDREW P. COLEFAX, MELINDA A. COLEMAN, BELINDA G. CURLEY, AND BRENDAN P. KELAHER)

UAVs are being increasingly used for cost-effective environmental monitoring and to collect data that is beneficial for coastal management. Recent improvements in affordability and accessibility of UAV systems have driven the development of a range of applications benefiting coastal management, such as wildlife monitoring, marine habitat mapping, fisheries compliance, topographic mapping, including modelling wave run-up and assessing storm damage and erosion. Compared to conventional manned aircraft, UAVs often provide financial, logistic, and safety benefits. Compared to satellite data, UAVs can provide higher resolution imagery and flexibility in frequency of data collection, as well as generate data in conditions where satellites are not particularly effective. Currently, UAVs bridge the gap between large-area low-resolution RS techniques and small-scale labour-intensive field sampling techniques, allowing for effective data collection in fine detail across relatively large areas. As the capabilities, benefits, and cost efficiencies of UAVs evolve with continual technological improvements and reforms to legislation, their application for coastal management will expand rapidly in the coming years.

CHAPTER 16 – FROM LAND TO SEA: MONITORING THE UNDERWATER ENVIRONMENT WITH DRONE TECHNOLOGY (DAVID R. GREEN AND BILLY J. GREGORY)

In the last 10 years, UAV or UAS technology has evolved very quickly to provide a number of low-cost platforms for numerous miniaturised sensors (e.g. RGB, Normalised Difference Vegetation Index (NDVI), thermal) that can be used to gather aerial photography and sensor imagery for numerous land-based monitoring, mapping, and modelling environmental applications. More recently, small affordable underwater platforms and sensors have emerged with the capability to see underwater and to gather different types of data and imagery below the waves in both fresh and saltwater. There are many examples of airborne platforms now in use that can survey the coast, wind turbines, flares, tank, and external ship inspections. This chapter explores some of the small affordable platforms and sensors now coming onto the market and some of their potential uses, advantages, and limitations.

Chapter 17 – A Question of UAS Ground Control: Frequency and Distribution (Jason J. Hagon)

Developments in miniaturised electronic components and SfM (SFM) photogrammetry software have led to the widespread use of UASs for multiple applications including topographical surveying. To date, much discussion around the accuracy UAS can achieve has been focused on data acquisition techniques and photogrammetric processing techniques. These factors are critical; however, this chapter aims to investigate the effects of GCP frequency and spatial distribution on Root Mean Square Error (RMSE) accuracy values. Three multi-rotor UASs were used to collect three aerial image datasets at two locations. The aerial images were processed using Pix4Dmapper Pro, where check points were used to determine the effects of changes in the frequency and spatial location of GCPs. Additionally, qProf – a plugin for QGIS – was used to analyse elevational differences in a variety of generated DSMs. Information deducted from the tests indicated that although there was a strong correlation between a reduction in GCPs and an increase in RMSE, other factors including the proximity to GCPs, their spatial distribution, GSD, and site topography all have a significant influence on RMSE values.

Chapter 18 – Launch and Recovery System for Improved Fixed-Wing UAV Deployment in Complex Environments (Alastair Skitmore)

This chapter details the design and implementation of an inexpensive fixed-wing UAV launch and recovery system, using a low-cost multi-rotor UAV as a versatile airlift and retrieval platform. A fixed-wing UAV is successfully airlifted to a pre-determined deployment altitude and launched, eliminating the need for a runway or costly rail launch system. The fixed-wing UAV is subsequently recovered by the multi-rotor in mid-air, using a net-based recovery approach. The finalised launch and recovery mechanisms are extremely inexpensive (~£40), easy to manufacture using widespread 3D printing methods, and readily adaptable to a number of multi-rotor platforms, allowing researchers to make use of their existing systems. Crucially, the developed system completely alleviates the need for a flat, open area to operate a fixed-wing UAV. Instead, the UAV is launched and recovered at altitude, clear of any obstacles on the ground. The system therefore has particular relevance to UAV operations in complex environments, such as areas of rugged terrain or dense vegetation, where the lack of a suitable take-off and landing space has traditionally prevented the use of fixed-wing UAVs. Finally, since the fixed-wing UAV is launched and recovered using the multi-rotor airlifter, its design can be optimised specifically for cruise flight, rather than take-off and landing, improving its range and endurance performance.

Chapter 19 – Epilogue (David R. Green, Billy J. Gregory, and Alex R. Karachok)

A few thoughts about the future of drones and applications are explored.

ACKNOWLEDGEMENTS

We would like to express our appreciation to all of the contributors to this volume who have made it possible to put this book together.

In particular, I would like to thank Ed Wedler who seeded my enthusiasm for this whole area of remote sensing with model aircraft; to Professor Jap van der Eyk (now deceased) and Roger Kirby (University of Edinburgh – Retired) for introducing me to aerial photography, remote sensing, and photogrammetry; to Professor David Rhind for his inspiration for geospatial technologies over the years and for kindly offering to write the foreword to this book; to Robert Wright, Mike Wood, and Bill Ritchie at the University of Aberdeen for giving me the opportunity to work at the University and on the MSc in GIS Degree Programme; to Norrie Kerr for his help in restarting an interest in model aircraft; and to many others who have been involved in the past 10 years in providing the basis for us to establish UCEMM (UAV/UAS Centre for Environmental Monitoring and Mapping) and DroneLite. I would also like to express my sincere thanks to my enthusiastic and supportive co-editors who have provided me with the energy and drive to complete this volume, to Billy for endless flights to acquire the imagery for many of our projects, and to Alex for designing the book cover.

REFERENCES

Green, D.R., April 1995. Preserving a fragile environment: integrating technology to study the Ythan Estuary. *Mapping Awareness*. Vol. 9. pp. 28–30.
Wedler, E., 1984. Experience with radio-controlled aircraft in remote sensing applications. *Proceedings 1984 ASP-ARSM Convention*, Washington, DC. March 11th–16th 1984. pp. 44–54.

2 From Radio-Controlled Model Aircraft to Drones

David R. Green
UCEMM – University of Aberdeen

Cristina Gómez
UCEMM – University of Aberdeen and
INIA-Forest Research Centre (CIFOR)

CONTENTS

INTRODUCTION

Both fixed-wing and multi-rotor platforms are increasingly being identified as potentially useful airborne platforms for a range of professional environmental remote sensing applications, including both expensive high-end specialist solutions and lower-cost, off-the-shelf commercial options. Small unmanned aerial vehicles (UAVs) are ideal for applications where aerial coverage requirements are small, flying experience and expertise is limited, and the operating budget is usually relatively small. With advances in battery technology, navigational controls, and payload capacities, many of the smaller UAVs are now capable of utilising several different sensors to collect photographic data, video footage, multispectral, thermal and hyperspectral imagery, as well as light detection and ranging (LIDAR). With the aid of specialist digital image processing (DIP) and soft-copy photogrammetry software, aerial data and imagery can be processed into a number of different products including ortho-photos, mosaics, and Digital Elevation Models (DEMs) and analysed to generate useful information for input to a Geographic Information System (GIS). This chapter begins by examining the history of small-scale aerial (SSA) platforms for aerial image acquisition as the context for the current growth in the popularity of UAVs, the associated technologies, and the range of small-scale sensors (e.g. GoPro Hero cameras) currently available for environmental monitoring, mapping, and modelling applications, together with image processing, analysis, and information extraction and soft-copy photogrammetry software such as Pix4D, AgiSoft, and AirPhotoSE. The past and present advantages and disadvantages of small aerial platforms are also considered, together with some of the emerging technologies and recent developments.

Aerial photography and video imagery have been acquired using small aerial platforms for over 30 years. Small-scale radio-controlled (RC) model aircraft and helicopters, using small 35 mm single-lens reflex (SLR) and video cameras, have all been used to acquire panchromatic, colour, colour infrared (CIR), and multispectral aerial photography for a wide range of environmental applications (Green, 2016).

Whilst not initially viewed as a serious aerial photography platform by many people, developments in miniaturised sensors, camera and battery technology, data storage, and small multi-rotor and fixed-wing aerial platforms known as UAVs in the past 5 years have served to reinvent the potential that such small platforms and sensors have for the low-cost acquisition of a wide range of aerial data and imagery.

Coupled with the development of **R**eady-**T**o-**F**ly (RTF) technology; low-cost digital cameras; multispectral, hyperspectral, thermal, and LIDAR sensors; Global Positioning System (GPS); image processing; and soft-copy photogrammetry software, UAVs now offer a sophisticated means for acquiring many different high-resolution photographic and 4K video datasets for small-area coverage studies.

With the aid of low-cost image processing and soft-copy photogrammetric software, photographic stills can easily be mosaicked and 3D models of the terrain and features constructed.

Beginning with a *reflective* examination of the early SSA RC platforms and cameras as a contextual setting for today's applications, this chapter sets out to explore the current potential of these small-scale platforms, the sensors, and some of the software that now provides the means to capture high-resolution imagery of the environment and to process it into spatial information. This chapter considers some of the current aerial platforms, sensors, developments in the technology and the means by which the data and imagery can now be processed into information. Focusing specifically on coastal applications, this chapter is illustrated with three examples of this airborne technology spanning some 30 years of small-scale aircraft systems that serve to show just how rapidly the technology has developed in recent years and evolved into the sophisticated systems now in use.

SMALL AIRBORNE PLATFORMS

The use of small aircraft to acquire aerial photography is not by any means new. Accounts in the literature, admittedly scattered across many academic journals, magazines, books, and the Internet, much of it not of an academic or scientific origin, reveal that a wide range of different aerial platforms have been successfully used over the years to acquire aerial photography of various different types and formats for many different environmental applications.

Long before the modern-day RTF Unmanned Airborne Vehicles or UAVs became widely available and so popular, there were already many examples of SSA platforms in existence known as RC model aircraft (both fixed-wing and helicopter) that were usually free-ranging, with some being tethered, and for the most part powered by fuel engines. These were generally known as radio-controlled (RC) aircraft or helicopters, whilst others were referred to as small or model aircraft (Green, 2016).

Similar in concept, aside from hobby and aerobatic uses, these platforms were also occasionally flown for aerial photography and were able to carry small cameras that could be triggered remotely to obtain aerial photographic coverage (Wedler, 1984).

Compared to the aerial platforms available today, model aircraft generally required specialist operation and experience (as they were quite difficult to fly), Do It Yourself (DIY) skills, and none had any of the modern attributes such as GPS, directional lights, and RTF capability associated with the modern platforms.

ENVIRONMENTAL REMOTE SENSING

Over the past 30 years, numerous studies have been reported in the literature citing the use of small-scale aircraft for field and environmental research where remotely sensed data needed to be collected. For example, Gregory et al. (1974); British Journal of Photography (1975); Hoer (1976); Lapp and Shea (1976); Kerr (1977); Taylor and Munson (1977); Lemeunier (1978); Miller (1979); Jonsson et al. (1980); Bowie (1981); Ellis et al. (1981); Clark (1982); North East River Purification Board (1982); Syms and Turner (1982); Velligan and Gossett (1982); Tomlins and Lee (1983); Canas and Irwin (1986); NASA (1996); Roustabout (n.d.).

A great deal of commercial interest also existed at this time with many small companies being established to take advantage of a wide range of small-scale RC platforms to acquire low-altitude aerial photography e.g. High Spy in the UK (www.highspy.co.uk).

According to Harding (1989), the earliest model aircraft photography was reportedly taken in 1939, although other researchers have even suggested the even earlier date of 1917. Significantly these are both during wartime periods when experimentation with unmanned aerial reconnaissance was already being undertaken. Wester-Ebbinghaus (1980) used a model helicopter to fly photography suitable for engineering applications. Tomlins (1983) used such platforms to acquire aerial photography for a variety of applications including forestry, pollution detection, wildlife habitat assessment, site mapping, publicity, wildlife inventories, and shoreline mapping. Tomlins and Manore (1984) report the use of aerial photography from model airborne platforms for assessing landfill operations, fisheries habitat, coastal studies, and environmental impact assessment (EIA). Wedler (1984) used a wide range of small platforms to monitor the ambient sediment load at pipeline crossings and to acquire vertical atmospheric temperature profiles. Thurling et al. (n.d.) used model aircraft to acquire crop biomass data and to investigate the extent of weed infestation in agricultural crops. Harding (1989) used model aircraft at Wessex Water to photograph sites for water quality surveys of reservoirs and publicity photographs of sewage treatment works. De Wulf and Goossens (1994) reported on the use of remotely piloted aircraft to acquire CIR photography to study winter wheat yield estimation. Fouche and Booysen (1994) wrote a number of papers about the application of remotely piloted aircraft (RCA) for the detection of nitrogen deficiency in crops and stress detection in tree crops. Mullins (1997) reported on the future development of so-called 'palmtop' planes or Micro Air Vehicles (MAVs) for remote sensing.

Wedler (1984) lists a wider range of environmental applications (Table 2.1). Some commercial scientific and non-scientific applications of the time are also shown in Table 2.2.

TABLE 2.1

A Range of Environmental Applications of SSA (Wedler, 1984)

- Monitoring ice conditions
- Observing coastal erosion and deposition features
- Recording ports and harbour sites
- Monitoring flood conditions at river ice jam sites
- Measuring storm-induced beach erosion/deposition
- Recording of engineering structures and facilities
- Detection of underground utilities
- Quality of rural highway decks
- Older dam sites and river embankment protection
- Preliminary site mapping
- Observations of domesticated farm animals and their environs
- Observations over tree nurseries and experimental agricultural farms
- Tree species identification and tree spacing
- Snowmelt behaviour
- Field drainage patterns

TABLE 2.2

Commercial Applications of SSA

- Commercial sites
- Real-estate site survey
- Company presentation and advertising brochures
- Local government municipal projects
- Local trades
- Legal work and law enforcement
- Construction industry
- Private properties
- Golf courses
- Plane to plane
- Insurance companies, accident sites
- Landscaping companies
- Mortgage companies
- Appraisal companies
- Land management
- Environmental impact assessment (EIA)
- Gifts e.g. aerial photographs of homes

PLATFORM NAMES

The literature available reveals a wide range of different types, sizes, and specifications for small aerial platforms with many different designs, including gliders and helicopters. Other airborne platforms are also mentioned such as airfoils, kites model rockets, and balloons (e.g. Johnson, 1978; Kendall and Clark, 1979; Levanon, 1979; Deuze et al., 1989; Marks, 1989). Whilst these platforms have been successfully used for many different applications over the years, they will not, however, be discussed further in this chapter.

The names given to some of these small platforms vary quite considerably in the literature. Some are clearly just different names for essentially the same thing, whilst others serve to distinguish between the platforms based on their size e.g. wingspan, or certain other characteristics e.g. the payload. The most common names used are shown in Table 2.3. A useful definition provided by Wedler (1984, p. 44) also states:

> Radio-controlled aircraft (RCA) … are small, retrievable aircraft, are remotely controlled directly and/ or indirectly, are built to carry a camera pay-load, and operate under specific photo mission requirements. They are not reduced-scale models, as in "model aircraft", but are full-size 1:1 scale aircraft (albeit small) with their own design and construction and flight characteristics.

On the whole, the terms RPA (remotely piloted aircraft), RPC (remotely piloted craft), RPV (remotely piloted vehicle), and photo-drone appear to correspond to the larger examples of these small aircraft. Photo-drones are generally the largest, with very large wingspans, and include more sophisticated and expensive RC platforms designed to carry large cameras and payloads. According to Wedler (1984), these examples also belong to one of the four classes of drones: target drones, harassment or decoy drones, weapon-carrier drones, and reconnaissance or surveillance drones and have largely been associated solely with military applications.

Examination of some example specifications also seems to suggest that, on the whole, the word 'model' (as in model aircraft) applies to the very small versions of these aircraft, with a small wingspan, the sort that are typically flown by model aircraft enthusiasts or hobbyists. These platforms are capable of carrying a small camera strapped to the underside of the aircraft. The word *model*, however, is perhaps an unfortunate label to have become associated with many of these aircraft; whilst they are a *model* in the sense that they are sometimes, but not always, small-scale replicas, they are far from being a model in many other respects. The authors of this chapter therefore consider the names 'toy' and 'model' inappropriate for use since they firstly provide an incorrect impression of the potential of these platforms, tending to conjure up an amateurish rather than a professional role,

---

TABLE 2.3

Common Names Given to SSA Platforms

- Toy aircraft
- Model aircraft
- Small-scale aircraft
- Small-scale model aircraft
- RCA (remotely controlled aircraft)
- RPV (remotely piloted vehicle)
- RPA (remotely piloted aircraft)
- Photo-drones
- Reconnaissance mini-drone
- Surveillance mini-drone
- Radio-controlled aircraft (RCA)
- Unmanned aerial or airborne vehicle (UAV)

---

and secondly are usually associated with hobbyists, where the aircraft is only capable of carrying a relatively small payload. Furthermore, by using such a heading, these platforms are often not considered to be a serious contender for aerial photographic data acquisition.

Alternative terminology often used is 'Small-Scale Aircraft' (SSA). This is perhaps the most appropriate since it helps to dispel the idea of the plane being a model and is one that is generally equally applicable to almost all of these aircraft, referring only to the size. SSA therefore seems to be the most appropriate terminology since it suggests that they are 'scaled-down' versions of a larger aircraft and will therefore be referred to as such in the remaining relevant parts of this chapter.

TYPES OF PLATFORMS

SSA fall into two distinct categories, the 'fixed-wing' and the 'rotary-wing', the latter more commonly being known as helicopters.

FIXED-WING

Fixed-wing platforms have been subdivided into high-wing monoplanes or biplanes, with both powered and power/powerless categories; the latter better known as 'gliders'. In practice, fixed-wing small-scale aircraft have tended to find more favour for applications than rotary wing in the past. The reason for this is that monoplanes have good stability and a slow flying capability. Biplanes, however, although more complicated to construct and more susceptible to damage, offer advantages of an increased wing area and therefore an increased capacity to carry equipment and to fly slowly. Seldom mentioned in the literature, gliders have also found uses where there is a need for quiet operation e.g. wildlife monitoring and habitat surveys, but they generally offer less of a practical solution than a powered plane.

ROTARY-WING

Helicopters appear to have been far less popular in the past because they were often more costly to buy, more difficult to fly, suffered from vibration, were unstable, had low-payload capacities, and, if they suffered from engine failure, could easily fall from the sky with the inevitable resultant loss of both the platform and the camera equipment. However, a quick search using *model and aircraft* as the search words revealed a surprising number of commercial companies using helicopters for aerial work. Benefits often cited were the ability to hover and the different photographic perspectives possible. The problem of vibration from the main rotor, tail rotor, and engine also seemed to be something of the past, being solved through the use of damping systems, and the larger ones can be flown in higher wind conditions and are more stable (personal communication and experience – Borich Aircams – www.borichaircams.co.uk).

CONSTRUCTION

SSA for aerial photography often originated as a standard model kit, purchased from a model hobby shop or alternatively were purpose built. Some kit aircraft had modifications to accommodate a camera mounting, with strengthened landing gear, and more powerful engines. More commonly, model enthusiasts were often responsible for custom-built designs specifically to carry out the aerial tasks.

Typically, wingspans for fixed-wing platforms used in the UK were between 1.8 and 3.7 m (Thurling, 1987). According to Thurling (1987), approximately one-third of the designs available included ailerons (a flap hinged to the trailing edge of the aircraft wing to provide lateral control as in a bank or roll) in addition to an elevator (control surface on the tail-plane of an aircraft to allow it to climb or descend) and rudder controls (a vertical control surface attached to the rear of the aircraft to steer it in conjunction with ailerons), and most included flaps (a movable surface attached to the trailing edge of an aircraft wing that increases lift during take-off and drag during landing).

Undercarriage configurations were usually of the tricycle-type, with two rubber wheels, plus either a nose wheel (fixed or movable) or a tail dragger. On the whole, the undercarriage had to be both flexible and forgiving, whilst also being very robust. Getting a fixed-wing SSA airborne was much easier to achieve than getting it down again on a level surface using an undercarriage that is strong enough to take the impact.

Most fixed-wing SSA used removable wings, tail-plane, and undercarriage for transportability. Elastic bands or nylon bolts were also used for fixings e.g. the wings, because they 'give' or 'break' on heavy impact and, in the event of an accident, minimise the damage to the rest of the aircraft and to the camera. Many of the SSA were constructed of an open balsa wood frame with plywood reinforcement and covered with lightweight nylon fabric. Some also used expanded polystyrene wings, which had the advantage of being easily repaired but the disadvantage of being heavier and easier to break. GRP (Glass Reinforced Plastic) construction was also used, as it had to be both strong and provide protection against damage but with the added penalty of weight.

Whilst there were a number of kit-based helicopters also available for purchase, in practice, most used for flying aerial photography appear to have been specialist equipment. Hi-Cam (www.hicam. com) in Australia cited a 1.5 m long platform flying at 0–300 feet, complete with a film camera, and a micro-video camera with 2.5 GHz microwave downlink for monitoring and flying. In addition, this platform also carried a mini-DV digital camcorder. Most helicopters available were made of lightweight materials using a combination of plastic, carbon fibre, and glass fibre.

ENGINES

Engine requirements varied markedly according to the size of the plane, the wing area, and the operating circumstances. Originally, most were either two- or four-stroke engines. For small planes, the two-stroke examples were powered by a methanol/castor oil mixture whilst the larger planes usually required a petrol engine. The position of the engine mounting on the plane also varied with options to be mounted at the front, the rear, or on top of the wings. Placement on the wing top or at the rear of the aircraft avoided the problem of oily exhaust fumes being blown back onto the camera lens. Mounting the engine at the front, however, required a pipe to take the oily mixture from the engine exhaust behind the camera. Most early helicopters also utilised gas engines. In all cases, to help avoid vibrations from the engine being transmitted to the camera, it was usual to rubber-mount the engine and pack the camera with plenty of latex foam.

OPERATIONAL CONSIDERATIONS

Practical operation of RC aircraft was well documented by Buckle (1985). Launching of SSA depended very much on their size and the environment in which they were to be used. Systems were designed to be launched by hand whereby the operator or operator's assistant supported the aircraft whilst it was prepared for take-off and at the appropriate time it was thrown gently upwards and forwards. Others were power-launched from flat ground e.g. roads, tracks, or a grass surface where available; the roof of a car or van; from water using small pontoons attached to the undercarriage; and from snow or ice with the aid of skis (Wedler, 1984). Some were even launched with the aid of a catapult and ramp. Providing the undercarriage was fairly robust and the wheels large enough, it was also possible to use ground that was not too densely vegetated and undulating as a runway.

Various opinions exist as to how difficult these aircraft were to launch and subsequently to fly and land. Discussions with experienced pilots suggested that considerable specialist experience and skills were required for overall control and operation; in many ways, flying a small-scale aircraft was actually considered much harder than flying the real thing! Certainly, the fixed-wing aircraft were easier to work with relative to the rotary-wing (helicopter), but both needed plenty of practice and experience. Given the cost of camera equipment and the platform then, it was advisable, perhaps even vital, to seek the advice and assistance of the expert.

Getting the aircraft off the ground often proved to be far less of a problem than getting it down again safely and especially in one piece. It was often advisable when launching the aircraft to ensure that the runway was level and had sufficient length for take-off. It was also advisable to take off into the wind and to avoid strong sidewinds. Landing the aircraft was sometimes more difficult, and it was usually best achieved in tall grass or in a bush to cushion the landing and to protect the camera that should in any case be well recessed into the underside of the fuselage. A hard landing, not necessarily in the exact location planned, often led to the aircraft and camera being badly damaged and even being written off, although a carefully designed aircraft would usually survive a crash landing quite well in practice. Windy conditions can also be quite hazardous for landing and take-off, as well as attempting to keep the camera vertical during the flight.

Some enthusiasts (e.g. Wedler, 1984) have also reported the use of waterborne aircraft mounted on floats. These presumably posed a completely different set of operational problems, mostly relating to ensuring that the aircraft can be retrieved and also that the camera and film remained dry.

Maintenance is also an essential part of successful operation, and in this context, it was frequently noted that inspection of the airframe regularly was required to maintain the aircraft and engine carefully and to avoid any problems. Nearly all operators also carried many spares including a second aircraft and camera. Some operators also reported the inclusion of a parachute or alternative recovery systems to prevent possible damage prior to landing or should a failure occur.

POSITIONING

The addition of a camera to an SSA's normal payload could prove a handful even for the most competent SSA pilot, as particular care was needed when launching, landing, and during low-level slow flying of the aircraft, judging distances, flight altitude, and attitude, especially when flying over long distances was equally difficult.

One of the problems identified when flying a SSA is positioning it at the desired height and judging its flight attitude; the yaw, pitch, and roll. For the experienced flyer, this was far less of a problem. Commercial operators suggested that image acquisition required both a pilot and a photographer to ensure correct locational positioning of the airborne platform and ultimately to aid in the acquisition of good aerial photography. Some made use of a GPS on helicopter platforms as well as a video downlink to provide a 'virtual experience' of being in the air. A number of alternatives for height estimation were also available including the use of an altimeter (unfortunately not always cost-effective for low-cost operations), painting a series of black bars or stripes on the underside of the wings that could be resolved at various distances by sighting and guesswork, by using optical range finders, or mathematically using trigonometry.

The problems associated with the attitude of the platform in relation to the position of the camera lens were often compensated for by mounting the camera in gimbals which ensured that the lens was always pointing directly downwards. Determining the field of view (FOV), a related problem, was also equally difficult. Once again, the experienced operator was able to ascertain fairly accurately when the plane was correctly positioned. Alternatives were to mount a video camera on the aircraft together with a downlink to a monitor (Fagerlund and Gunnershed, 1975) to provide the pilot with a visual positioning of the aircraft. However, this also was not always practical because of the cost of the additional equipment, the requirement for the video downlink, the size of the aircraft, and its overall potential payload limitations.

A similar problem was that of being able to judge how far away the aircraft was from the operator, and whether or not the camera was covering the desired area. Various solutions were proposed. One was to fly the area in stages, using colleagues positioned in the field to signal when the aircraft was overhead. Another alternative, suggested by Wedler (1984), allowing long-distance operation of up to 15 km, was to place the operator in a pick-up truck travelling along the flight line. This was only practical, however, in certain terrain that was accessible to a vehicle by road and not too rough.

Particular care needed to be exercised in some areas to avoid other low-flying aircraft. In highland areas of Britain, it is vital to contact the relevant authorities to establish the flight plans of military jets prior to planning a flight. Part of the problem for SSA pilots is that it is difficult to gain an idea of the spatial position (X, Y, Z) of the small-scale aircraft relative to another airborne vehicle at a distance from the operator. To improve the potential visibility of the small-scale aircraft, solutions such as painting the small-scale aircraft in a bright colour e.g. fluorescent green/orange have been used.

Considerable care has also to be exercised when flying these aircraft to avoid other people, power cables, as even aircraft of this size can be *lethal* and can cause quite a lot of damage in a collision, and a comprehensive insurance policy is important. In some areas, it is not permitted to fly e.g. near airports and because of the noise factor. To ensure complete and successful control of the aircraft, radio transmitters must be free from interference as loss of control can also be a possibility because of the dependency on a radio signal for control. Similar issues can arise when flying near to model aircraft clubs if other people are flying on the same frequency.

FLYING REGULATIONS

In most countries, there have been flight regulations in place for SSA for some time. In the UK, for example, the Civil Aviation Authority (CAA) regulations restricted the total flying weight of an SSA to 7 kg and below (various communications with the CAA, 1988/1989). Most, if not all, countries also had some legislation in place.

PHOTOGRAPHY

Photographic scales acquired, using flying altitudes of between 30 and 300 m, have been used to provide contact scales of between 1:800 and 1:70,000 (Wedler, 1984). Both panchromatic and colour films were often used, for either oblique or vertical photographs, most frequently taken with a film speed of 100 ASA (American Standards Association). Most of the cameras used were 35 mm SLR and Polaroid cameras, although some researchers have reported the use of 8 mm movie cameras and small video systems. A typical camera system would have been a Contax 137D 35 mm camera with an electric film wind and a 50 mm f1.7 Zeiss Planar lens (Boddington, 1989). Another would have been a Konica FS-1/FT-1 fitted with a Konica 22 or 40 mm lens (Boddington, 1989; Harding, 1989). Shutter speeds of 1/500 second or faster were generally used (Boddington, 1989). Wester-Ebbinghaus (1980) used both Hasselblad and Rolleiflex 60×60 format cameras. Others report the use of Kodak 110 cameras (Harding, 1989; Wedler, 1984). Researchers have also mentioned film speeds of 400 ISO using a 28 mm lens at f/5.6, with the use of a Kodak Portra 400 VC colour negative film because of the fine grain structure. By being closer to the 'subject', a telephoto lens was not needed which allowed slower shutter speeds and more light to the film and slow, fine grain, high-contrast film to be used.

ADVANTAGES

SSA were soon shown by a number of researchers e.g. Wedler (1984) to have a number of distinct advantages over more conventional light aircraft, autogyros, and micro-lights, the platforms most often used for acquiring small-format aerial photography. Beyond the low cost of these small platforms and the imagery, the operation and maintenance was comparatively cheap, all of which were deemed important for small-area studies. Another advantage was that the photography could be acquired at virtually any time (except under extreme weather conditions e.g. high wind and rain). The availability of a small, easily transported aircraft also provided relative freedom for image acquisition; no booking was needed for flying time, the aircraft, or the risk of flight cancellation due to poor weather conditions. Flying regulations for light aircraft also prevent low-altitude flying (<200 m) for the acquisition of large-scale photography, something that was easily achievable with an SSA. Furthermore, when special types of aerial photography, taken at one or more specific times of the day/year, are needed, the SSA platform was ideal as it allowed complete flexibility. In addition, taking aerial photography at a lower altitude reduced the effects of atmospheric haze and improved the clarity of the imagery.

TECHNOLOGICAL DEVELOPMENTS

Interestingly, towards the end of the initial popularity of SSA for aerial work, Warner et al. (1996) mentioned rapid technological advances in lightweight optics, electronic equipment, strong composite materials, high-performance engines, and digital cameras as some of the later developments in SSA that were considered to greatly improve their future potential for aerial imagery. In addition, some operational examples started to use video cameras to aid in SSA navigation, and there was growing use of small videocams on helicopter platforms.

TODAY

Reflecting briefly on the SSA platforms and sensors used in the past for aerial photography provides a very useful and important contextual setting for the current platforms that have become so popular today. Whilst commonly seen in the past as *toys for the boys* with little professional and commercial role for them in a serious aerial data acquisition role, it is clear from the relatively few published examples cited in the literature, that many of the original SSA were in fact very sophisticated and practical airborne data acquisition platforms even then that utilised, what were at the time, very *up-to-date* technologies to fly the aircraft and to acquire the aerial imagery. Although all of these technologies have all clearly evolved considerably since then, providing considerable improvements in the opportunities to acquire and process imagery, interestingly many of the practical, operational, and flying constraints today are still remarkably similar to those faced in the past, albeit perhaps now more relevant than ever because of the rapid growth in the end-user community now able to access and use RTF platforms, the lower costs associated with these technologies, and the ease with which the data and imagery acquired can now be processed into information.

Today, unmanned systems are still associated with a host of terms in the literature and the media: Unmanned Aerial Systems (UAS), drones, RPA, Unmanned Vehicle Systems (UVS), and unmanned airborne or aerial vehicles (UAVs) all reflecting the variety of system configurations and fields of application in use. Different sources use UAV or UAS as the preferred term, and although UAV is the term adopted by the UK CAA, others suggest that UAS is more correct.

An UAV is flown without a pilot on board and is either remotely and fully controlled from another place (e.g. ground, another aircraft, space) or programmed and fully autonomous (ICAO, 2011). A UAV comprises the flying platform – an aircraft designed to operate without human pilot on board, the elements necessary to enable and control its navigation, including taxiing, take-off, and launch; flight and recovery/landing; and the elements needed to accomplish mission objectives: sensors and equipment for data acquisition and transfer of data, including devices for precise location when necessary.

As noted earlier, aerial and remotely controlled systems for surveillance and the acquisition of earth surface data have a relatively long history, many typically originating with military activities. Photogrammetry and remote sensing technologies identified the potential of UAV-sourced imagery acquired at low altitudes with high spatial resolution, more than 30 years ago (Colomina and Molina, 2014). However, civilian research on UAVs only began in the 1990s (Skrypietz, 2012). Currently, the rapid emergence of UAVs in many civilian applications has once again raised awareness of the vast potential of these aerial systems.

In the UK, UAVs are usually classified by their size and weight, from small and lightweight (less than 2.7 kg) with a relatively short distance range, up to systems with more than 20,000 km range and weight of approximately 12,000 kg. To date, only small platforms (<20 kg) can be used for civilian applications in the UK.

MULTI-ROTOR UAV

One category of UAV currently available, and perhaps the most popular for recreational flying and latterly aerial photography and video work, has been the multi-rotor UAV platform (Figure 2.1). Multi-rotors usually have four, six, or eight rotors powered by electric motors. Similar in many ways to single-rotor helicopter platforms, that are still also widely available, multi-rotors now provide a more stable aerial platform for a range of cameras, including cameras on mobile phones, the GoPro Hero series, and many DSLR (digital single-lens reflex) cameras e.g. Panasonic GH4. Popular examples of multi-rotors include the DJI Phantom and DJI Inspire (Quadcopter – 4R), the DJI S900 (Hexacopter – 6R), and the DJI S1000 (Octacopter – 8R) (www.dji.com).

As this technology has developed, so too has the range of platforms from toy to specialist and from small to large. Developments in the technology are now also providing nanodrones, miniature UAVs able to carry small still and video cameras. The palm-size Micro Drone 2 (http://www.micro-drone.co.uk) weighs 0.034 kg and has a flying range of 120 m and an endurance of 6–8 minutes. Other small drones can now be flown as tethered aerial vehicles to circumvent the risks associated with free flying; e.g. the Pocket Flyer by CyPhy Works (http://cyphyworks.com/pocket-flyer) is a 0.080 kg tethered platform that can fly continuously for 2 hours or more.

Compared to early SSA, modern rotary-wing aircraft have very complex mechanics to allow them to be flown at low speeds. Among their main strengths, rotary-wing UAVs can fly vertically, take-off and land in a very small space, and hover over a fixed position and at a given height. This makes rotary-wing UAVs well suited for applications that require manoeuvring in tight spaces and the ability to focus on a single target for extended periods (e.g. inspections). Disadvantages of rotary-wing UAVs are that they can be less stable than fixed-wing aircraft under some conditions and also more difficult to control during flight. Single-rotor and co-axial rotor platforms (with two counter-rotating rotors on the same axis) are very similar to conventional helicopters, with a single lifting rotor and two or more blades. These platforms maintain directional control by varying blade pitch via a servo-actuated mechanical linkage. Single-rotor and co-axial rotor UAVs are typically RC and powered by electric motors, although some of the heaviest examples use petrol engines.

Multi-copters have an even number of rotors and utilise differential thrust management of the independent motor units to provide lift and directional control. As a general rule, the more rotors, the higher the payload they can take and are functional in strong wind conditions, as the redundant

FIGURE 2.1   A multi-rotor UAV.

lift capacity provides for increased safety, and more control in the event of a rotor malfunction or failure. Rotary-wing UAVs are commonly used to capture oblique aerial photographs and video and may be used for mapping tasks. Some commercially available vehicles are also equipped with GPS/IMU sub-systems and are capable of autonomous flights that significantly improve the capability to undertake repeat aerial video and photography to cover the ground in a systematic manner for mapping applications.

FIXED-WING UAV

Fixed-wing UAVs are an alternative to multi-rotor platforms and have some distinct advantages and disadvantages. Whilst larger and with a capacity to carry a large payload, fixed-wing aircraft are able to carry out aerial sorties for a longer duration of time and can cover larger areas under autonomous flight. Disadvantages include complicated launch and retrieval requirements and limitations to linear and areal coverage. Fixed-wing UAVs are characterised by a relatively simple structure, making them reasonably stable platforms that are relatively easy to control during autonomous flights. Their efficient aerodynamics enables longer flight duration and higher speeds. This makes fixed-wing UAVs ideal for applications such as aerial survey which require the capture of georeferenced imagery over large areas. On the down side, fixed-wing UAVs need to fly forward continuously and need space to both turn and land. These platforms are also dependent on a launcher (person or mechanical) or a runway to facilitate take-off and landing, which can have implications on the type of payloads they carry. Typical lightweight fixed-wing current commercial platforms have a flying wing design (Figure 2.2) with wings spanning between 0.8 and 1.2 m and a very small fin at both ends of the wing. In-house vehicles tend to have slightly longer wings to enable carrying the required heavier sensors (Petrie, 2013). A second type of design is the conventional fuselage. The dimensions are around 1.2–1.4 m length for the fuselage and 1.6–2.8 m wing length. In the UK, there are around 20 companies operating commercial airborne imaging services using fixed-wing UAVs (Petrie, 2013).

COMBINED MULTI-ROTOR AND FIXED-WING PLATFORMS

A relatively recent development has been that of an aerial platform that combines the best characteristics of a rotary-wing (take-off and landing, hovering) and a fixed-wing platform (long-distance flight). An example of this technology is the VTOL Flying Wing (http://www.vtol-technologies.com).

FIGURE 2.2   A fixed-wing UAV.

SENSORS

Rapid developments in digital and microprocessor technology have led to a wide range of low-cost and miniaturised digital cameras and sensors that are suitable for mounting on UAVs. These include panchromatic, colour, and CIR, multispectral, hyperspectral, thermal, and LIDAR sensors. This also includes the possibility to make use of digital cameras and attachments on mobile phones. Whilst not all sensors have been developed specifically for UAVs, as UAV technology has evolved so too there has been a demand for specialist sensors for various applications. Different sensors are limited to different platforms by the payload and lift capacity of the aerial platform. A growing range of both passive and active sensors measuring naturally occurring radiation reflected or emitted by the target objects are becoming available. Some are off-the-shelf whilst others are now being developed specifically for UAVs.

PASSIVE SENSORS

Passive optical sensors measure radiation in the visible (0.4–0.7 μm) and infrared (IR) (0.7–14 μm) part of the electromagnetic spectrum and rely on the sun as the illumination source, which makes them only suitable in daylight conditions. They are also limited by atmospheric effects such as clouds, haze, or smoke. There are an increasing number of optical sensors now available for UAVs, ranging from small cameras capable of still photography and video (e.g. GoPro series and iLook (http://www.walkera.com/index.php/Goods/info/id/37.html) cameras), to both small and large DSLR cameras, stereo cameras, multispectral and hyperspectral cameras, high-resolution cameras on smartphones, or low-cost developments like the HackHD camera (www.hackhd.com). The range of opportunities provided by optical sensors is constrained by various issues concerning the digital frame cameras that can be deployed on lightweight UAVs. These include the camera weight relative to the available UAV payload; the very small format of the camera images; the numerous non-metric characteristics of many of the lower-cost cameras; the lenses and resolution; photographic intervals; the need for very short exposure times to help combat the effects of platform instability (i.e. due to speed, roll, pitch, and yaw); the requirements for high framing rates arising from the speed of the UAV platform over the ground from a very low altitude; and the very large longitudinal and lateral overlaps (i.e. percentage endlap and sidelap) that need to be employed for mapping purposes.

Multispectral and Near Infrared (NIR)

Multispectral imagery is produced by sensors that measure reflected energy within several specific bands of the electromagnetic spectrum. They usually have three or more different band measurements in each pixel of the images they produce. Examples of bands in these sensors typically include visible green, visible red, and near infrared (NIR). Simultaneous measurement of multiple spectral wavelengths provides information that can be visually or automatically interpreted. For a given location, algebraic combinations of values in various spectral wave bands can be very useful to aid in the detection of environmental features; e.g. multispectral imaging of vegetation is very useful in the identification of plant stress, disease, and nutrient or water status. Marcus UAV (http://www. marcusuav.com) has prototyped a custom payload manifold for the housing of a Tetracam ADC Lite MS (http://www.tetracam.com/Products-ADC_Lite.htm) camera system on a UAV. Using IR camera images collected at specific time intervals, overlapped and aligned, and some spectral filtering software (e.g. Pixel Wrench (http://www.pixelwrench.co.uk)), NDVI (Normalised Difference Vegetation Index) images can be derived to provide information about plant condition and status.

Short-Wave Infrared (SWIR)

Radiation in the short-wave infrared (SWIR) (typically 0.9–1.7 μm) is not visible to the human eye but can be sensed by dedicated indium gallium arsenide (InGaAs) sensors. Images from an InGaAs camera are comparable to visible images in their resolution and detail, making objects easily recognisable (as opposed to thermal imagery). One of the main benefits of SWIR imaging is its low power consumption, as it uses a thermoelectric cooler or no cooler if the dark current is low enough, while still providing good-enough imagery in low light conditions. InGaAs sensors can be made extremely sensitive, literally counting individual photons. Thus, when built as focal plane arrays with thousands or millions of tiny sensor pixels, SWIR cameras will work in low light conditions.

Only a few commercial companies are making SWIR cameras and even fewer are making the detector material, indium gallium arsenide (InGaAs). Sensors Unlimited (www.sensorsinc.com) and Teledyne Judson (www.judsontechnologies.com) are the only two US developers of SWIR technology, subject to strongly controlled exporting regulations. Xenics (www.stemmer-imaging. co.uk) in Belgium, Allied Vision Technologies (www.alliedvision.com) in Germany, and Chunghwa (www.leadinglight.com.tw) in Taiwan are other providers. Sensors Unlimited – UTC Aerospace Systems (utcaerospacesystems.com) – has introduced the smallest SWaP (size weight and power) SWIR camera for unmanned vehicles. The 640×512 pixel (25 µm pitch) camera weighs 27 g. The $25.4 \times 25.4 \times 25.3 \, mm^3$ total volume allows it to easily fit on board most UAS or unmanned ground vehicle systems (UGS).

Hyperspectral

Hyperspectral imaging samples a wide variety of bandwidths in the light spectrum to provide a rich dataset and detect objects of interest not visible to single-bandwidth imaging sensors. With a larger number of fine spectral bandwidths, the identification of specific conditions and characteristics are greater. Sensors with hundreds of bands (e.g. 255) are increasingly being used for many applications. Developers of hyperspectral sensors provide flexible and customisable options for the number and resolution of spectral bands in the visible and IR (e.g. Rikola Ltd. (www.rikola.fi)). Headwall Photonics (www.headwallphotonics.com) e.g. now specialises in hyperspectral imaging sensors that are small and rugged enough to fit on relatively small UAVs. Adding hyperspectral imaging as a standard element in an electro-optical (EO) and IR sensor suite for UAVs has, however, until recently has presented a difficult engineering challenge.

Thermal

Some surfaces and features have been found to show up well in thermal infrared (TIR) imagery because of temperature differences between the different surfaces. Lightweight thermal cameras have been adapted or specifically developed for use in UAVs. The FLIR Quark 640 (www.flir-tau-buy.com/product/flir-quark-2-thermal-camera/) long-wave IR (7.5–13.5 µm) thermal sensor was incorporated by Sky-Watch (Denmark) into one of its drones. FLIR Quark 640 is a very small ($22 \times 22 \times 12 \, mm^3$ without lenses) sensor with a flexible configuration of lenses (6–35 mm), scene range from −40°C to 160°C, and sensitivity of 50 mK. Its weight depends on configuration but is less than 30 g. Tamarisk 320 developed by DRS Technology (www.drsinfrared.com) is a long-wave IR (8–14 µm) thermal sensor of similar size (~$30 \times 30 \times 30 \, mm^3$). It can also be configured with lenses (7–35 mm) and has a scene range from −40°C to 67°C and sensitivity of 50 mK. Its weight depends on the configuration, ranging from 30 to 135 g.

Fluorescence

Fluorescence spectroscopy has also proven to be useful for some applications. Estimates of fluorescence (F) can be derived from multispectral and hyperspectral radiance sensors, exploiting the Fraunhofer line and decoupling F from the reflected flux. Furthermore, optical indices related to F can be derived from reflectance sensed by multispectral sensors. However, the quantitative estimation of F from the air is complicated by the absorption of the atmosphere en route to the sensor, and approaches to deal with atmospheric effects have yet to be developed (Meroni et al., 2009). For the estimation of F, very high spectral resolution (0.05–0.1 nm) sensors are recommended (Meroni et al., 2009) to allow resampling and application of estimation methods, from multispectral to hyperspectral.

Active Sensors

Active sensors emit radiation and measure the fraction reflected by the target objects as well as the difference in the time between emission and reception. Active sensors require power supplied by a source that inevitably adds some considerable weight to the aerial system. For this reason, active equipment is less versatile for use on UAVs when compared with passive equipment. Radar Synthetic Aperture Radar (SAR) is a type of radar using relative motion between an antenna and its target region, to provide distinctive coherent-signal variations. SAR pulses radio waves at wavelengths of 0.002–1 m repeatedly towards a target region. The many echo waveforms received successively at the different antenna positions are coherently detected and stored and then post-processed together to resolve elements in an image of the target region. SAR is used for a wide variety of environmental applications. Oil spills in the ocean or water bodies can be detected using SAR imagery because the oil changes the backscatter characteristics of the water. As with thermal imagery, differential imaging is necessary for the detection of oil leaks. All-weather, night and day capacity for data collection makes radar technology appealing and convenient for surveillance in difficult environments.

Radar

Radar systems have been integrated into large UAV systems by the military, but there is still a need for small, low-cost, high-resolution radar systems specifically designed for operation on small UAVs. Brigham Young University (BYU) in Utah, USA, has developed some compact, low-cost, low-power SAR systems, including a series of microSAR systems designed for operation on small UAVs. The microSAR design represents a trade-off between coverage and precision versus cost and size. It is an ultra-low-power system (16 W) designed for operation on a UAV with ~2 m wingspan. The system records data continuously for over an hour on a pair of compact flash disks then loaded onto a laptop for processing data into images using SAR image formation and auto-focusing software. Image downlink capacity and real-time processing are being developed. The microSAR system consists of a stack of circuit boards ($7 \times 8.5 \times 7 \text{cm}^3$) and two flat microstrip antennas ($0.1 \times 0.5 \text{m}^2$). Minimal enclosures reduce the flight weight to less than 1 kg. Unlike conventional SAR in which short pulses are transmitted and received separated by an interval, microSAR transmission and reception occur simultaneously via continuous-wave linear-frequency modulation, enabling low-power operation. To optimise performance, microSAR uses bi-static operation, in which transmission and receipt occur via different antennas. Designed for operation at 130–800 m height and speeds of 20–50 m/s, microSAR has a swath width of 200–900 m with a nominal one-look spatial resolution of $0.1 \times 0.6 \text{m}^2$, which is multi-look averaged to $1 \times 1 \text{m}^2$ in processed imagery. The averaging reduces the *speckle noise* inherent in SAR images. The first microSAR operated in the C-band (5.56 GHz), but microSAR systems using other bands have also been built.

ImSAR (Utah, USA) has developed a NanoSAR (http://www.imsar.com/pages/products.php?name=nanosar) series improving from the first NanoSAR-A (0.5 m resolution operating at 500 m height) to NanoSAR-C (0.3 m resolution at 2000 m height). NanoSAR-C weighs less than 1 kg. ImSAR radar has printed circuit board technology in place of the heavy metal tubes that serve as radio wave guides in standard SARs. Etched on fiberglass boards, the radar circuits are similar to the lightweight circuits used in laptop computers and cell phones. NanoSAR-C has a range of 1–16 km and power consumption between 25 and 70 W.

LIDAR

Over the last few years, airborne LIDAR surveys have become popular for many applications. Typically, LIDAR surveys are carried out using manned aircraft flying at 1500 m or above. At this height, they are susceptible to atmospheric conditions and have poor image resolution or a small image footprint. In general, the main constraint to the use of LIDAR on a UAV has been the size and weight of the sensor. UAV imaging systems are generally much lighter than an equivalent LIDAR system. This is due in part to two factors: LIDAR sensors are not nearly as small and light as cameras and a LIDAR system depends entirely upon an accurate inertial navigation system (INS). The imaging system in contrast uses traditional photogrammetry and does not require any INS. As a result, UAV imaging systems can cover a larger area in less time at less initial expense.

Riegl (www.riegl.com/products/unmanned-scanning/ricopter) has developed a LIDAR system for UAVs, the VUX-1, beaming 500k shots per second of NIR radiation; it has an estimated accuracy of 10 mm. Its dimensions are $22.7 \times 18.0 \times 12.5 \, cm^3$ and weighs 3.6 kg. The RIEGL VUX-1 is designed to be mounted in any orientation and has a 330° FOV. Its maximum range is 900 m. Yellowscan has developed a LIDAR sensor for UAVs with 2.2 kg beaming 800k shots per second with multi-echo technology (three echoes per shot) in the 905 nm wavelength. It has power for 2 hours of autonomy, a maximum range of 100–150 m depending on conditions and a 100° FOV. Velodyne (velodynelidar.com/vlp-16.html) has developed a new HDL-32E LIDAR sensor, with a scanning rate of 700k shots per second. At less than 1.5 kg and smaller than $15 \times 9 \, cm^2$, this LIDAR scanner is therefore ideal for UAV applications. Velodyne has also announced a new LIDAR sensor with a scanning rate of 300k shots per second, 0.6 kg of weight, and dimensions $10 \times 6.5 \, cm^2$. Both LIDAR systems work with radiation of 905 nm and have a range of ~100 m. Hokuyo (www.hokuyo-aut.jp/02sensor/07scanner/utm_30lx.html) has developed a number of small, lightweight LIDAR devices for UAVs, e.g. UTM-30LX. The UTM-30LX is $60 \times 60 \times 87 \, mm^3$ size and 0.37 kg, has a scanning rate of 40k shots per second, and a range up to 30 m. Interface to the sensor is through USB 2.0 with an additional synchronous data line to indicate a full sweep. The FOV is 270°, and a 12 V power source is required. Characteristics such as range, flying height, and power consumption are trade-offs to consider for specific applications and platforms systems.

To accommodate LIDAR sensors on UAVs, Phoenix Aerial has recently developed the Scout, designed to host Velodyne LIDAR sensors and a combination of LIDAR and photogrammetric equipment. The Scout has dimensions $12.5 \times 22.4 \times 18.5 \, cm$ and weighs 2.5 kg. LIDAR sensors designed for UAV and specific equipment designed to host sensors on UAVs.

Most recently, Riegl has developed a specialised UAV platform and bathymetric sensor – Bathycopter – as a means to survey water depths (bathymetry) in shallow water areas.

(http://www.riegl.com/uploads/tx_pxpriegldownloads/BathyCopter_at_a_glance_2015-09-11.pdf)

VIDEO AND STILL CAMERAS

The GoPro Hero series of cameras and competing ultra-compact and lightweight cameras (e.g. Walkera iLook) have proven themselves to professionals and amateurs alike as ideal sensors for recording high-resolution digital still photographs and video on UAV platforms such as the DJI Phantom series. Some UAV systems now also take advantage of smartphones and computer tablets instead of traditional screen viewers to monitor the aerial view of the surface. With the addition of video downlink hardware and view screens (first-person view (FPV)), aerial flights on UAVs can be monitored in real time (although it is necessary to acquire the help of an additional person to the UAV pilot to monitor the flight using the FPV screen) and even on smart devices such as the phones and computer tablets.

High-resolution imagery (e.g. 12+ MP) provides detailed colour and filtered imagery from low- to medium-altitude flights on platforms such as the DJI Phantom 1, Phantom 2, and 3D Robotics IRIS (3dr.com). Using traditional tools and techniques and DIP software (e.g. AgiSoft, MosaicMill, 2d3 (insitu.com), or Pix4D), the imagery acquired can then be interpreted and analysed manually or on-screen or information extracted using semi-automated approaches. 3D ortho-photo imagery and Digital Terrain Models (DTMs) can also be generated.

Larger platforms can of course carry larger payloads and even multiple cameras. This means that much better cameras e.g. the Panasonic GH4 DSLR can be mounted on a UAV with the end result being much higher resolution digital stills and video (e.g. 4 K) that moves the imagery from amateur to the professional quality required for commercial applications. In addition, the capability to carry a larger payload means that a better gimbal can be used, as well as numerous other sensors to provide complementary imagery.

STEREO CAMERAS

Stereo cameras have been utilised for a number of roles on UAVs. Firstly, for capturing stereo imagery: taking two photographs simultaneously from two slightly different viewpoints provides a basis for generating 3D imagery or views. Cameras such as the Fuji Finepix Stereo have been successfully mounted on small N-copters for this purpose (Haubeck and Prinz, 2013). Secondly, stereo cameras can be used as the basis for UAV navigation systems. Stereo imagery can be used to work out the distances to any obstacles, such as any planes, buildings, or mountains ahead; detect the nearest object in the FOV, to enable a warning to be issued, if the UAV is about to collide with the obstacle; and enable the UAV to use the distance-to-obstacle feature, to calculate a flight path and to avoid the obstacle. Stereo imaging can be accomplished through the use of two imaging Charge-Coupled Device (CCD) cameras and suitable software, such as the Stanford Research Institute (SRI) Small Vision System software. One image can be fitted with a vertical polariser and the other with a horizontal polariser. The difference between the images from the two imagers can be used to detect the presence of water, since only light with a horizontal polarisation is reflected from a water surface.

ADVANCES IN THE TECHNOLOGY

Advances in UAV and related technologies have been very rapid in recent years. Hazel and Aoude (2015) e.g. list the following areas that are in various stages at present: *advanced manufacturing techniques, batteries and other power, communication systems, detect, sense, and avoid capabilities, GPS, lightweight structures, microprocessors, motors, engines, and sensors.* Within a very short space of time, many of these areas of research are already finding their way into off-the-shelf products e.g. the DJI Phantom now has Flight Limit and No-Fly Zone technology built in, and for the Phantom 4, this includes a collision avoidance system, the obstacle sensing system (Fisher, 2016).

Platforms

There are now many manufacturers of both multi-rotor and fixed-wing aerial platforms of varying types and configurations. These include quadcopters, hexacopters and octa/okta-copters, of various different sizes and types, low cost to high cost, from toys to professional kits. Alongside the standard platforms, there are now also a number of waterproof platforms that not only allow the UAV to land and take-off from water bodies but also to prevent loss of the platform should it ditch in water owing to a power or physical malfunction e.g. Aquacopter Bullfrog (www.aquacopters.com) and Splashdrones (www.fpvfactory.com). These developments now provide operators with a wide choice of platforms which range from the simple and easy to maintain to the professional platform that requires a fair amount of expert maintenance. Today the technology is still evolving very rapidly, and there are a number of ongoing areas of development that will significantly enhance the future potential of UAV technology.

Multiple Drone Configurations

Although most applications of drones to date have involved only a single platform, there has been a lot of very sophisticated ongoing research into the development and deployment of multiple platforms or so-called *swarms* flying in configuration. According to Madey (2013), *Swarms offer numerous advantages over single UAVs, such as higher coverage, redundancy in numbers, and reduced long-range bandwidth requirements.* Already, SenseFly e.g. has developed software for controlling multiple drones for the mapping of remote areas (eijournal.com/news/industry-insights-trends/swarm-technology-for-drone-mapping-released).

Ready to Fly (RTF)

The growing number of RTF drones now available is allowing virtually anyone to fly and capture airborne data and imagery as the basis for environmental monitoring, mapping, and modelling applications. Although some of the larger platforms require more skills, knowledge, and understanding of the platform and sensor combination, as well as a certain level of training required to fly safely and within the current legislation (e.g. CAA), the freedom to acquire high-resolution data and imagery as and when needed is really unprecedented. The growing ease of use is one of the factors that is driving the popularity of these platforms, especially when compared to many earlier small-scale aircraft platforms.

Batteries

Since most of the platforms and their attached systems are now mostly reliant upon batteries e.g. LiPo (lithium polymer), developments in battery technology have become essential to allow for longer flight duration and reliability. There is currently considerable effort being put into developing battery technology with rapid advancements; for instance, the ratio capacity to size is increasing which facilitates more power-consuming activities, like longer flights and heavier payloads (Gómez and Green, 2015). Various new battery technologies are also emerging that are more compact and more lightweight alongside their extended life. Metal–air battery technologies (e.g. zinc–air, aluminium–air, lithium–air) are being developed. These batteries have valuable qualities that will benefit the UAVs industry, e.g. high-density power. Graphene cells, although not yet commercially available, also have potential for UAVs. For the moment though, LiPo batteries of varying sizes and capacities seem to dominate the market, although lithium–sulphur technology (Li-STM) is viewed as the next development (www.barnardmicrosystems.com/UAV/engines/batteries.html). Within a very short space of time battery life, even for smaller UAVs (e.g. the DJI Phantom) the flight time has increased from 5–8 minutes to 15–20 minutes and even longer with some larger and enhanced LiPo batteries. Naturally larger UAVs with more payload and lift capacity can extend this flight time further using multiple on-board battery packs. Other sources of energy used by a few UAVs are fuel cells and solar power. Some even use new technologies such as micro-generators, micro-turbines, and chemically powered systems that may replace batteries in the future. Nevertheless, battery technology is far from becoming obsolete and new designs introduced by battery manufacturers now offer relatively light, high capacity, and reliable sources of power (Gómez and Green, 2015).

Autonomous Navigation, GPS, and Collision Avoidance

Autonomous flight capability is well developed for the larger and more sophisticated drone platforms and covers the ability of the platform to navigate amongst other things, obstacles in the flight path. This is clearly advantageous for systems that may be expected to fly in enclosed spaces (Gómez and Green, 2015). Although many small UAVs are still flown manually, with each new release of UAV platform model (e.g. DJI Phantom 3 and 4), new technology has made it increasingly easy for a pilot to fly the aircraft. With the aid of improved GPS units and mobile phone and tablet apps e.g. Drone Deploy (www.dronedeploy.com), Ground Station (www.dji.com/product/pc-ground-station), and Litchi (flylitchi.com), it is now possible to pre-plan a UAV flight – either before going on-site – or on-site – and let the software and UAV do the work e.g. flying coverage of a field or specified area. This may include auto-take-off and landing, operation of the camera to acquire stereo imagery, altitude settings, and so on. Low-cost Real-Time Kinematic (RTK) GPS units are now also becoming available for UAVs to provide more accurate positional locations to aid in e.g. aerial survey applications; e.g. Piksi (www.swiftnav.com/piksi.html).

With the unfortunate number of incidents involving *drones* covered by the media and the press, as well as the desire to accomplish more complicated flight paths and commercial uses of drones for surveying and product delivery though has been the growing emphasis on including as standard built-in collision avoidance systems and software that prevents pilots from flying in certain areas or within a certain distance of e.g. an airport (Fisher, 2016).

Sensors

Many of the early aerial sensors were off-the-shelf digital cameras of varying makes and sizes used without much modification for use on UAVs. One of the most popular has been the GoPro Hero (gopro.com) series of small, waterproof, and physically robust digital sports cameras capable of taking photographic stills and video footages. Although many applications have used these cameras *as is*, recent developments have seen a number of modifications to the camera sensor wavelengths and the lens types. Manufacturers have developed low-cost specialist modifications for digital cameras for off-the-shelf platforms that extend the capabilities for environmental monitoring e.g. MapIR (www.mapir.camera) and HackHD (www.hackhd.com). Further developments lie with the higher end digital SLR and video cameras that can provide very high-resolution imagery from slightly larger airborne platforms with higher payloads. The development of UAV gimbals by companies such as DJI e.g. their ZenMuse (www.dji.com/product/zenmuse-z15) range has provided the means to carry off-the-shelf digital cameras that are recognised to be highly suited to aerial photography from small-aircraft platforms.

A growing demand for additional aerial remote sensors to extend the wavelength sensing capabilities beyond the visible and the NIR include thermal cameras. One example is the range of modifications offered by IRPro (www.ir-pro.com) in the USA to extend the sensitivity of GoPro cameras to include multispectral images, NDVI, as well as a range of flat lenses that remove the often-cited disadvantages of the GoPro namely the semi-fisheye lens. Another advantage of these customised popular sensors is the end product that results, that in the case of the GoPro NDVI modification, offers the end user, client, or customer a service with a product that is presented in the form of information that can then be used in situ planning and decision-making. The Flir Tau 2 (www.flir.com) is a small thermal sensor designed specifically for the small UAV platform e.g. DJI Phantom and DJI Inspire. Likewise, hyperspectral cameras from Headwall Photonics (www.headwallphotonics.com) and LIDAR sensors e.g. LidarPod (www.routescene.com) and Riegl's Bathycopter LIDAR (www. riegl.com) are all products specifically designed to fit the payload and size constraints of small-scale aircraft. The miniaturisation of these sensors offers the potential to gather new and unique data and environmental information from small airborne platforms. The major constraint for many operators at present though is the price associated with these particular sensors, and whilst the costs are dropping, they are still generally outwith the budget of most organisations who may only use the equipment infrequently.

Software

An increasing demand to be able to process the digital still and video imagery in various different ways has also grown rapidly. Initially, many hobbyists made use of commercially available or Opensource (opensource.org) DIP software and occasionally the soft-copy photogrammetric modules within these. Whilst some people made use of specialist environmental remote sensing software e.g. Erdas Imagine (www.hexagongeospatial.com) and ENVI (www.harrisgeospatial.com), others utilised graphic software such as Adobe Photoshop to undertake lens and light corrections and to stitch together multiple photographic stills into mosaics. Not surprisingly within the past 5 years, various software vendors have seen the opportunity to develop a number of relatively low-cost commercial and educational soft-copy photogrammetry and image processing software specifically tailored to imagery acquired from UAV platforms and common digital cameras such as the GoPro. Commercially available products such as AgiSoft (http://www.agisoft.com), Pix4D (https://pix4d.com), and most recently ENVI-UAV (http://www.harrisgeospatial.com) have emerged to allow rapid processing of UAV imagery into a range of products such as mosaics and 3D visual surface models. With each new release of the software, these products are rapidly improving, and like RTF drones, the interfaces are becoming easier to use for the non-specialist. Higher-end soft-copy photogrammetric software solutions are also offered by MosaicMill (http://www.mosaicmill.com), Pix4D, and others. Other products are GeoApp UAS from Hexagon Geospatial (www.hexagongeospatial.com). Opensource software such as AirPhotoSE (http://www.uni-koeln.de/~al001/airphotose.html) also provides the basis for researchers to utilise UAV imagery, in this case specifically targeting archaeology.

SUMMARY AND CONCLUSIONS

SSA platforms and sensors have clearly come a considerable way in a relatively short period of time, going from model aircraft with fuel-based engines and SLR cameras that needed to be flown by a specialist to the current plug-and-play RTF platforms with custom-designed UAV sensors. In a relatively short period of time, UAV technology has advanced from being a hard to use, often novelty toy, to a very serious aerial platform that offers considerable research and commercial potential, in whatever form, to gather unique data and imagery from a range of altitudes and coverages for small-area, large-scale requirements for a very wide range of environmental applications.

The reduced costs and ease of image acquisition, processing, analysis, and generation of usable products have grown phenomenally in the last 5 years. Recent developments also show that the technologies are still evolving very quickly, and with the ongoing developments in the technology, this is set to continue, grow, and expand in the next few years. The significance of these small aerial platforms is perhaps best summed up by Wyman (2015) who observes: *by 2035, the number of unmanned aerial vehicle operations per year will surpass that of manned aircraft.*

ACKNOWLEDGEMENTS

I would like to extend special acknowledgements to two people who have contributed to my long-term interest in RC models, and now UAVs, over the years:

1. Ed Wedler – whose office I shared in the Ontario Centre for Remote Sensing (OCRS) in the mid/late 1980s when I was a postgraduate student studying at the University of Toronto in Canada – as the person who was first responsible for cultivating my interest in SSA all those years ago.
2. Norrie Kerr of Aberdeen in Scotland, UK, for his considerable practical knowledge and expertise and for willingly spending a lot of his free time flying the various model aircraft he designed and built over the Ythan Estuary to the north of Aberdeen.

An interest that has endured to the present day, leading finally to the establishment of my UAV Centre for Environmental Monitoring and Mapping (UCEMM) at the University of Aberdeen in the Department of Geography and Environment, Scotland, UK, with a 'squadron' of multi-rotor UAV hardware, accompanying software and applications.

REFERENCES

Boddington, D., 1989. Cyclops: Aerial photography of field experiments using an RPV. *Radio Control Models and Electronics*. pp. 448–450.

Bowie, P., 1981. Eye in the sky. *RC Model Builder*. pp. 38–40.

British Journal of Photography, 1975. Remote aerial photography from a model helicopter. Vol. 122.

Buckle, B., 1985. Practical safe operation of very small RPVs for civil users. *Proceedings of the 5th International Conference on RPVs—Remotely Piloted Vehicles*. Bristol, UK, 9–11 September 1985. pp. 4.1–4.9

Canas, A.A.D., and Irwin, D.A., 1986. Airborne remote sensing from remotely piloted aircraft. *International Journal of Remote Sensing*. Vol. 7(12). pp. 1623–1635.

Clark, A.S., 1982. Canadair rotary-wing RPV technology development—part II. *Proceedings of the 3rd International Conference on Remotely Piloted Vehicles.* Bristol, UK, 13th–15th September 1982. Paper II.

Colomina, I., and Molina, P. 2014. Unmanned aerial systems for photogrammetry and remote sensing: A review. *ISPRS Journal of Photogrammetry and Remote Sensing.* Vol. 92. pp. 79–97.

De Wulf, R.E., and Goossens, R.E., 1994. Winter wheat yield estimation with CIR imagery from a remotely piloted aircraft. *Proceedings 1st International Airborne Remote Sensing Conference and Exhibition. Applications, Technology, and Science: Today's Progress for Tomorrow's Needs.* Strasbourg, France, 12–15 September 1994.

Deuze, J. L., Devaux, C., Herman, M., Santer, R., Balois, J. Y., Gonzalez, L., Lecomte, P., Verwaerde, C., 1989. Photopolarmetric observations of aerosols and clouds from balloon. *Remote Sensing of Environment.* Vol. 29. pp. 93–109.

Ellis, R.M., Totten, J.A., and Fuller, A.R., 1981. The use of radio controlled aircraft in pollution studies. *Proceedings 2nd International Conference on Remotely Piloted Vehicles.* Bristol, UK. Paper 11.

Fagerlund, E., and Gunnershed, N., 1975. Systems analysis and development of a mini-RPV for reconnaissance. The 'Skatan' Project. FDA Report D30021-t1. National Defence Research Institute, Stockholm. p. 50.

Fisher, J., 2016. DJI adds collision avoidance system to Phantom 4 drone. PC Mag UK. http://uk.pcmag.com/dji-phantom-3-professional/75655/news/dji-adds-collision-avoidance-system-to-phantom-4-drone

Fouche, P.S., and Booysen, N., 1994. Low altitude surveillance of agricultural crops using inexpensive remotely piloted aircraft. *Proceedings 1st International Airborne Remote Sensing Conference and Exhibition.* Strasbourg, France, 11th–15th September 1994. pp. III-315–III-326.

Gómez, C., and Green, D.R., 2015. Small-scale airborne platforms for oil and gas pipeline monitoring and mapping. Unpublished AICSM Interface Report for Redwing Ltd. p. 54.

Green, D.R., 2016. Acquiring environmental remotely sensed data from small-scale aircraft for input to Geographic Information Systems (GIS). Unpublished Paper. p. 34.

Gregory, T.J., Bailey, R.O., and Nehms, W.P., 1974. RPVs—Exploring Civilian Applications. *Astronautics and Aeronautics.* pp. 38–47.

Harding, B., 1989. Model aircraft as survey platforms. *Photogrammetric Record.* Vol. 13(74). pp. 237–240.

Haubeck, K., and Prinz, T., 2013. A UAV-based low-cost stereo camera system for archaeological surveys—experiences from Doliche (Turkey). *International Archives of the Photogrammetry, Remote Sensing and Spatial Information Sciences*, Vol. XL-1/W2, Rostock, Germany. pp. 195–200.

Hazel, B., and Aoude, G., 2015. *In Commercial Drones, the Race Is On: Aviation's Fastest-Growing Sector Outpaces US Regulators.* Oliver Wyman Aviation, Aerospace and Defense. Marsh and MacLennan Companies, New York, NY. p. 16.

Hoer, J., 1976. Aerial photography from model aircraft. *British Journal of Photography.* Vol. 123(8). p. 156.

ICAO, 2011. ICAO circular 328, Unmanned Aircraft Systems (UAS). Technical Report. International Civil Aviation Authority. Montreal, Canada.

Johnson, G.W., 1978. Balloon photography for archaeological exploration and mapping. *ASP Proceedings 43rd Annual Meeting.* 27th February–5th March 1978.

Jonsson, I., Mattsson, J.O., Okla, L., and Stridsberg, S., 1980. Photography and temperature measurements from a remotely piloted vehicle. *Oikos.* Vol. 35. pp. 120–125.

Kendall, D.J.W., and Clark, T.A., 1979. Balloon borne for infrared Michelson interferometer for atmospheric emission studies. *Applied Optics.* Vol. 18(8). pp. 346–353.

Lapp, H.S., and Shea, E., 1976. A night photo system for remotely piloted vehicles. *Proceedings Society of Photo-Optical Instrumentation Engineers*, Reston, Virginia, 24th–25th March 1976. pp. 174–178.

Lemeunier, P., 1978. Electronic reconnaissance with Pilotless aircraft. *Electronic and Applications Industrial.* Vol. 247. pp. 41–43.

Levanon, N., 1979. Ice elevation map of Queen Maud Land, Antarctica from balloon altimetry. *Nature.* Vol. 278(507). pp. 842–845.

Madey, A.G., 2013. Unmanned Aerial Vehicle swarms: The design and evaluation of command and control strategies using agent-based modeling. *International Journal of Agent Technologies and Systems.* Vol. 5(3). pp. 1–3.

Marks, A.R., 1989. Aerial photography from a tethered helium filled balloon. *Photogrammetric Record.* Vol. 13(74). pp. 257–261.

Meroni, M., Rossini, M., Guanter, L., Alonso, L., Rascher, U., Colombo, R., and Moreno, J. 2009. Remote sensing of solar-induced chlorophyll fluorescence: Review of methods and applications. *Remote Sensing of Environment.* Vol. 113. pp. 2037–2051.

Miller, P., 1979. Aerial photography from radio-controlled aircraft. *Aerial Archaeology.* Vol. 4. pp. 11–15.

Mullins, J., 1997. Palmtop planes. *New Scientist.* (2076). p. 9.

NASA, 1996. *Workshop on Remotely-Piloted Aircraft for U.S. Global Change Research*, Williamsburg, VA, USA, 12–15 November 1996.

North East River Purification Board, 1982. Aerial photography using a model aircraft—An evaluation. Unpublished Report. p. 16.

Petrie, G., 2013. Commercial operation of lightweight UAVs for aerial imaging and mapping, *GEOInformatics*, Vol. 1. pp. 28–38.

Roustabout, n.d. A magazine for people in the North Sea oil industry. Aerial Surveying in Miniature. Issue No. 133. p. 35.

Skrypietz, T., 2012. Unmanned aircraft systems for civilian missions. Brandenburg Institute for Society and Security (BIGS). Policy Paper No. 1. p. 28.

Syms, P., and Turner, P.S., 1982. ASAT: The UK's First Turbojet RPV. *Proceedings 3rd International Conference Remotely Piloted Vehicles*, Bristol, UK, 13th–15th September 1982. Paper 12.

Taylor, J.W., and Munson, K., 1977. *Jane's Pocket Book of Remotely Piloted Vehicles: Robot Aircraft Today.* Collier Books, New York, NY. p. 239.

Thurling, D.J., 1987. Design and operation of low-cost remotely-piloted aircraft for scientific field research. *Proceedings of 6th International Conference on Remotely Piloted Vehicles*. Bristol, UK, 6th–8th April 1987.

Thurling, D.J., Harvey, R.N., and Butler, N.J., n.d. *Aerial Photography of Field Experiments Using Remotely Piloted Aircraft.* Unpublished Manuscript. p. 7.

Tomlins, G.F., 1983. Some considerations in the design of low-cost remotely-piloted aircraft for civil remote sensing applications. *The Canadian Surveyor.* Vol. 37(3). pp. 157–167.

Tomlins, G.F., and Lee Y.J., 1983. Remotely piloted aircraft—An inexpensive option for large-scale aerial photography in forestry applications. *Canadian Journal of Remote Sensing.* Vol. 9. pp. 76–85.

Tomlins, G.F., and Manore, M.J., 1984. Remotely piloted aircraft for small format aerial photography. *Proceedings 8th Canadian Symposium on Remote Sensing.* pp. 127–135.

Velligan, F. A., and Gossett, T.D., 1982. Extended use of the aquila RPV system. *Proceedings 3rd International Conference on Remotely Piloted Vehicles.* Bristol, UK, 13th–15th September 1982. Paper 6.

Warner, W.S., Graham, R.W., and Read, R.E., 1996. *Small Format Aerial Photography.* Whittles Publishing, Dunbeath. p. 347.

Wedler, E., 1984. Experience with radio-controlled aircraft in remote sensing applications. *Proceedings 1984 ASP-ARSM Convention.* Washington, DC, 11th–16th March 1984. pp. 44–54.

Wester-Ebbinghaus, W., 1980. Aerial photography by radio-controlled model helicopter. *Photogrammetric Record.* Vol. 10(55). pp. 85–92.

Wyman, O., 2015. www.oliverwyman.com/our-expertise/insights/2015/apr/in-commercial-drones-the-race-is-on.html

3 Aquatic Vegetation Monitoring with UAS

Hamish J. Biggs
National Institute of Water and Atmospheric Research (NIWA)

CONTENTS

INTRODUCTION

Aquatic vegetation is a key component of rivers, lakes, wetlands, estuaries, and other ecosystems. It performs ecologically important roles such as providing habitat for invertebrates and fish (Shupryt and Stelzer, 2009; Figueiredo et al., 2015). It also helps to improve water quality by stabilising sediments, promoting the deposition of fine sediment, and decreasing turbidity (Madsen et al., 2001). Aquatic vegetation is generally considered to be beneficial at low to medium biomass; however, at high biomass, it causes a multitude of problems. In rivers, the hydraulic resistance of aquatic vegetation slows flows, raises water levels, increases flooding risks, and causes excessive accumulation of fine sediments (Butcher, 1933; Gurnell et al., 2006). The accumulated fine sediments can destroy habitat for invertebrates, molluscs, and fish spawning (Wood and Armitage, 1997; Österling et al., 2010; Kemp et al., 2011). Water bodies choked with aquatic vegetation also experience large variations in dissolved oxygen, which can lead to changes in fish abundance and community structure (Miranda and Hodges, 2000; Killgore and Hoover, 2001).

Management of aquatic vegetation can be necessary when it reaches high biomass. Techniques include: manual clearing and physical control (Wade, 1990), flushing flows (Tena et al., 2017), biological control with herbivorous fish (Van der Zweerde, 1990), and chemical control (Murphy and Barrett, 1990). Active management of aquatic vegetation is very expensive, with global costs running in the billions of dollars annually (Dawson, 1989; Rockwell, 2003; Madsen and Wersal, 2017). To efficiently manage aquatic vegetation, it is critical to base management decisions on quality survey data (Murphy, 1990). Surveys may be bank-based assessments, distributed point samples (Madsen and Wersal, 2017), or aerial deployments to cover larger spatial extents (Murphy, 1990; Marshall and Lee, 1994; Kirkman, 1996; Silva et al., 2008). From river banks, it is difficult to derive quantitatively useful information about aquatic vegetation abundance and species composition (Figure 3.1a). However, from aerial imagery, it is straightforward to quantify vegetation cover, species composition, and demarcate the boundaries of individual specimens to investigate size distributions (Figure 3.1b).

Recent advances in UAS technology have enabled leaps forward in low-cost, high-resolution surveying of aquatic vegetation (Watts et al., 2012; Klemas, 2015). UAS-based remote sensing of aquatic vegetation has been applied in wetlands (Husson et al., 2014; Boon et al., 2016), rivers (Visser et al., 2013; Flynn and Chapra, 2014; Biggs et al., 2018), and coastal environments (Dugdale, 2007; Klemas, 2015; Parsons et al., 2018). There are even opportunities for UAS-based detection and spraying to control invasive weeds (Göktoğan et al., 2010). UAS surveying of aquatic vegetation is well suited for shallow areas with good optical transparency. For deep areas or turbid water, other techniques such as scuba diving or surveying with depth sounders may be required (Capers, 2000; Ibáñez et al., 2012; Madsen and Wersal, 2017).

This chapter provides readers with enough information to undertake aquatic vegetation monitoring with UAS. Section 2 covers field deployments to obtain RGB aerial imagery, samples of vegetation biomass, and basic hydraulic conditions. Section 3 covers processing the RGB aerial imagery

FIGURE 3.1 Monitoring aquatic vegetation in the River Urie, UK, with (a) river bank imagery and (b) aerial imagery. The red arrows indicate the flow direction of the river.

and analysis techniques to determine vegetation cover and size distributions. Section 4 covers lab-work to quantify the biomass of aquatic vegetation samples. Section 5 combines the processed aerial imagery, biomass samples, and hydraulic data to quantify distributions of aquatic vegetation at larger spatial scales. Section 6 covers emerging technologies in aquatic vegetation surveying such as multispectral and hyperspectral imagery. This chapter uses case studies from rivers; however, most methods are general and can be used for aquatic vegetation surveying in other environments.

AQUATIC VEGETATION UAS FIELD DEPLOYMENTS

SITE SELECTION AND MISSION PLANNING

UAS surveying is suitable for shallow sites with clear water and good visibility of aquatic vegetation. Sites should have adequate open spaces to place Ground Control Points (GCPs). Consideration also needs to be made for riparian vegetation, bridges, power lines, and large trees that may interfere with aerial imagery, flight paths, and visual line of sight from the UAS pilot to the aircraft. A laser range finder (such as the Nikon Forestry Pro) can be very useful to determine the height of trees and other obstacles. The rules of UAS operations from the relevant aviation authority need to be checked during mission planning. Common considerations are to not fly over people without their consent, to obtain land owner/administrator permission before flying, and not to fly within restricted airspace (e.g. near an airfield). During mission planning, flight altitude should be selected to provide the required pixel resolution to demarcate vegetation boundaries. As a rule of thumb, 1 cm resolution is suitable for most aquatic vegetation surveying. Flight paths should be selected that enable sufficient overlap between images (e.g. 60%–80%) to use structure from motion software to generate a geo-referenced orthomosaic. A georeferenced orthomosaic is essentially a high-resolution map derived from the aerial images and is a precursor to classification of aquatic vegetation. Consideration for battery life, emergency landing locations, and battery swap locations should be made during mission planning. For surveying large rivers or water bodies, it can be convenient to walk along the bank while the UAS flies a zigzag path orthogonal to the bank. This maintains proximity of the operator and UAS (at the end of each zigzag) for battery swaps. For these missions, it may be useful to carry a mobile landing pad (such as rolled-up carpet) for safe landing on long grass, rocky, or uneven terrain. Finally, it is worth considering the possibilities of what may go wrong and make contingency plans for the most likely scenarios. Poor mission planning, large trees, and inadequate altitude can be expensive mistakes for UAS operators (Figure 3.2).

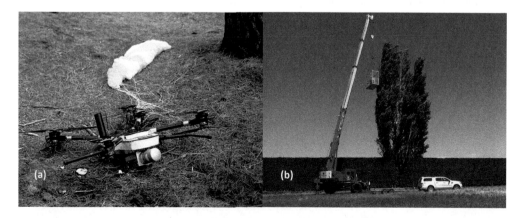

FIGURE 3.2 (a) UAS loses radio contact and returns to home by the shortest path (through a large tree) and (b) incorrect flight path and inadequate prior surveying of tree height.

UAS Hardware

There is a wide variety of UAS hardware to choose from, and selecting the optimal system depends on the deployment application (Watts et al., 2012; Klemas, 2015). For low-altitude aquatic vegetation surveying multirotor UAS are convenient and commonly used. Selecting an appropriate model of multirotor UAS should be based on the required payload and flight time. For high-performance RGB, multispectral, or hyperspectral cameras, a heavy-lift UAS (such as a DJI Matrice 600 Pro) may be needed. However, for most RGB applications, the camera on a small-surveying UAS is sufficient (e.g. the 20 MP camera on the DJI Phantom 4 Pro). These small systems are generally easy to use, relatively inexpensive, have decent flight time, and pose lower safety risks than heavy lift systems. For extensive surveying missions (e.g. lakes and wetlands), fixed-wing UAS can be used (Husson et al., 2014).

Ground Control Points (GCPs)

GCPs are generally needed to accurately georeferenced orthomosaics (e.g. for scaling, rotation and translation). At least eight GCPs should be used and distributed throughout the survey site. GCPs should be placed in open areas where they are easily visible in multiple overlapping images (e.g. away from trees and bushes). GCPs can be manufactured from canvas for durability and secured with tent pegs or rocks (Figure 3.3). GCP designs should have a well-defined centre that can be easily located when surveying them with a Real-Time Kinematic (RTK) Global Positioning System (GPS). For repeated surveys, it is convenient to install more permanent GCPs or use defined landmarks (such as the corner of a building). In some situations (e.g. if camera origin is known with RTK or Post Processing Kinematic (PPK) GPS accuracy), it may be possible to avoid the use of GCPs altogether.

Lighting, Cloud, and Wind

Lighting conditions, cloud cover, and wind are all important considerations when surveying aquatic vegetation (and for through water imagery in general). In the author's experience, flying at solar noon, with no cloud cover, and minimal wind is optimal (Figure 3.4). If sunglint off the water surface is problematic (more likely at latitudes closer to the equator), then survey time of day, camera angle, and flight direction can be adjusted (Overstreet and Legleiter, 2017). Polarising filters are useful for through water imagery taken from river banks (Figure 3.1a); however, the author has found little benefit for UAS-mounted cameras with nadir (vertical) orientation. In this case, polarising filters simply reduce the total amount of light entering the lens, necessitating tradeoffs with shutter speed and aperture.

FIGURE 3.3   Easily visible GCPs with a well-defined centre.

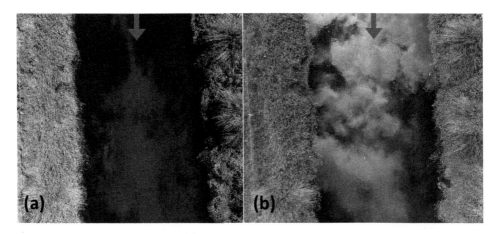

FIGURE 3.4   Comparison of aerial imagery with (a) blue sky and (b) low cloud. Surface reflections of low cloud yield unusable imagery for aquatic vegetation surveying.

CAMERA SETTINGS

Camera settings are critical for obtaining quality aerial imagery. Fast shutter speed is the most important factor to avoid blurry images. The author recommends: shutter priority mode with 1/1,000 shutter (or faster in good lighting conditions), focus (infinity or auto), aperture (usually auto), and ISO (auto). Image blur can be estimated from UAS velocity multiplied by shutter speed. For example, a UAS travelling at 3 m/s (or 3,000 mm/s) with a shutter speed of 1/1,000 seconds will have 3 mm of blur. If the altitude of the UAS provides 24 mm pixels, then this blur is not significant; however, if the shutter speed was only 1/100 seconds, then problematic blur of 30 mm occurs. It is recommended to adjust shutter and flight speed to account for blur, rather than stopping and hovering at each image capture location. This is because more significant blur arises from changes in camera orientation than camera origin. A UAS flying a programmed route at constant speed is flying through clean air and is generally more stable than one trying to hover in one location where it sits above its own rotor wash. Stopping at each location to take an image is also very battery hungry and will greatly reduce spatial coverage. The best solution is to provide plenty of light for the camera (e.g. time of day, atmospheric conditions, large camera sensor, and fixed focal length lens with wide aperture), which enables a fast shutter speed without exceeding 400 ISO. For most aerial surveying applications (over 20 m altitude), imagery depth of field is not an issue and wide open (or auto) lens aperture can be used.

FIELDWORK FOR VEGETATION BIOMASS

The most common ways to quantify aquatic vegetation abundance are: cover (e.g. planform area), volume, and biomass (Wood et al., 2012). Cover is readily quantified from aerial imagery (Biggs et al., 2018); however, it doesn't provide any information about the three-dimensional abundance of aquatic vegetation and how it fills the water column. Volume is useful to quantify vegetation density and is discussed later in this chapter. Vegetation biomass is generally the most informative way to quantify abundance but is difficult to measure directly at large spatial scales. To get the best of both worlds, it is possible to combine spot samples of vegetation biomass with cover, then extrapolate the data to larger spatial scales. In practice, an initial UAS flight is used to record natural vegetation abundance (Figure 3.5), then aquatic vegetation specimens/samples can be removed and UAS flights repeated. Samples/specimens should be distributed throughout the study site and be representative of the natural size distribution of aquatic vegetation. The area of each specimen/sample is found by comparing the initial and final georeferenced orthomosaics (e.g. the area of specimen

FIGURE 3.5 Manual image segmentation in the River Urie, UK. Classes are *Ranunculus penicillatus* (green), *Myriophyllum alterniflorum* (black), and river boundaries (blue).

C01_P0001 in Figure 3.5 if it was removed). Before specimens/samples are removed, it is useful to survey their location with GPS or even to photograph them from above with a GPS camera to help identify them in the georeferenced orthomosaics. Vegetation biomass that is removed from the river can either be bulked together if only total biomass is needed or kept separate if the properties of each specimen/sample are to be investigated. Vegetation biomass should be transported to the lab in tanks or heavy-duty polythene bags with the same water from where it was collected.

FIELDWORK FOR HYDRAULIC MEASUREMENTS

To understand the spatial distribution of aquatic vegetation, it is critical to quantify the hydraulic conditions where they live (Franklin et al., 2008; Biggs et al., 2018). In streams and rivers, the key parameters are volumetric discharge, slope, and bathymetry. Discharge data can be obtained from a continuous flow gauging station if one is close to the study site. These stations are commonly operated by local councils or environmental agencies and provide a discharge time series. This data is useful for investigation of flow variability (floods) that may remove aquatic vegetation and influence spatial distributions. If there is no flow data available or it is of dubious quality (which occurs at gauging station where rating curves are not updated as aquatic vegetation grows), then local discharge measurements should be performed. Discharge may be obtained by using techniques such as salt dilution gauging, Acoustic Doppler Current Profiler (ADCP) cross-sections (Figure 3.6), and distributed point velocity measurements. When selecting an appropriate technique for discharge gauging in vegetated rivers, it is prudent to remember that: (i) the bottom tracking feature of ADCPs does not work over submerged vegetation (as it looks like mobile bed materials) and (ii) logarithmic velocity profiles do not apply in vegetated rivers so depth averaged velocity cannot be obtained from one point measurement at 0.6 of depth (instead the average of a velocity profile must be used). For ADCP discharge gauging, it is convenient to select a cross-section free of vegetation (or clear it first) to obtain reliable results without having to manually record velocity profiles. River slope can be obtained using equipment such as a piezometer, manometer, or total station. Cross-sections of bathymetry can be obtained with an ADCP or depth sounder (if free of vegetation) or manually with a ruler. Data on discharge, slope, and bathymetry is then used to derive information

FIGURE 3.6   Discharge gauging with a Teledyne StreamPro ADCP in the River Urie, UK.

such as: cross-sectional mean velocity, Froude number, stream power, and roughness coefficients (Biggs et al., 2018).

PROCESSING AQUATIC VEGETATION IMAGERY DATA

GEOREFERENCED ORTHOMOSAIC

Aerial images are captured with at least 60%–80% overlap on all sides and GCPs distributed throughout the study site. Georeferenced orthomosaics or Digital Elevation Models (DEM) are generated using Structure-from-Motion (SfM [SFM]) software such as AgiSoft Photoscan or Pix4D. The usual workflow is to: load images, align images, place GCP markers, input the coordinates of GCP markers, build a dense point cloud, build a mesh, and build texture. Then a georeferenced orthomosaic (or a DEM) can be generated and exported. There are many tutorials online, and the exact process is software specific, so only the general steps are discussed here. Georeferenced orthomosaics should be output with sufficient resolution (Figure 3.7) for image segmentation and classification of aquatic vegetation.

MANUAL IMAGE SEGMENTATION AND CLASSIFICATIONS

Manual segmentation of RGB imagery into many classes of aquatic vegetation can achieve overall accuracy of 95% (Husson et al., 2014), whereas automatic techniques struggle to achieve 65% overall accuracy (Husson et al., 2016). Manual image segmentation also suits demarcation of a wider range of classes, analysis of the geometry of individual specimens, and is convenient for biomass conversions. There are multiple software packages that can be used for manual image segmentation; however, a simple workflow for ArcMap is provided here:

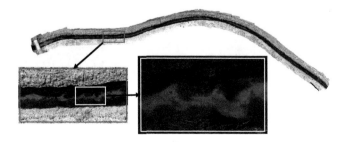

FIGURE 3.7   High-resolution UAS survey of aquatic vegetation (*Potamogeton crispus* and *Elodea canadensis*) in a 430 m reach of the Halswell River, New Zealand.

- Load the georeferenced orthomosaic.
- Adjust the brightness, contrast, and gamma of the georeferenced orthomosaic to better resolve aquatic vegetation.
- Decide how many classes are needed and create a polygon shapefile for each. For example:
  - Classification to species level (Figure 3.5),
  - Classification for total vegetation only (Figure 3.8).
- Draw the boundaries of objects within each class.
- Save the shape files.
- Load the shape files in a software of choice (e.g. MATLAB®) to compute the area and geometric properties of each of the polygons.

AUTOMATIC IMAGE SEGMENTATION AND CLASSIFICATIONS

Automatic image segmentation and classifications are useful for routine surveying of visually distinctive vegetation over large spatial extents. However, automatic classifications of RGB imagery are less accurate than manual classifications (Husson et al., 2016). It can also be very time consuming to develop robust classification rules (Visser et al., 2018) and often substantial manual classification data is required to assess the accuracy of the automatic classifications. Automatic classifications are also unable to distinguish the boundaries between neighbouring specimens of the same class, making them unsuitable for resolving size distributions within a population. If automatic classifications are deemed appropriate for the classification task, there are many techniques to choose from (Richards, 2013), for example: Adaptive Cosine Estimator (ACE) and Spectral Angle Mapper (SAM) (Flynn and Chapra, 2014); maximum likelihood techniques (Biggs, 2017); object-based image analysis techniques (Visser et al., 2018); feature learning algorithms (Hung et al., 2014); support vector machines (SVMs) (Mountrakis et al., 2011); and random forests (Husson et al., 2016). Ground truth data is mandatory for automatic classifications and can be comprised of georeferenced field measurements/observations or obtained from manual classifications (if classes are easily distinguishable). Ground truth data and automatic classification data can then be used to compute the user and producer accuracy (Congalton, 1991; Richards, 2013) to assess how well the classification performed. To perform repeatable automatic classifications, it may be useful to convert RGB data to another colour space where brightness and colour are separated (e.g. Hue Saturation Value (HSV)

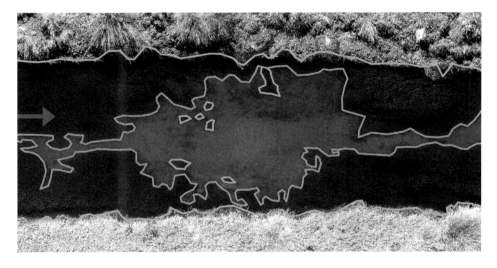

FIGURE 3.8 Manual image segmentation in the Halswell River, New Zealand. Classes are: river bed (orange), isolated vegetation (green), and river boundaries (blue). The river is dominated by *Potamogeton crispus* and *Elodea canadensis*. Total vegetation area = river area – bed area+isolated vegetation area.

or CIE-Lab) then only use colour channels for classification. This step can help to develop classification rules that are not specific to the lighting conditions on the day when measurements occurred.

GEOMETRIC PROPERTIES OF INDIVIDUAL PLANTS

Common geometric properties are area, length, width, aspect ratio, and orientation (Biggs et al., 2018). Length can be defined as the maximum separation of any two points on the plant. Orientation can be defined as a unit vector parallel with direction of length and originating at the plants centre of area. Width can be defined as the maximum span of the plant orthogonal to the orientation vector. Aspect ratio is simply length divided by width. These parameters can be calculated from the polygons in the shapefiles (e.g. in MATLAB), alternatively area, length, and width can be measured directly in software such as ArcMap.

LABWORK FOR AQUATIC VEGETATION BIOMETRICS

SAMPLE HEALTH AND HANDLING

Aquatic vegetation samples should be transported to the laboratory and processed as soon as possible. If it is not possible to process samples the same day they are cut, then they should be stored overnight in tanks of water. Storage water should be taken from the same site that the aquatic vegetation was removed from or should have similar properties (e.g. temperature and water chemistry). Storage water tanks must also be aerated with a bubbler system (Vettori, 2016) to reduce vegetation deterioration that rapidly occurs in anoxic conditions.

SAMPLE DRYING TECHNIQUES FOR FRESH BIOMASS

Aquatic vegetation is wet when it is removed from its habitat or laboratory storage tanks. To determine its fresh biomass, surface water must be removed without vegetation losing internal moisture (Westlake, 1965; Madsen, 1993). Drying techniques include 'vigorous shaking and hand-squeezing' (Neil et al., 1985), air drying, drying with paper towels (Biggs, 2017), and centrifugal drying with a salad spinner (Neil et al., 1988; Rooney and Kalff, 2000 and 2003; Garrison et al., 2005; Vermaire et al., 2011; Bickel and Perrett, 2015). Centrifugal removal of surface water is arguably the most defensible technique (Westlake, 1965). However, most implementations of this technique lack detailed information on the spin regime and are in no way standardised or repeatable (Bickel and Perrett, 2015). The following section of this chapter provides a way to report standardised drying data based on the physics of the spin regime.

STANDARDISED CENTRIFUGAL DRYING FOR FRESH BIOMASS

Water removal during centrifugal drying is dependent on the centripetal acceleration profile $a_c(t)$ over spin time T. The centripetal acceleration on a body moving in a circle of radius r and linear velocity v (or angular velocity ω) is $a_c = v^2/r = r\omega^2$ (Halliday et al., 2005). Angular velocity is easily measured and has units of radians per second, with conversion from revolutions per minute (RPM) as $\omega_{rad/s} = 2\pi\omega_{RPM}/60$. Vegetation centrifuges can either be built from scratch or large-volume top-loading washing machines can be commandeered (Edwards and Owens, 1960) (Figure 3.9a). The angular velocity profile during the washing machine spin cycle must be quantified or a prescribed angular velocity profile can be implemented by modifying the washing machine speed controller. While the latter approach is advantageous for scientific accuracy and standardisation, care must be taken around maximum angular velocities and unbalanced/unstable loads, which can cause serious safety risks if the standard safety features of the washing machine have been circumvented. Angular velocity can be quantified as $\omega = 2\pi/\Delta t$, where Δt is the time for one revolution

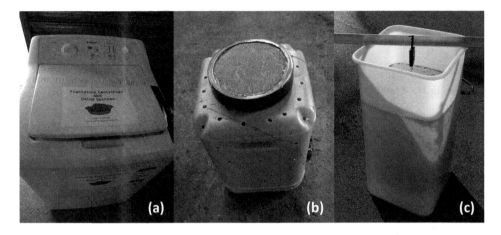

FIGURE 3.9 (a) Centrifugal vegetation dryer using the spin cycle of a 9.5 kg top-loading washing machine and (b) and (c) determining aquatic vegetation volume by Archimedes principle and the buoyancy force from displaced water.

of the washing machine. There are many ways to measure the time for one revolution, but in the implementation in Figure 3.9a, a magnet, Hall effect sensor, and high-frequency analogue to digital converter (ADC) are used (model LabJack U3-HV).

DRY BIOMASS

It is also common practice to dry aquatic vegetation to constant weight at 105°C then report biomass (Edwards and Owens, 1960; Westlake, 1965; Madsen, 1993; Wood et al., 2012). This technique is convenient if large ovens or drying rooms are available but suffers from uncertainties in the amount of non-organic matter (e.g. silt, sand, and mineral deposits) that remains within vegetation samples. A technique that circumvents this problem is known as 'ash-free dry mass' or 'ash-free dry weight' (Westlake, 1965) and is commonly used for small samples of periphyton. However, this technique is impractical and usually unnecessary for aquatic vegetation samples that comprise tens of kilograms of fresh biomass (Madsen, 1993). For most applications, thoroughly rinsing aquatic vegetation then drying to constant weight provides a suitable metric for biomass. Vegetation must be weighed immediately after drying or it may absorb water from the atmosphere yielding 5%–10% error (Westlake, 1965).

VEGETATION VOLUME AND DENSITY ESTIMATION

Measurements of volume are used to determine the density and porosity of aquatic vegetation. In situ field measurements of volume are reported by some authors (Wood et al., 2012), but results are highly dependent on the survey technique/resolution used (Westlake, 1965). In flowing waters, vegetation volume also changes due to dynamic reconfiguration. As such, vegetation field volume is not a standardised metric for vegetation abundance and is mainly used to obtain information on porosity. Lab measurements of vegetation volume provide a useful metric for the abundance of plant material and can be extrapolated the same way as biomass to estimate the total volume of aquatic vegetation tissue within a water body. Vegetation volume V and fresh biomass B_f are also used to determine vegetation density ρ_v, where $\rho_v = B_f / V$. Vegetation density is important for flow veg-etation interactions and to determine whether biomass will be concentrated near the bed or water surface. The volume of aquatic vegetation can be determined by submerging vegetation in water and measuring volumetric displacement. This method is simple to perform but lacks precision. Another approach is to use Archimedes principle (Hughes, 2005) where the vegetation is suspended in water and the weight force of the displaced liquid is measured. This can be implemented by placing vegetation inside a container such as a barrel with holes in it (Figure 3.9b) or a mesh bag and submerging it in a tank of water Figure 3.9c). Heavy weights can be added to the container to ensure negative buoyancy. The procedure for determining vegetation volume from the mass of displaced water is shown in Figure 3.10.

BIOMETRICS AND BIOMECHANICS

Distributions of plant biomass or volume throughout a study site address the questions of 'where does it grow?' and 'how much is present?', but to address questions such as 'why does it grow there?' may require further information on vegetation biometrics and biomechanics. Biometric measurements cover parameters such as density, average stem diameter, stem length per unit mass, and leaf counts per unit mass (Biggs, 2017). Biomechanical measurements comprise tension and bending tests to determine parameters such as Young's modulus, breaking stress, breaking strain, breaking force, flexural rigidity, tensile toughness, degree of elasticity, and energy ratio (Niklas and Spatz, 2012; Miler et al., 2012; Vettori, 2016; Biggs, 2017). Care should be taken that aquatic plant stems do not dry out during testing, and biomechanical tests can even be performed underwater (Łoboda et al., 2018).

AQUATIC VEGETATION AT LARGER SPATIAL SCALES

COVER

To assess aquatic vegetation abundance at larger spatial scales, it is useful to quantify percentage cover. Percentage cover is simply calculated as: $Cover = 100\% \times \dfrac{A_{VT}}{A_R} = 100\% \times B_{SA}$, where A_{VT} is the total planform area of aquatic vegetation, A_R is the river planform area, and B_{SA} is the commonly reported parameter of surface area blockage. In practice, the total planform area A_c of a class c at a study site is $A_c = \sum_{p=1}^{Pc} a_{c,p}$, where $a_{c,p}$ is the area of each polygon in class c, and P_c is the total number of polygons in class c. The area of the river reach A_R can be found by demarcating the boundaries of the river with polygons (Figures 3.5 and 3.8). For short reaches, the boundaries of the whole reach can be demarcated by one polygon; however, for longer reaches (Figure 3.7), it can be convenient to use multiple adjoining polygons. The total area of the river reach is, therefore, $A_R = \sum_{p=1}^{P_R} a_{R,p}$, where

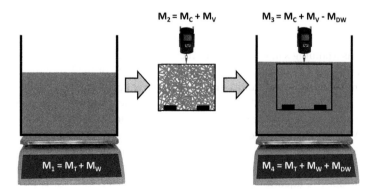

FIGURE 3.10 Procedure to determine vegetation volume by the mass of displaced water, where M_T is the tank mass, M_W is the water mass, M_C is the mass of the container and weights, M_V is the vegetation mass, and M_{DW} is the displaced water mass. The volume of the displaced water V_{DW} is M_{DW} divided by the water density. The volume of the vegetation V_V is V_{DW} minus the volume of the container and weights V_C. A benchtop scale or a hanging scale can be used, but the vegetation and container must always be suspended.

p is the polygon index, P_R is the number of polygons in the river class, and $a_{R,\,p}$ is the area of polygon p in the river class. For the total area of aquatic vegetation, there may be more than one species of aquatic vegetation, so it may be necessary to sum across multiple classes. $A_{VT} = \sum_{v=1}^{V} \sum_{p=1}^{P_v} a_{v,p}$, where v is the vegetation class index, V is the number of vegetation classes, p is the polygon index, P_v is the number of polygons in vegetation class v, and $a_{v,p}$ is the area of polygon p in vegetation class v. When calculating vegetation cover by this method (Figure 3.5), it is important that there is no overlap between vegetation polygons during segmentation or the total planform area of aquatic vegetation will be incorrect. If overlapping polygons have been used, then the aquatic vegetation layers must be flattened/combined before calculating total vegetation area. This can be achieved by converting from polygons to a high-resolution raster map for each class, then performing an OR operation on all the raster maps.

An alternative approach to obtain total cover in water bodies dominated by aquatic vegetation is to focus on demarcation of non-vegetated areas (Figure 3.8). The total area of aquatic vegetation is then $A_{VT} = A_R - A_B + A_{IV}$, where A_B is the sum of polygon areas in the non-vegetated river bed class, and A_{IV} is the sum of polygon areas of isolated vegetation that are within the boundaries of polygons of A_B. This approach can be an efficient way to determine total cover comprised of multiple indistinguishable species of aquatic vegetation (Figure 3.8).

BIOMASS

Large-scale estimates of aquatic vegetation biomass can be achieved by combining aerial surveys with field samples. The total biomass M_c of class c at the study site can be estimated as $M_c = A_c F_c$, where A_c is the total planform area of class c, and F_c is an average biomass per unit area (kg/m^2) conversion factor. Calculation of the biomass conversion factor requires samples of aquatic vegetation with known planform area to be collected from throughout the study reach and processed in the laboratory for biomass (as previously discussed). The biomass conversion factor is then calculated as $F_c = \dfrac{\sum_{n=1}^{N_c} m_{c,n}}{\sum_{n=1}^{N_c} a_{c,n}}$, where N_c is the total number of samples/specimens collected from class c, $m_{c,n}$ is the mass of each sample/specimen in class c, and $a_{c,n}$ is the planform area of each sample/specimen in class c. F_c can be obtained using any of the biomass metrics (e.g. fresh biomass, dry biomass, or ash-free dry biomass); however, which biomass metric is used and how samples were processed must be explicitly stated.

HYDRAULIC INTERACTIONS

Aquatic vegetation can increase flow resistance in rivers by an order of magnitude or more (Chow, 1959; Nikora et al., 2008). Flow resistance is usually parameterised with resistance (or roughness) coefficients, such as Manning's n, Chezy's C, or the Darcy–Weisbach friction factor f (Nikora et al., 2008). These coefficients can be calculated from measurements of the hydraulic characteristics of rivers; however, they are difficult to predict directly from river geometry and aquatic vegetation abundance (Arcement and Schneider, 1989). Often the most practical method is to survey a river reach and compare it to a set of reference reaches to estimate resistance coefficients (Chow, 1959; Barnes, 1967; Hicks and Mason, 1991). Another approach in vegetated rivers is to parameterise aquatic vegetation blockage (e.g. cross-sectional blockage B_X, surface area blockage B_{SA}, or volumetric blockage B_V), then look up published relationships to estimate resistance (Green, 2005; Nikora et al., 2008). The most detailed study of aquatic vegetation hydraulic resistance was that of Nikora et al. (2008) who found that the reach averaged cross-sectional blockage was a good predictor of resistance. Cross-sectional blockage can be defined as $B_X = \dfrac{w_v h_v}{WH} = \alpha \dfrac{h_v}{H}$, where w_v is the width of the cross-section that is occupied by vegetation, W is the total width of the cross-section, h_v is the average height of vegetation in the cross-section, H is the average depth of the cross-section, and α is the proportion of the cross-section's width occupied by vegetation. Cross-sectional blockage is usually reported as an average throughout a study reach, such that $B_X = \frac{1}{N}\sum_{i=1}^{N}\frac{w_{v,i}h_{v,i}}{W_i H_i} - \frac{1}{N}\sum_{i=1}^{N}\alpha_i\frac{h_{v,i}}{H_i}$, where angle i is the cross-section index, and N is the number of cross-sections. B_X is practical to obtain from field measurements for short river reaches or with few cross-sections that are spaced far apart. However, to accurately characterise vegetation in a river reach, it is important to have a high number of cross-sections and to have them spaced relatively closely together. As the number of cross-sections increases, the cross-sectional blockage will tend to the volumetric blockage B_V.

The volumetric blockage can be defined as $B_V = \frac{V_{VT}}{V_R} = \frac{h_v A_{VT}}{H A_R} = \frac{h_v}{H}B_{SA} \approx \beta B_{SA}$, where V_{VT} is the volume of vegetation in the study reach (vegetation field volume), V_R is the volume of the study reach, h_v is the average height of vegetation throughout the study reach, H is the average water depth throughout the study reach, and β is a conversion parameter from B_{SA} to B_V that approximates $\frac{h_v}{H}$ based on an experimentally realistic sampling campaign of river bathymetry and vegetation heights. If the study reach is morphologically diverse (e.g. comprised of pools, riffles, and runs), then it is important that average water depth is derived from measurement points distributed throughout the study site, not just above aquatic vegetation. Since $B_X \approx B_V \approx \beta B_{SA}$, it is feasible to obtain a close approximation for B_X from an aerial survey for B_{SA} and ground truth measurements for β. Relationships between B_X and resistance coefficients can then be found from Nikora et al. (2008) (e.g. Manning's $n = 0.025e^{3.0B_X}$ with $r^2 = 0.89$). Alternative equations have been proposed by Green (2005); however, their use is not recommended as they were obtained by a simple linear fit (e.g. $n = 0.0043B_X - 0.0497$ with $r^2 = 0.66$) through field data that was clearly non-linear (Green, 2005; Figure 3.4a). The resistance equations of Green (2005) are also physically invalid as they predict a negative resistance coefficient when there is no vegetation in the river (i.e. when $B_X = 0$ then $n = -0.0497$). By contrast, the equations of Nikora et al. (2008) provide a Manning's n value of 0.025 when there is no vegetation present, which is a realistic base resistance for natural rivers (Chow, 1959).

The approach detailed above will provide a workable estimate of aquatic vegetation resistance in rivers; however, it should be mentioned that the exact relationship between hydraulic resistance and aquatic vegetation is extremely complicated and difficult to parameterise. For example, spatial distributions of aquatic vegetation, biomechanics, dynamic reconfiguration, porosity, total surface area, and many other factors influence hydraulic resistance. The dependence of B_X and B_V on water depth can also be problematic, as they are therefore discharge dependent. Further research is needed to determine whether the depth invariant metrics of B_{SA} or aquatic vegetation biomass are good predictors of hydraulic resistance.

Aquatic vegetation also shows interesting variations in tolerances (or preferences) for hydraulic conditions amongst species. For example, Biggs et al. (2018) found that the highest abundance of *Ranunculus penicillatus* occurred in river reaches with specific hydraulic characteristics (notably stream power per unit area). Feedbacks between aquatic vegetation growth and hydraulics also become important at larger spatial scales. For example, as aquatic vegetation grows, hydraulic resistance increases, mean velocity decreases, and depth increases. These feedbacks may adjust hydraulic conditions to an optimum for peak vegetation biomass, above which self-limiting factors (such as sedimentation and vegetation removal during floods) could limit further biomass accrual. At larger spatial scales river morphology is also affected by flow interactions with aquatic vegetation. For example, flow redirection around dense clusters of aquatic vegetation can cause localised bank erosion, while fine sediments accumulate within and downstream from vegetation clusters (Figure 3.11).

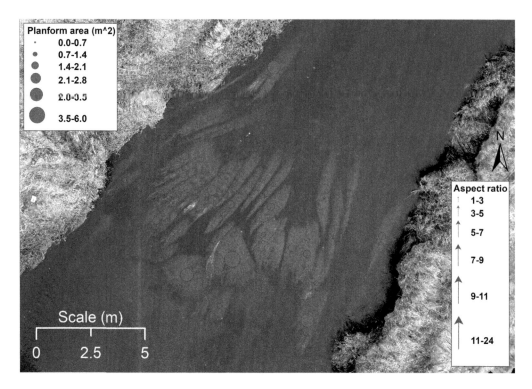

FIGURE 3.11 Dense clusters of *Ranunculus penicillatus* in the River Urie, UK. Aquatic vegetation plan-form area can be used to identify areas of high biomass. Orientation scaled by aspect ratio provides an indication of local flow direction and magnitude.

Targeted Removal of Aquatic Vegetation

Aquatic vegetation at low to medium biomass provides many ecological and environmental benefits. However, current management practices often entail either leaving or removing all aquatic vegetation. Neither of these options is optimal and a better solution is required. The use of UAS to survey aquatic vegetation (Figure 3.11) provides river managers with high-resolution maps to plan management campaigns and only target sites with problematic levels of biomass.

EMERGING TECHNOLOGIES IN AQUATIC VEGETATION SURVEYING

Hydraulic Measurements from UAS

The 'Fieldwork for Hydraulic Measurements' section detailed standard methods for obtaining hydraulic data; however, in a book on UAS for Environmental Applications, it is worth mentioning some exciting new ways to obtain hydraulic data from the air. In clear water, bathymetry can be obtained from through water imagery corrected for surface refraction (Woodget et al., 2015; Dietrich, 2017). For terrestrial areas, it is possible to obtain a DEM from aerial imagery using SfM (SFM) techniques with comparable accuracy to aerial LIDAR (Fonstad et al., 2013). For long river reaches with significant slope (such as braided gravel bed rivers), it is possible to derive river slope from the SfM (SFM) DEM. It may also be possible to obtain surface velocities or even discharge from UAS-based imagery in the future (Detert et al., 2017). These developments would greatly improve the efficiency of gathering complementary hydraulic data during UAS surveys of aquatic vegetation.

MULTISPECTRAL IMAGERY

Standard RGB cameras have wide bands of sensitivity in three colour channels. This information can then be used to quantify the colour of reflected light based on the relative contribution of the three colour channels. This provides sufficient resolution to distinguish significant differences in reflectance spectra (e.g. a red car vs a green plant) but struggles to distinguish subtle spectral differences between vegetation types or species. For this purpose, multispectral cameras can be very useful. They are usually comprised of six or more cameras, with band-pass filters that only allow transmission of light in a narrow wavelength band (e.g. 10 nm bandwidth). These band-pass filters can be user selected to target reflectance or absorption peaks of pigments in the aquatic vegetation they are surveying. For example, to survey algae and cyanobacteria, published absorption peaks of chlorophyll A, chlorophyll B, phycoerythrin, and phycocyanin may be selected (Rowan, 1989).

Multispectral cameras have few spectral bands, compared to the hundreds of hyperspectral cameras; however, they are cheaper and easier to use. They are also able to capture a 2D image from each of their sub-cameras at one instant in time (compared to a hyperspectral scan) which enables imagery of moving objects. The disadvantage of multispectral cameras being comprised of sub-cameras is that the origin and alignment of each camera is not the same. Therefore, post processing is required to align imagery before generating georeferenced multispectral orthomosaics. For Airphen brand multispectral cameras, there is a plugin for AgiSoft Photoscan that takes care of this. Other disadvantages of multispectral cameras are that targeted wavelengths must be known in advance and the limited spectral bands may provide insufficient information to separate vegetation types or species. For routine monitoring of total aquatic vegetation, multispectral imagery can be very effective; however, for classification to species level, hyperspectral imagery is more suitable (Richardson, 1996).

HYPERSPECTRAL IMAGERY

Recent advances in hyperspectral cameras have resulted in high-performance, low-weight systems that can be flown on UAS (e.g. the Resonon Pika L Airborne System or the Headwall Nano-Hyperspec). These systems are used to generate georeferenced hyperspectral data cubes with hundreds of spectral channels and provide exciting opportunities for automated classification of aquatic vegetation. Hyperspectral data is well suited to supervised classification approaches, such as maximum likelihood (Shafri et al., 2007) and SAM algorithms (Hirano et al., 2003; Hestir et al., 2008), or machine learning approaches, such as SVMs (Mountrakis et al., 2011; Parsons et al., 2018), and random forest classifiers (Abdel-Rahman et al., 2014). Training data can be provided by selecting pixels corresponding to specific vegetation classes (Hirano et al., 2003) or by using hand-held spectroradiometers for ground-truthed reflectance spectra (Penuelas et al., 1993). If properly normalised for incident lighting and collected with standardised techniques (Pfitzner et al., 2006; Hueni et al., 2017), then this data can be used to develop a library of spectral signatures (Williams et al., 2003; Hueni et al., 2009; Rossel et al., 2016) covering a wide variety of aquatic vegetation species. The price of hyperspectral camera systems currently limits their widespread use; however, as technology improves and mass production drives down costs, hyperspectral remote sensing of aquatic vegetation from UAS is likely to become routine.

ACKNOWLEDGEMENTS

The author would like to thank David Plew, Jochen Bind, and Murray Hicks for advice and contributions that improved this chapter. The author would also like to thank Shaun Fraser, Christopher Gibbins, Konstantinos Papadopoulos, David R. Green, and Vladimir Nikora for assistance with fieldwork and valuable advice. Funding for this book chapter and the research contained within it was provided by the following grants, which are gratefully acknowledged:

- 'Drone Flow Research Programme', MDIE Smart Ideas Grant (C01X1812).
- 'Flow-Macrophyte-Sediment Research Programme', NIWA Grant (CDPD1706) under the 'Sustainable Water Allocation Research Programme' (SWAP).
- 'Hydrodynamic Transport in Ecologically Critical Heterogeneous interfaces' (HYTECH), under the European Union's Seventh Framework Programme (Marie Curie FP7-PEOPLE-2012-ITN), European Commission grant agreement number (316546).

REFERENCES

Abdel-Rahman, E., Mutanga, O., Adam, E. and Ismail, R. (2014). Detecting Sirex noctilio grey-attacked and lightning-struck pine trees using airborne hyperspectral data, random forest and support vector machines classifiers. *ISPRS Journal of Photogrammetry and Remote Sensing*, 88(1), 48–59.

Arcement, G. and Schneider, V. (1989). Guide for selecting Manning's roughness coefficients for natural channels and flood plains (No. 2339). USGS, Washington.

Barnes, H. (1967). Roughness characteristics of natural channels (No. 1849). USGS, Washington.

Bickel, T. and Perrett, C. (2015). Precise determination of aquatic plant wet mass using a salad spinner. *Canadian Journal of Fisheries and Aquatic Sciences*, 73(1), 1–4.

Biggs, H. (2017). Flow-vegetation interactions: from the plant to the patch mosaic scale (Doctoral dissertation). University of Aberdeen, Aberdeen, UK.

Biggs, H., Nikora, V., Gibbins, C., Fraser, S., Green, D., Papadopoulos, K. and Hicks, M. (2018). Coupling Unmanned Aerial Vehicle (UAV) and hydraulic surveys to study the geometry and spatial distribution of aquatic macrophytes. *Journal of Ecohydraulics*, 3(1), 45–58.

Boon, M., Greenfield, R. and Tesfamichael, S. (2016). Wetland assessment using unmanned aerial vehicle (UAV) photogrammetry. *The International Archives of the Photogrammetry, Remote Sensing and Spatial Information Sciences*, XLI-B1, 781–788.

Butcher, R. (1933). Studies on the ecology of rivers: I. On the distribution of macrophytic vegetation in the rivers of Britain. *The Journal of Ecology*, 21(1), 58–91.

Capers, R. (2000). A comparison of two sampling techniques in the study of submersed macrophyte richness and abundance. *Aquatic Botany*, 68(1), 87–92.

Chow, V. (1959). *Open channel hydraulics*. McGraw-Hill, New York, NY.

Congalton, R. (1991). A review of assessing the accuracy of classifications of remotely sensed data. *Remote sensing of environment*, 37(1), 35–46.

Dawson, F. (1989). Ecology and management of water plants in lowland streams. *Freshwater Biology Annual Report*, 57(1), 43–60.

Detert, M., Johnson, E. and Weitbrecht, V. (2017). Proof-of-concept for low-cost and non-contact synoptic airborne river flow measurements. *International Journal of Remote Sensing*, 38(8–10), 2780–2807.

Dietrich, J. (2017). Bathymetric Structure-from-Motion: extracting shallow stream bathymetry from multi-view stereo photogrammetry. *Earth Surface Processes and Landforms*, 42(2), 355–364.

Dugdale, S. (2007). An evaluation of imagery from an Unmanned Aerial Vehicle (UAV) for the mapping of intertidal macroalgae on Seal Sands, Tees Estuary, UK (Masters dissertation), Durham University, Durham, UK.

Edwards, R. and Owens, M. (1960). The effects of plants on river conditions: I. Summer crops and estimates of net productivity of macrophytes in a chalk stream. *The Journal of Ecology*, 48(1), 151–160.

Figueiredo, B., Mormul, R. and Thomaz, S. (2015). Swimming and hiding regardless of the habitat: prey fish do not choose between a native and a non-native macrophyte species as a refuge. *Hydrobiologia*, 746(1), 285–290.

Flynn, K. and Chapra, S. (2014). Remote sensing of submerged aquatic vegetation in a shallow non-turbid river using an unmanned aerial vehicle. *Remote Sensing*, 6(12), 12815–12836.

Fonstad, M., Dietrich, J., Courville, B., Jensen, J. and Carbonneau, P. (2013). Topographic structure from motion: a new development in photogrammetric measurement. *Earth Surface Processes and Landforms*, 38(4), 421–430.

Franklin, P., Dunbar, M. and Whitehead, P. (2008). Flow controls on lowland river macrophytes: a review. *Science of the Total Environment*, 400(1), 369–378.

Garrison, P., Marshall, D., Stremick-Thompson, L., Cicero, P. and Dearlove, P. (2005). *Effects of pier shading on littoral zone habitat and communities in Lakes Ripley and Rock, Jefferson County, Wisconsin*. Wisconsin Department of Natural Resources, PUB-SS-1006, Madison, WI.

Göktoğan, A., Sukkarieh, S., Bryson, M., Randle, J., Lupton, T. and Hung, C. (2010). A rotary-wing unmanned air vehicle for aquatic weed surveillance and management. *Journal of Intelligent and Robotic Systems*, 57(1–4), 467.

Green, J. (2005). Comparison of blockage factors in modelling the resistance of channels containing submerged macrophytes. *River Research and Applications*, 21(6), 671–686.

Gurnell, A., Van Oosterhout, M., De Vlieger, B. and Goodson, J. (2006). Reach-scale interactions between aquatic plants and physical habitat: river Frome, Dorset. *River Research and Applications*, 22(6), 667–680.

Halliday, D., Resnick, R. and Walker, J. (2005). *Fundamentals of Physics* (7th ed.). Wiley, New York.

Hestir, E., Khanna, S., Andrew, M., Santos, M., Viers, J., Greenberg, J., Rajapakse, S. and Ustin, S. (2008). Identification of invasive vegetation using hyperspectral remote sensing in the California Delta ecosystem. *Remote Sensing of Environment*, 112(11), 4034–4047.

Hicks, M. and Mason, P. (1991). *Roughness Characteristics of New Zealand Rivers: A Handbook for Assigning Hydraulic Roughness Coefficients to River Reaches by the "Visual Comparison" Approach*. Water Resources Survey, DSIR, Wellington, New Zealand.

Hirano, A., Madden, M. and Welch, R. (2003). Hyperspectral image data for mapping wetland vegetation. *Wetlands*, 23(2), 436–448.

Hueni, A., Nieke, J., Schopfer, J., Kneubühler, M. and Itten, K. (2009). The spectral database SPECCHIO for improved long-term usability and data sharing. *Computers and Geosciences*, 35(3), 557–565.

Hueni, A., Damm, A., Kneubuehler, M., Schläpfer, D. and Schaepman, M. (2017). Field and airborne spectroscopy cross-validation—some considerations. *IEEE Journal of Selected Topics in Applied Earth Observations and Remote Sensing*, 10(3), 1117–1135.

Hughes, S. (2005). Archimedes revisited: A faster, better, cheaper method of accurately measuring the volume of small objects. *Physics Education*, 40(5), 468–474.

Hung, C., Xu, Z. and Sukkarieh, S. (2014). Feature learning based approach for weed classification using high resolution aerial images from a digital camera mounted on a UAV. *Remote Sensing*, 6(12), 12037–12054.

Husson, E., Hagner, O. and Ecke, F. (2014). Unmanned aircraft systems help to map aquatic vegetation. *Applied Vegetation Science*, 17(3), 567–577.

Husson, E., Ecke, F. and Reese, H. (2016). Comparison of manual mapping and automated object-based image analysis of non-submerged aquatic vegetation from very-high-resolution UAS images. *Remote Sensing*, 8(9), 724.

Ibáñez, C., Caiola, N., Rovira, A. and Real, M. (2012). Monitoring the effects of floods on submerged macrophytes in a large river. *Science of the Total Environment*, 440(1), 132–139.

Kemp, P., Sear, D., Collins, A., Naden, P. and Jones, I. (2011). The impacts of fine sediment on riverine fish. *Hydrological Processes*, 25(11), 1800–1821.

Killgore, K. and Hoover, J. (2001). Effects of hypoxia on fish assemblages in a vegetated waterbody. *Journal of Aquatic Plant Management*, 39(1), 40–44.

Kirkman, H. (1996). Baseline and monitoring methods for seagrass meadows. *Journal of Environmental Management*, 47(2), 191–201.

Klemas, V. (2015). Coastal and environmental remote sensing from unmanned aerial vehicles: an overview. *Journal of Coastal Research*, 31(5), 1260–1267.

Łoboda, A., Przyborowski, Ł., Karpiński, M., Bialik, R. and Nikora, V. (2018). Biomechanical properties of aquatic plants: the effect of test conditions. *Limnology and Oceanography: Methods*, 16(4), 222–236.

Madsen, J. (1993). Biomass techniques for monitoring and assessing control of aquatic vegetation. *Lake and Reservoir Management*, 7(2), 141–154.

Madsen, J., Chambers, P., James, W., Koch, E. and Westlake, D. (2001). The interaction between water movement, sediment dynamics and submersed macrophytes. *Hydrobiologia*, 444(1–3), 71–84.

Madsen, J. and Wersal, R. (2017). A review of aquatic plant monitoring and assessment methods. *Journal of Aquatic Plant Management*, 55(1), 1–12.

Marshall, T. and Lee, P. (1994). Mapping aquatic macrophytes through digital image analysis of aerial photographs: an assessment. *Journal of Aquatic Plant Management*, 32(1), 61–66.

Miler, O., Albayrak, I., Nikora, V. and O'Hare, M. (2012). Biomechanical properties of aquatic plants and their effects on plant–flow interactions in streams and rivers. *Aquatic Sciences*, 74(1), 31–44.

Miranda, L. and Hodges, K. (2000). Role of aquatic vegetation coverage on hypoxia and sunfish abundance in bays of a eutrophic reservoir. *Hydrobiologia*, 427(1), 51–57.

Mountrakis, G., Im, J. and Ogole, C. (2011). Support vector machines in remote sensing: a review. *ISPRS Journal of Photogrammetry and Remote Sensing*, 66(3), 247–259.

Murphy, K. (1990). Survey and monitoring of aquatic weed problems and control operations. Aquatic Weeds. In: Pieterse, A., Murphy, K. (Eds.), *The Ecology and Management of Nuisance Aquatic Vegetation*. Oxford University Press, Oxford, pp. 228–237.

Murphy, K. and Barrett, P. (1990). Chemical control of aquatic weeds. In: Pieterse, A., Murphy, K. (Eds.), *The Ecology and Management of Nuisance Aquatic Vegetation*. Oxford University Press, Oxford, pp. 136–173.

Neil, J., Kamaitas, G. and Robinson, G. (1985). Aquatic plant assessment Cook's Bay, Lake Simcoe. Distribution, biomass, tissue nutrients and species composition. Report prepared for the Ontario Ministry of the Environment, Rexdale.

Neil, J., Graham, J. and Warren, J. (1988). Growth of macrophytes in Cook's Bay, Lake Simcoe. Report prepared for the Ontario Ministry of the Environment, Rexdale.

Niklas, K. and Spatz, H. (2012). *Plant Physics*. University of Chicago Press, Chicago.

Nikora, V., Larned, S., Nikora, N., Debnath, K., Cooper, G. and Reid, M. (2008). Hydraulic resistance due to aquatic vegetation in small streams: field study. *Journal of Hydraulic Engineering*, 134(9), 1326–1332.

Österling, M., Arvidsson, B. and Greenberg, L. (2010). Habitat degradation and the decline of the threatened mussel *Margaritifera margaritifera*: influence of turbidity and sedimentation on the mussel and its host. *Journal of Applied Ecology*, 47(4), 759–768.

Overstreet, B. and Legleiter, C. (2017). Removing sun glint from optical remote sensing images of shallow rivers. *Earth Surface Processes and Landforms*, 42(2), 318–333.

Parsons, M., Bratanov, D., Gaston, K. and Gonzalez, F. (2018). UAVs, hyperspectral remote sensing, and machine learning revolutionizing reef monitoring. *Sensors*, 18(7), 2026.

Penuelas, J., Gamon, J., Griffin, K. and Field, C. (1993). Assessing community type, plant biomass, pigment composition, and photosynthetic efficiency of aquatic vegetation from spectral reflectance. *Remote Sensing of Environment*, 46(2), 110–118.

Pfitzner, K., Bollhöfer, A. and Carr, G. (2006). A standard design for collecting vegetation reference spectra: implementation and implications for data sharing. *Journal of Spatial Science*, 51(2), 79–92.

Richards, J. (2013). *Remote Sensing Digital Image Analysis* (5th ed.). Springer, Berlin.

Richardson, L. (1996). Remote sensing of algal bloom dynamics. *BioScience*, 46(7), 492–501.

Rockwell, W. (2003). Summary of a survey of the literature on the economic impact of aquatic weeds. Report to the Aquatic Ecosystem Restoration Foundation, AERF Publications, Marietta, Georgia.

Rooney, N. and Kalff, J. (2000). Inter-annual variation in submerged macrophyte community biomass and distribution: the influence of temperature and lake morphometry. *Aquatic Botany*, 68(4), 321–335.

Rooney, N. and Kalff, J. (2003). Submerged macrophyte-bed effects on water-column phosphorus, chlorophyll A, and bacterial production. *Ecosystems*, 6(8), 797–807.

Rossel, R., Behrens, T., Ben-Dor, E., Brown, D., Demattê, J., Shepherd, K., Shi, Z., Stenberg, B., Stevens, A., Adamchuk, V. and Aïchi, H. (2016). A global spectral library to characterize the world's soil. *Earth-Science Reviews*, 155(1), 198–230.

Rowan, K. (1989). *Photosynthetic Pigments of Algae*. Cambridge University Press, Cambridge.

Shafri, H., Suhaili, A. and Mansor, S. (2007). The performance of maximum likelihood, spectral angle mapper, neural network and decision tree classifiers in hyperspectral image analysis. *Journal of Computer Science*, 3(6), 419–423.

Shupryt, M. and Stelzer, R. (2009). Macrophyte beds contribute disproportionately to benthic invertebrate abundance and biomass in a sand plains stream. *Hydrobiologia*, 632(1), 329–339.

Silva, T., Costa, M., Melack, J. and Novo, E. (2008). Remote sensing of aquatic vegetation: theory and applications. *Environmental Monitoring and Assessment*, 140(1–3), 131–145.

Tena, A., Vericat, D., Gonzalo, L. and Batalla, R. (2017). Spatial and temporal dynamics of macrophyte cover in a large regulated river. *Journal of Environmental Management*, 202(2), 379–391.

Van der Zweerde, W. (1990). Biological control of aquatic weeds by means of phytophagous fish. In: Pieterse, A., Murphy, K. (Eds.), *The Ecology and Management of Nuisance Aquatic Vegetation*. Oxford University Press, Oxford, pp. 201–221.

Vermaire, J., Prairie, Y. and Gregory-Eaves, I. (2011). The influence of submerged macrophytes on sedimentary diatom assemblages. *Journal of Phycology*, 47(6), 1230–1240.

Vettori, D. (2016). Hydrodynamic Performance of Seaweed Farms: an Experimental Study at Seaweed Blade Scale (Doctoral dissertation). University of Aberdeen, Aberdeen, UK.

Visser, F., Wallis, C. and Sinnott, A. (2013). Optical remote sensing of submerged aquatic vegetation: opportunities for shallow clearwater streams. *Limnologica-Ecology and Management of Inland Waters*, 43(5), 388–398.

Visser, F., Buis, K., Verschoren, V. and Schoelynck, J. (2018). Mapping of submerged aquatic vegetation in rivers from very high-resolution image data, using object-based image analysis combined with expert knowledge. *Hydrobiologia*, 812(1), 157–175.

Wade, P. (1990). Physical control of aquatic weeds. In: Pieterse, A., Murphy, K. (Eds.), *The Ecology and Management of Nuisance Aquatic Vegetation*. Oxford University Press, Oxford, pp. 93–135.

Watts, A., Ambrosia, V. and Hinkley, E. (2012). Unmanned aircraft systems in remote sensing and scientific research: Classification and considerations of use. *Remote Sensing*, 4(6), 1671–1692.

Westlake, D. (1965). Some basic data for investigations of the productivity of aquatic macrophytes. In: Goldman, C. (Ed.), *Primary Productivity in Aquatic Environments*, University of California Press, Berkeley, pp. 231–248.

Williams, D., Rybicki, N., Lombana, A., O'Brien, T. and Gómez, R. (2003). Preliminary investigation of submerged aquatic vegetation mapping using hyperspectral remote sensing. *Environmental Monitoring and Assessment*, 81(1–3), 383–392.

Wood, P. and Armitage, P. (1997). Biological effects of fine sediment in the lotic environment. *Environmental Management*, 21(2), 203–217.

Wood, K., Stillman, R., Clarke, R., Daunt, F. and O'Hare, M. (2012). Measuring submerged macrophyte standing crop in shallow rivers: a test of methodology. *Aquatic Botany*, 102(1), 28–33.

Woodget, A., Carbonneau, P., Visser, F. and Maddock, I. (2015). Quantifying submerged fluvial topography using hyperspatial resolution UAS imagery and structure from motion photogrammetry. *Earth Surface Processes and Landforms*, 40(1), 47–64.

4 Unmanned Aerial Vehicles for Riverine Environments

Monica Rivas Casado
Cranfield University

Amy Woodget
Loughborough University

Rocio Ballesteros Gonzalez
Castilla-La Mancha University

Ian Maddock
University of Worcester

Paul Leinster
Cranfield University

CONTENTS

INTRODUCTION

The key role of rivers in society has been recognised for centuries, from the provision of fertile valleys (e.g., Nile) to their recreational value (e.g., Danube). However, rivers and freshwater ecosystems in general have always been at the receiving end of anthropogenic activity, from intense pollution to obstruction caused by hard engineering structures. In the last decades, regulatory efforts across the world have focused on the protection and even improvement of the ecological quality of rivers, with some countries granting rivers the same rights as human beings [1]. Perhaps the most relevant regulatory example is the European Union's Water Framework Directive (WFD) [2], which aims to achieve good ecological status or potential of inland and coastal waters. The implementation of such regulatory frameworks requires robust monitoring programmes that provide information concerning the chemical, physico-chemical, biological, and hydromorphological characteristics, amongst others.

The chemical and physico-chemical characteristics describe the quality of the water and the existence of any source of pollution, such as lead from industrial activity or sediments from agricultural practice. The biological elements inform us about the particular species (fauna and flora) that inhabit the river, and the hydromorphological characteristics describe the quantity and dynamics of water flow and connection to groundwater bodies (hydrology), river depth and width variation (morphology), structure and substrate of the river, structure of the riparian zone, and river continuity.

From a broader perspective, monitoring programmes are also required to better inform planning and management decisions. This is of paramount importance when it refers to flood management and water abstraction licensing. Lessons learnt from past events can inform the strategies to be followed in future and perhaps even during more challenging events. These monitoring programmes are also required from a scientific perspective too, for example, to improve our understanding of erosion and deposition processes [3] to develop accurate flood prediction tools. Our knowledge of the relation between hydromorphology conditions and biology is still rather limited. The implementation of strategic monitoring programmes can also contribute to address multiple gaps in knowledge within that domain; from the identification of the key parameters that drive fish pass attractiveness [4] to the recognition of the drivers for eutrophication [5].

The range of monitoring approaches available is vast and varied, ranging from purely observational in situ methods to remote sensing techniques (e.g., satellite imagery). The choice of a suitable technique depends upon the specific parameters under investigation. Chemical and physicochemical properties rely on strict protocols for data collection, whereas multiple methods are available for hydromorphological characterisation. For example, the REFORM project [6] summarises multiple methods, models, and tools available for hydromorphological assessment across Europe. In brief, these methods rely on the identification of patterns (e.g., hydraulic units or habitat units (Figure 4.1)) within a river reach that describe areas of homogeneous hydraulic or habitat characteristics (Table 4.1). These in turn are used as proxy indicators of the suitability of the reach to hold specific species (fauna and flora) and the overall river reach quality.

Regarding methods used for hydromorphological monitoring, a trade-off exists between resolution and wide-area coverage. Typically, remote sensing methods can provide wide-area but low-resolution information whereas in situ methods provide localised but generally high-resolution data. For example, the Environment Agency in England uses satellite imagery and aerial imagery from manned aircraft to broadly determine flood extent and impact [7]. In contrast, scientists have relied on high-resolution (≈ 2 cm) Unmanned Aerial Vehicle (UAV) aerial imagery to determine the volume of debris flows [8] or different types of aquatic plant species [9]. The uptake of UAVs in recent years for environmental surveying tasks has enabled wide-area data capture with unprecedented detail (Figure 4.2).

There can be resistance amongst practitioners in adopting new technologies. Developers and promoters think it is self-evidently a better way of doing things. In contrast, practitioners may think it is nothing more than a toy looking for a grown-up application, which may be of interest for researchers but could never have a role in routine applications. On occasions, this may be because innovative ideas are not marketed properly. Developers should think through and articulate the benefits of the new approach and how this will facilitate the achievement of outcomes that are important to the person they are trying to convince to adopt their idea.

The starting point for effective environmental management is an appropriate level of understanding of the current state of the environment. This has typically required on-site assessment or investigation. Such activities, including walkover surveys, which provide vital information, can be very resource- and or labour-intensive. This then reduces the resources available to implement measures to remediate, protect, and improve the environment.

In this context, there are several current and potential applications of UAVs that will help in the management of river catchments to better protect and improve these environments for people and wildlife. These could be used routinely by practitioners to increase their effectiveness when carrying out their environmental assessment and management work. UAVs can also be used in combination with other remote sensing technologies, Geographical Information Systems (GIS), modelling, and visualisation approaches to further extend their usefulness. DeBell et al. [12] review the potential

FIGURE 4.1 Examples of selected habitat and hydraulic units.

TABLE 4.1
Examples of the Most Common Habitats and Hydraulic Units Based on [10,11]

Unit	Description
Cascade/rapid	Highly turbulent series of short falls and small scour basins, frequently characterised by very large substrate and a stepped profile
Chute	Narrow steep slots or slides in bedrock
Run	Moderately fast and shallow gradient with ripples on the water surface. Deeper than riffles with little, if any, substrate breaking the surface
Riffle	Most common type of turbulent fast water mesohabitat in low-gradient alluvial channels. Substrate is finer than other fast turbulent mesohabitats. Less white water, with some substrate breaking the surface
Glide	Smooth 'glass-like' surface or with some minor boundary turbulence, with visible flow movement along the surface. Relatively shallow compared to pools
Pool	Relatively deep and slow flowing (compared to glides), with fine substrate. Distinct deepening of the channel compared to upstream and downstream sections. Usually little surface water movement visible

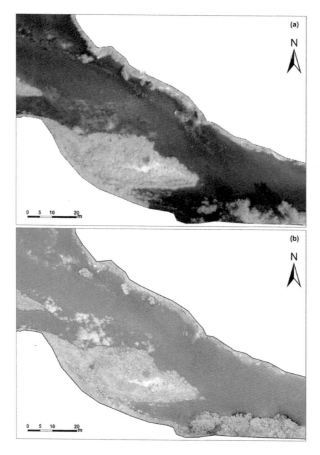

FIGURE 4.2 Comparison of UAV imagery of different resolutions (a) 25 cm and (b) 2.5 cm for a reach along the River Dee, UK.

applications of UAVs for large-scale water resource management which include erosion control, failure of water resources infrastructure, control of invasive species, and pollution control, amongst others. Some of these applications that are already feasible and potential developments are outlined below.

- **Catchment flyover surveys:** the information gathered from UAVs, such as bankside conditions, areas of erosion, land-use and management practices, and the extent of invasive plants, will be similar to that obtained through walkover surveys. This information can then be used to assess in-river and bankside habitat quantity and quality.

    UAVs should enable more frequent repeat surveys to be carried out than existing approaches allow, for example, to see if there are seasonal impacts or changes in land management practices that are affecting the catchment and river environments. In addition, UAVs can operate in conditions that other remote sensing systems, such as satellites and planes, cannot function (e.g., cloud cover). It may also be possible to use UAVs in combination with other remote-controlled surveillance systems, such as radio-controlled boats, to extend their monitoring capabilities and improve the position of the data collected.
- UAVs offer new possibilities for fluvial surveillance by providing georeferenced imagery with high spatial and temporal resolution in near real time. These platforms have evolved rapidly becoming more flexible than the traditional remote sensing platforms.

- **Hydromorphology flyover surveys:** these are a subset of the surveys described above. Good agreement has been found between the habitat and ecology classifications derived from UAV collected data and the walkover approach typically currently used [13–16].
- **Investigations:** UAVs can potentially help locate where pollutants are entering a river and the extent of any pollution [12]. The pollution pathway can then be tracked back to the source such as spillages and land management practices giving rise to sediment. UAVs can be used to identify occurrences of fly-tipping alongside rivers or within channels, and they also have a potential application in dealing with poachers and other illegal fishing activities.

  Pollution within rivers can be tracked if there is a visible indicator and the extent of fish and other species in distress can be assessed. It may also soon be possible to use UAVs to take water samples from the suspected area of pollution in the river channel for subsequent laboratory analysis [17].
- **River management:** the weed growth within the channel and the condition of the banks can readily be assessed and used to guide maintenance activities [12]. UAVs can subsequently be used to check on the effectiveness of the interventions and as part of routine inspections.
- **Flood risk management:** UAVs can assist with maintenance, inspection, and repair activities. They can be used to check on access to locations where equipment is going to be deployed to ensure it is clear and of the correct width. They can also be used to survey flood risk management assets, in particular, the condition of embankments. Prior to flooding events, they can also be used to identify any build in blockages or build-up of debris on structures such as bridges and screens to determine whether any clearance activities are required.
- UAVs already have a role to play in helping to determine the extent of flooding during and after an event [18,19].They also have a potential role in checking on the functioning of natural flood risk management interventions and in identifying the extent of damage after a flooding event. This would include providing information on the scale of any breaches or other damage. They can also provide information as breaches are being repaired or levels of banks including shingle are being restored.

More work must be carried out before the applications outlined above become an integral part of routine operations. However, UAVs clearly have the potential to fulfil an important range of functions in practical environmental operations, increasing the effectiveness of surveillance, assessment, remediation, protection, and improvement activities.

Within this chapter, we aim to give an insight into the use of UAVs for fluvial applications. In Section 2, we review current data collection and processing considerations. In Section 3, we explore examples of UAV use for riverine feature detection and mapping, and in Section 4, we demonstrate the quantification of physical river parameters from UAV data.

DATA COLLECTION AND PROCESSING CONSIDERATIONS

UAV surveys require careful dedication to field work planning and flight design (Figure 4.3). These considerations range from ensuring the collection of high-quality data to complying with national and international aviation regulation. Lack of planning in any of these aspects could result in incomplete datasets, failure to deploy the UAV, or even jail sentences if the regulation is not followed. In recent years, there has been an increase in the number of near misses and drone accidents (e.g., [20,21]), which may result in more strict regulatory and flight planning requirements. The following paragraphs aim at informing the reader on the key considerations to be taken into account for successful UAV imagery collection and processing within riverine environments.

FIGURE 4.3 Workflow summarising the key steps for UAV aerial imagery data collection. GCP, GPS, and DTM stand for ground control point, global positioning system, and digital terrain model, respectively.

DATA COLLECTION

The workflow showing the key steps and considerations to be followed to collect high-resolution UAV aerial imagery is summarised in Figure 4.3. The process starts with flight project designing after identifying the objective of data collection. Within Section 1 of this chapter, the potential objectives for which data are collected in river environments have been described. General objectives, such as catchment flyover surveys, require large area but coarse resolutions, whereas surveys falling under the investigations categories will require smaller areas but finer resolutions. The objective will therefore determine the type of survey products required, which in turn will inform the choice of UAV platform, sensor type (RGB, multispectral, hyperspectral or thermal cameras), image resolution, flight altitude (i.e., ground sampling distance (GSD) and flight area). In brief, the more detailed (finer resolution) the product is required to be, the lower the flying altitude. The relationship between the sensor type, the flying altitude, and the GSD is summarised in Equation 4.1 and Figure 4.4. The configuration of camera characteristics, such as quality, zoom, focal length,

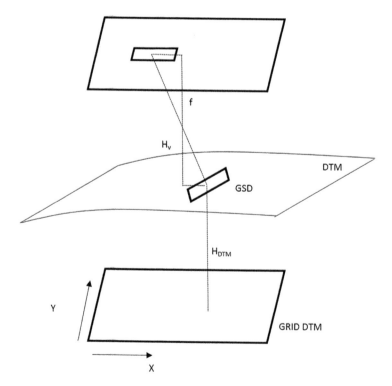

FIGURE 4.4 Spatial relationships between the sensor and the ground cover where f is the focal length, H_v is the altitude, DTM is the digital terrain model, GSD is the ground sample distance, and HDTM is the terrain model altitude [22].

ISO, and image stabilisation, also need to be taken into account to ensure the greatest field of view with the highest resolution and absence of movement is covered.

$$\frac{f}{H_v} = \frac{p}{\text{GSD}}$$
(4.1)

where f is the focal length, H_v is the altitude, p is the pixel size, and GSD is the ground sample distance.

Once the flight project is defined, the next step is to plan the flight (Figure 4.3) to generate a navigation file that guides the UAV to automatically capture the images at specific waypoints, with an appropriate level of overlap and considering all requirements desired by the user (mainly GSD) and those required to perform the data analysis (i.e., photogrammetric workflow).

Field inspections should be performed at this stage to avoid obstacles during the flight, such as buildings or electrical wires. This step also requires the delineation of the area of interest, the specification of the flight direction, the selection of a coordinate system of reference, the identification of the take-off and landing points that could or not be the same, the location of the waypoints, and the selection of the flying speed (Figure 4.5). Map-based or online resources are checked prior to the field inspection to assess the overall ability to fly the area safely (i.e., the area is not in a no-fly zone such as near airports, prisons, and power stations) or whether it has a significant number of other potential hazards (e.g., power lines, bird sanctuary, wind farms).

Users should specify a target level of overlap between consecutive images along and across the flight path. Overlapping ensures that key features are identifiable by the image processing software so that it is possible to accurately stitch all the images together into a single orthoimage. Common practice is to establish the overlap between consecutive images of 60% and the side lapping in 80%. Different over and side lapping levels are required by the photogrammetric process. Therefore, the flight planning software estimates the images' footprint. Recent research also advocates that images should be collected at convergent view angles, rather than solely at nadir [23]. This helps to avoid introducing systematic doming or dishing errors occurring within output digital elevation

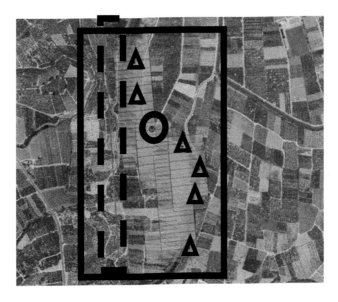

FIGURE 4.5 Example of delineated flight area. The line (within the dashed line rectangle) shows the direction of the flight, the point within the circle is the landing and take-off location, the small circles within the triangles are the GCPs, and the squares (within the solid line rectangle) show the footprint of each image collected.

models (DEMs) as a result of inadequate automated camera lens calibrations within the subsequent Structure-from-Motion (SfM [SFM]) image processing phase [14].

The endurance of the UAV batteries (or alternative power source) will determine the number of flights to be undertaken to collect the required imagery. Planning for successive flights requires attention to (i) the number of waypoints completed within each flight and (ii) the selection of multiple take-off and landing areas.

Another consideration is whether ground control points (GCPs) are to be deployed within the survey area, and if so, how many of them are required. The GCPs enable for the orthoimage to be georeferenced during the photogrammetric process. For that purpose, the coordinates of the centroids of the GCPs are measured with high accuracy using a Real-Time Kinematic (RTK) GPS or total station. These coordinates are then used to georeference the aerial images during the photogrammetric process (Figure 4.6). The number of GCPs to be used depends upon the extent of the area to be surveyed, the accuracy required in the final data products, and access to the site. Some authors have looked in detail at estimating the right number of GCPs for freshwater ecosystem mapping. Perhaps the work by Vericat et al. [24] and James et al. [25] are the most comprehensive. In Vericat et al. [24], a minimum of 58 GCPs are estimated to be required within one hectare to reduce the registration error down to 10 cm when using a third-order polynomial geometric transformation model. That level of location accuracy is perhaps necessary for investigative UAV surveys where the intent is to precisely characterise the surveyed area, for example, to detect the change in erosion and deposition patterns over time. In James et al. [25], a Monte-Carlo approach is proposed to considerably reduce the number of GCPs, with minimal effect on the survey quality and to minimise field effort. Their work highlights the need to take into account datum alignment to gravity, the presence of vegetation at the scale of the GCPs, and absolute 3D positioning. The number of GCPs deployed depends upon the project accuracy required, the network geometry, and the quality of image observations [25]. The work by James et al. [25] uses two examples to explain the relationship between GCP and quality of the geomatic products obtained (DTM). For a survey precision of approximately 50 mm, imaged from 100 m, at a spatial resolution of 23 mm per pixel, a total of 15

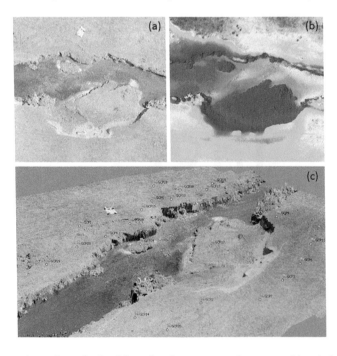

FIGURE 4.6 Geomatic products obtained from the photogrammetric process; (a) orthoimage, (b) DSM, and (c) 3D model.

GCPs are required for one of the case study areas presented. The number of GCPs can be substantially reduced if 'a camera that is sufficiently accurately pre-calibrated that self-calibrating bundle adjustment is not required' [25]. Significant research has gone into developing software that can accurately stitch the imagery without the need for GCPs [26,27].

Specific considerations apply to data collection in riverine environments. The flight needs to be designed to maximise data capture. To our knowledge, the majority of studies to date have focused on the collection of bed-channel imagery and adjacent banks (10 m both sides), with a small number of studies focusing on the use of UAVs to characterise the river floodplain. Within this context, the flights need to be performed via multi-passes that follow the direction of the river and accounting for the oblique angle of the UAV at turning points that allow for changes in orientation (upstream vs. downstream). The flight plan should also assess flight timing and season. Flying at solar noon ensures that the undesirable effects of the shadows are minimised. Flying in winter as opposed to summer ensures the river bed is not occluded by overhanging vegetation.

Further data collection considerations also include planning for safety in both the air and on the ground. Air safety is covered when complying with the aviation authorities of the country where the flight will take place. In the specific case of UK, this relates to the Civil Aviation Authority (CAA) and their regulations CAP393 and CAP722. In brief, the regulations specify the conditions under which UAV flights are authorised and the types of notifications/checks that have to be issued before flying to ensure the safety of all airspace users. The fast pace of technological development within the field requires constant scrutiny and review of existing legislation by national and international aviation authorities. For example, the Federal Aviation Administration (FAA) has recently updated the USA regulatory framework for UAVs [28], the CAA in UK is planning to change the regulation [29] in the near future, and the European Aviation Safety Agency (EASA) in Europe will open a consultation on the topic in the not too distant future [30]. It is therefore of paramount importance for UAV operators to constantly review regulatory changes to ensure safety in their operations.

Ground safety considerations focus on making sure that the survey area is safe to operate and include, for example, making sure the area is free of cattle, all footpaths are under control, no pedestrians interfere with the flight, the UAV is not flying above roads, the weather conditions are favourable, and ensuring the platform is fit to fly. Frequent weather checks are essential to ensure safety when flying and during take-off and landing. Weather conditions, such as wind speed, and rain probability, are decisive as they affect the stability of the UAV and its endurance, in addition to the quality of the imagery collected (e.g., blurred images).

DATA PROCESSING

The independent frames collected with the UAV are processed with photogrammetric techniques to obtain a mosaicked image of the overall area surveyed. Photogrammetry allows the generation of three different geomatic products: dense 3D point clouds, digital surface models (DSMs), and orthoimages (Figure 4.6). These geomatic products are generated automatically by specialised software that combines computational vision techniques with photogrammetry methods based on self-calibration of the camera lens and orientation of the images to increase the accuracy of the final products. The photogrammetric workflow (Figure 4.7) includes: (i) inspection of the imagery, (ii) image alignment, (iii) calibration and pose estimation, (iv) dense matching, (v) mesh and texture building, and (vi) the generation of the geomatic products.

Currently, different commercial software are available (e.g., ImageMaster from Topcon, AgiSoft Photoscan, Pix4D, and Apero-MicMac). The differences between these software programs include the processing time, the quality of the orthoimage, the automatisation of photogrammetric workflow, and whether the software is open source or commercially available.

The photogrammetric workflow (Figure 4.7) initiates with the inspection of the imagery collected. Blurred images (Figure 4.8) should be removed from the set of images to be processed to obtain high-quality geomatic products. Manual selection of non-blurred images is a tedious and a

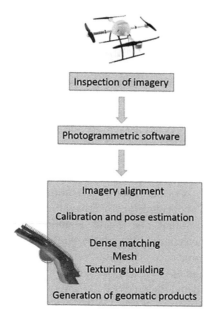

FIGURE 4.7   Photogrammetric workflow.

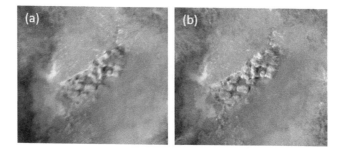

FIGURE 4.8   Example showing (a) blurred and (b) non-blurred images.

highly time-consuming procedure. As a result, recent work has focused on automating the detection of low-quality images [26]. Once the quality of the input imagery has been assessed, the photogrammetric analysis focuses on the alignment of the imagery. This is achieved through the identification and matching of common points on consecutive images. The majority of commercial software packages usually find the position of the camera for each image and define the camera calibration parameters. This is of paramount importance for the generation of accurate geomatic products such as the orthoimage, where all the images need to be corrected for a consistent matching based on the distortion generated during data collection.

Further photogrammetric steps focus on (i) the calibration and pose estimation, (ii) the dense matching, and (iii) mesh and texture building. The calibration and pose estimation module requires the number and specific location of GCPs placed in the surveyed area [31] based on the exact coordinates of the GCP centroids measured in the field. The majority of commercial software programs recommend at least ten GCPs across the area to be reconstructed to achieve results of the highest quality, both in terms of the geometrical precision and georeferencing accuracy, although as mentioned above, some studies have suggested a greater number are required for high precision studies [24,25].

Within the (ii) dense matching and (iii) mesh (geometry) and texture building, the dense point cloud is built from the camera positions and the imagery collected. Most of the commercial software

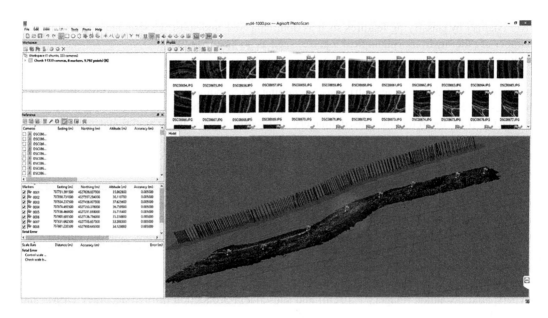

FIGURE 4.9   General view of AgiSoft Photoscan application window.

programs reconstruct a 3D polygonal mesh representing the surveyed area (Figure 4.9) and some
3D editor tools can be used for more complex editing. The texturing step is required for the generation of the orthomosaic, the DSM, and the orthoimages. In brief, the texturing step assigns different
colour tones to each of the elements that constitute the orthoimage. Once the DSM has been generated, the orthoimages from every aerial image are combined in a complete mosaic.

The advantage of using commercial software over other alternatives is that, commercial options
allow for the generation of pdf reports that contain all the key information on survey implementation (e.g., camera location and image overlap, cameras specification details, camera calibration) and
data processing (e.g., error of GCP in planimmetry and altimmetry, details about the resolution and
point density, and the selected processing parameters). The geomatic products generated are associated to specific quality indicators that are useful for the interpretation of the outputs generated.

Overall, high-resolution UAV aerial images together with photogrametric analysis and post-
processing algorithms are a useful tool to monitor riverine ecosystems. For example, the orthoimage generated allows for the identification of habitat and hydraulic untis (Table 4.1), amongst other
hydromorphological parameters that enable the assessement of quality of the reach. In addition, the
3D model of the sampled reach is a powerful tool to visualise key metrics and characteristics of
the riverine environment. The following sections explain how the generated geomatic products can
be post-processed for riverine feature detection and mapping.

RIVERINE FEATURE DETECTION AND MAPPING

Classification systems have often played a key role in mapping physical features of the fluvial environment and subsequently facilitating our understanding and management of rivers. Such classification systems are numerous and include, amongst others, the definition and mapping of biotopes,
surface flow types (SFTs), channel geomorphic units (CGUs), substrate size [32], and other hydromorphological features. These classification schemes offer relatively rapid and consistent methods
of summarising the detail which exists at the fluvial microscale over mesoscale extents of channel.
Traditionally, the practical application of these schemes takes the form of walkover surveys and
sometimes includes field mapping or direct measurements of hydraulic or geomorphic conditions.
For example, the UK's River Habitat Survey assesses physical habitat quality over 500 m reaches of

river channel by classifying features such as dominant SFT and substrate size and quantifying the number of other hydromorphological features such as pools and riffles.

Remote sensing approaches have also been explored as a means of applying these kinds of classification schemes. This has included the use of multispectral imagery (e.g., [33–35]), hyperspectral imagery (e.g., [36,37]), and terrestrial laser scanning (TLS) point clouds (e.g., [38,39]) for mapping biotopes and SFTs. Most of these studies report good agreement between the SFT/biotope mapping conducted from the remotely sensed imagery and those mapped using traditional methods from the bankside. In some cases, the remote mapping is thought to provide a more accurate and precise representation of the spatial arrangement of physical habitat, although quantitative evidence is not always provided [34–36]. Remote sensing approaches offer permanent, digital, spatially continuous, and spatially explicit datasets, against which change over time can be assessed and other types of spatial analyses performed. Unfortunately, however, the additional cost, logistical issues, and difficulties associated with surveying at an appropriate spatial scale have sometimes made remote sensing options prohibitive for physical habitat surveying in comparison to traditional walkover approaches.

The development of small UAVs and SfM (SFM) photogrammetry has recently provided an alternative remote sensing approach (referred to here as UAV-SfM (SFM). Wider environmental applications of this method have demonstrated the rapid and flexible acquisition of unprecedented hyperspatial resolution imagery (<10 cm) and subsequent production of high-accuracy DEMs and very detailed 3D datasets (e.g., [40,41]). Within this section, how a UAV-SfM (SFM) approach can be used specifically for the detection and classification of riverine features is demonstrated, including the hydromorphology, substrate size, bankside and in-channel vegetation, and woody debris. Examples are used from the published literature, which have only begun to emerge in the last few years and from our own experiences with UAV data acquisition and analysis for riverine feature detection (Table 4.2). These examples can be broadly categorised into those which (i) use automated methods of UAV image classification (either supervised or unsupervised) and (ii) rely on manual delineation of features of interest.

TABLE 4.2

Summary of Recent Papers on the Use of UAVs for Feature Detection and Quantification

Aim	Feature	Analysis Method	UAV Type	Camera (Image Resolution)	Reported Accuracy	
Feature detection	Hydromorphology	Automated – ANNs	Rotary wing (Astec Falcon 8) 1.9 kg	Sony Alpha 6000 RGB (0.025 m)	Classification accuracies: Erosion 8% Submerged veg 29% Riffles 81% Emergent veg 82% Deep water 93% Vegetated bars 100%	[15]
		Manual mapping	Unspecified	Unspecified (0.11–0.16 m)	Not assessed	[45]
	Intermittent streams	Automated – maximum likelihood supervised classification	Fixed-wing (Aeromapper)	Sony NEX-5T RGB 16 MP (0.02 m)	Classification accuracy: kappa = 0.82	[42]

(Continued)

TABLE 4.2 *(Continued)*
Summary of Recent Papers on the Use of UAVs for Feature Detection and Quantification

Aim	Feature	Analysis Method	UAV Type	Camera (Image Resolution)	Reported Accuracy	
	Large woody debris	Automated – supervised classification	Motorised paraglider (ABS-Aerolight Pixy)	Canon Powershot G5 5 MP RGB (0.04–0.19 m)	Majority of wood not detected at pixel sizes >0.10 m False detections greater and missed detections lower at finer resolutions LWD length estimations do not improve but diameter estimations do improve at finer resolutions	[43]
		Manual mapping	Rotary wing (Aeryon Scout)	Photo 3S RGB (0.05 m)	Not assessed	[52]
	Vegetation (riparian)	Automated – Object-Based Image Analysis (OBIA) and supervised classification	Fixed-wing (Gatewing X100)	Ricoh GR3 10 MP (0.05–0.10 m)	Best classification accuracies by species (overall accuracy/ kappa): *Impatiens Glandultfera* = 68% / 0.45 *Heracleum mantegazzianum* = 97% / 0.91 Japanese knotweed = 69% / 0.38	[44]
		Automated – pixel-based and object-oriented approaches for single UAV images and mosaics	Motorised paraglider (ABS-Aerolight Pixy)	Canon Powershot G5 5 MP RGB (0.07–0.22 m) Canon EOS 5D 12 MP RGB (0.03–0.12 m)	Species mapping results: Single image accuracies (overall accuracy / kappa): Pixel-based = 84% / 0.66 Object-based = 91% / 0.79 Image mosaic accuracies (overall accuracy / kappa): Pixel-based = 71% / 0.60 Object-based = 63% / 0.47	[58]
	Vegetation (emergent and floating-leaved)	Manual mapping on paper print outs at 1: 800	Fixed-wing (SmartOne) 1.5 kg	Canon Ixus 70 7 MP RGB (0.056 m)	Overall species mapping accuracy = 80.4% (rivers)	[9]
	Algae	Automated – Supervised classification (ACE & SAM algorithms)	Rotary wing (DJI Phantom) 1.2 kg	GoPro Hero 3 RGB 12 MP	Overall accuracy/Tau co-efficient: ACE = 90% / 0.82 SAM = 92% / 0.84	[59]

(Continued)

TABLE 4.2 (*Continued*)

Summary of Recent Papers on the Use of UAVs for Feature Detection and Quantification

Aim	Feature	Analysis Method	UAV Type	Camera (Image Resolution)	Reported Accuracy	
Quantification	Topography (exposed)	SfM (SFM) to derive a DEM	Helikite (Allsopp Skyshot)	Canon A480 10 MP	Mean vertical error = 0.07 m Standard deviation of vertical error = 0.15 m	[41]
		SfM (SFM) to derive a DEM	Rotary wing (Draganflyer X6) 1.5 kg	Panasonic Lumix DMC-LX3	Mean vertical error = 0.005–0.111 m Standard deviation of vertical error = 0.019–0.203 m	[14]
	Topography (submerged) and water depth	SfM (SFM) to derive a DEM and simple refraction correction (RC)	Rotary wing (Draganflyer X6) 1.5 kg	Panasonic Lumix DMC-LX3	Mean vertical error (after RC) = −0.008 to 0.053 m Standard deviation of error (after RC) = 0.064–0.086 m	[14]
		SfM (SFM) to derive a point cloud and complex refraction correction (CRC)	Rotary wing (DJI Phantom 3 Advanced)	Not specified	Mean vertical error (after CRC) = −0.011 to 0.014 m	[49]
	Topographic change	SfM (SFM) to derive DEMs and orthophotos for change detection	Rotary wing (MikroKopter Hexa XL)	Canon EOS 500D	RMSE topographic change = 0.05 m	[50]
		SfM (SFM) to derive point clouds, change detection using M3C2 algorithm [60]	Rotary wing (DJI Phantom 2)	Canon IXUS/ Powershot 16 MP	RMSE topographic change = 0.30–0.45 m (dependent on surface characteristics)	[51]
	Substrate size	SfM (SFM) to derive a point cloud and correlation of cloud roughness with grain size	Rotary wing (make/model unspecified)	GoPro Hero3+ Silver edition 5 MP	Slope of observed versus predicted = 0.94	[61]
		SfM (SFM) to derive a point cloud and correlation of cloud roughness with grain size	Rotary wing (Draganflyer X6) 1.5 kg	Panasonic Lumix DMC-LX3	Mean residual error of grain size = −0.0001 m Slope of observed versus predicted = 0.77	[53]
	Velocity	Large-scale particle imaging velocimetry (LSPIV)	Rotary wing (DJI Phantom 2)	GoPro Hero 3	Maximum velocities obtained using the drone are closer to benchmarks than alternative Large-scale particle image velocimetry (LSPIV) configurations	[55]

Of the automated approaches, the work of Rivas Casado et al. [15] provides perhaps the most comprehensive example of the use of a UAV-SfM (SFM) approach for mapping a variety of fluvial hydromorphological features. A small, rotary-winged Falcon 8 UAV mounted with a Sony Alpha 6000 RGB camera was flown at c. 100 m over the River Dee in Wales, to obtain c. 2.5 cm resolution imagery over a 1.4 km stretch of channel. The UAV imagery was processed using SfM (SFM)-photogrammetry software (AgiSoft's PhotoScan Pro), including the use of GCPs for georeferencing and the removal of geometric distortions to produce an orthophoto. A type of supervised image classification called an artificial neural network (ANN) was then used to classify the hydromor-phological features of interest within the imagery using their spectral characteristics; including side bars; areas of erosion, riffles, glides, pools; areas of shallow water; and areas of vegetation includ-ing overhanging cover, vegetated banks and bars, submerged and emergent floating vegetation, and grassed areas. Areas of shading were also identified by the ANN process. An example of the clas-sification output is given in Figure 4.10, and readers are referred to Rivas Casado et al. [15] for full details of this method. The success of the classification procedure was assessed using ground truth data collected during a walkover survey. Ground truth data comprised a series of points arranged in a regular 2×2 m grid covering the entire study area, each of which was assigned a hydromor-phological feature class. These points were accurately located using RTK GPS and documented in the field using colour photographs. A confusion matrix was used to derive the overall classifica-tion accuracy of the ANN approach (81%) and the accuracy of mapping individual hydromorphic features, of which a number showed accuracies of greater than 85%. The features most difficult to classify included areas of erosion, where the steep or vertical banks were not visible from the 2D planar aerial view of the UAV. Submerged vegetation was usually misclassified as shallow water or riffles, or patches of shadow where the vegetation had been covered with deposits of dark brown sediments. Some confusion between features was noted in transition zones, and this was thought to relate to small errors in the spatial positioning of the datasets. The darkening of spectral signa-tures introduced by shadowing also caused some misclassifications. Careful UAV flight planning to minimise the presence of shadowing is recommended for ameliorating this problem. Despite these issues, the classification of most other hydromorphological features was highly successful

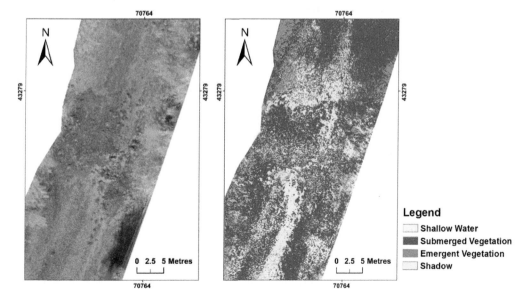

FIGURE 4.10 Example of output of classified habitat units obtained based on the work by Rivas et al. [15] in the River Jucar, Spain.

(accuracies 81%–100%), thereby showcasing a method which is objective and unbiased and which has the potential to be easily transferred to other river sections or indeed whole river networks.

Others have used a variety of automated mapping approaches to focus on the delineation of particular riverine features, as detailed in Table 4.2. For example, Spence and Mengistu [42] used fixed-wing UAV imagery map, the presence of intermittent streams in the Canadian prairie lands of Saskatchewan, with a high classification accuracy (kappa = 0.82) and significantly out-performing a comparable assessment using coarse resolution (10 m) SPOT5 satellite imagery. Early work by MacVicar et al. [43] used imagery from a motorised paraglider to map the position of large woody debris using a supervised classification. Mapping success was found to deteriorate at coarser image spatial resolutions, and at pixel sizes greater than 0.10 m, the majority of woody debris was not detected at all given their typical diameters of 0.12 m or less. In other examples, highly variable results were found in the mapping invasive riparian vegetation species using object-based image analysis and subsequent supervised classifications (kappa = 0.38–0.91) [44]. The poor accuracy values for some species are thought to relate to species mixing within stands and variable luminosity and shading during the UAS surveys.

Riverine feature identification using UAV imagery has also been conducted using simpler, manual techniques, as summarised in Table 4.2. Rather than assess classification or mapping accuracy, these examples instead tend to illustrate the possibility for exceptionally detailed mapping made possible by the high spatial resolution of UAV imagery and how this mapping may be used within applied science and management scenarios. For example, Cheek et al. [45] manually mapped boulders, large woody debris, and mesohabitat types (e.g., runs, pools, riffles) from fixed-wing UAV imagery. The mapping was used to assess associations between physical habitat conditions and fish assemblages at a variety of spatial scales. Habitat variability at the microscale was the most important in structuring fish assemblages within their study site and concluded that the UAV method offers the opportunity to produce '...*an inventory of instream habitats and substrates far more comprehensive than could be generated at the same cost using more traditional approaches*' ([45], p. 9).

As these examples demonstrate, the data provided by classifications of UAV imagery have permitted spatially continuous classifications of river features and have the potential to contribute to river science and management in a variety of ways. As with all classification systems however, choices must be made by the user in terms of the specific scheme of classification, the nature of the sampling procedure, and the method for interpolation or extrapolation of evidence. These choices may be defined by the project aims but are also typically determined by funding, resource allocation, and available time and by the prevailing practices within particular geographical regions or subject disciplines. In partial consequence of this, classification methods may be subjective or appropriate only at particular scales or on particular rivers. As a result, it can be suggested that the use of quantitative data continuums, which are now possible to obtain using a UAV-SfM (SFM) approach, rather than qualitative classifications, offer a new method for capturing and understanding river systems [46]. This is discussed further in Section 4.

QUANTIFYING RIVERINE ENVIRONMENTS

River systems comprise of numerous quantitative continuums, including those which describe the spatial variation in topographic elevation, water depth, grain size, and flow velocity, for instance. Remotely sensed datasets have permitted researchers to work with such continuums of data in recent decades. For example, traditional aerial photography has facilitated fluvial grain size mapping through the use of image texture algorithms (e.g., [47]), and airborne LIDAR has allowed us to quantify riverine topography more quickly and accurately than has previously been possible (e.g., [48]). Yet few, if any, of these methods can quantify multiple continuums simultaneously and at those spatial scales which have most value for assessments of river habitat in particular. In contrast, a number of initial studies have demonstrated the potential for data acquired from a UAV-SfM (SFM) approach to fill this gap. This section demonstrates how a UAV-SfM (SFM) approach can be used specifically for quantifying physical river habitat parameters. Examples from the published literature and from the authors' experiences in recent years are presented.

As demonstrated in Table 4.2, the majority of UAV-SfM (SFM) river research to date has focused on the quantification of topography, particularly in the exposed parts of the channel. This work was pioneered by Fonstad et al. [41] who derived a DEM from imagery acquired using a Helikite platform over the Pedernales River in Texas, USA. Comparison with airborne LIDAR suggested this novel method could produce DEMs with centimetre-range vertical precision and at relatively low costs. Since 2012, further research has shown the UAV-SfM (SFM) approach capable of producing DEMs in exposed areas with vertical mean error and precision as good as 0.005 and 0.02 m, respectively [14]. Reconstruction of submerged areas has also been demonstrated in relatively shallow (<1 m deep), clear flowing rivers. Here, the implementation of refraction correction algorithms is necessary to improve mean errors and standard deviations. Woodget et al. [14] used a correction process on the UAV-SfM (SFM) DEMs to reduce errors in submerged areas. This method assumes the use of nadir imagery and applies a simplified version of Snell's law, which effectively increases water depth estimates by the refractive index of clear water (1.34). Results indicate that errors can be reduced by up to c. 50% to obtain elevation accuracy and precision values as good as 0.008 and 0.064 m, respectively [14]. More recently, Dietrich [49] developed a more complex, iterative refraction correction procedure which estimates the refractive error for each and every point within the UAV-SfM (SFM) point cloud. A direct comparison with the simpler method of Woodget et al. [14] found mean and standard errors to be comparable, but a notable improvement in precision-flying height ratios was observed [49] (Figure 4.11). These papers demonstrate the potential of UAV-SfM

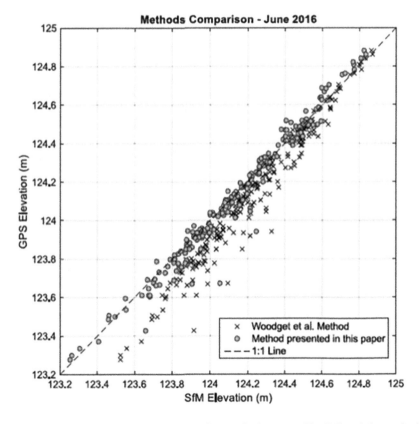

FIGURE 4.11   Comparison of the refraction correction method proposed by [14] and the method presented in [49] for the June 2016 White River dataset.

(SFM) surveys for generating continuums of topographic data in both exposed and submerged areas, with exceptionally high spatial resolutions (1–2 cm), high accuracies and precisions (Table 4.2). Studies which use the UAV-SfM (SFM) approach for quantifying change in topography have also started to emerge in recent years, using both 2.5D DEMs [50] and 3D point clouds [51].

Very recently, the topographic data derived from a UAV-SfM (SFM) approach has also been used to quantify continuums of fluvial grain size by employing one of two different methods. The first of these relies on the development of an empirical correlation between grain size and the image texture within the UAV-SfM (SFM) output orthophotos. For example, Tamminga et al. [52] apply the texture methods which were initially developed by Carbonneau et al. [47] for application to traditional aerial photography. Whilst they find a strong calibration relationship (R2 = 0.82) between texture and grain size, their study does not include any validation data, and therefore, the accuracy and precision of their grain size predictions are unknown. Alternative grain size estimation methods have made use of a calibration relationship between the roughness of UAV-SfM (SFM)-derived point clouds and grain size. Validation results suggest that continuums of fluvial grain size can be derived with accuracies of <1 mm and precisions of c. 2 cm [53,54]. However, it is not yet clear how applicable this method will be at different scales and at different sites. For example, this approach is likely to be less successful where grains are well packed or imbricated, such that their size is not expressed in their 3D relief. In such instances, the image texture approach [47], which relies on the 2D pattern of grains, may be a better predictor of grain size, but this has yet to be proven. An overview of the outputs from these different methods is presented in Figure 4.12.

To date, the use of UAVs and SfM (SFM) for quantification of continuums for other physical river parameters has been lacking. Some initial, proof-of-concept studies have used particle imaging velocimetry to measure flow velocities [55,56], but comprehensive error assessments have not yet been presented.

As a direct result of the technological developments in UAVs and digital photogrammetry in the last few years, clear progress has been made in capturing data continuums of the fluvial environment. These continuums hold great potential for providing exceptionally high-resolution, high-accuracy, high-precision, quantitative measurements of the physical environment over a range of spatial and temporal scales. Until now, the luxury of such datasets has not been available to river scientists and practitioners [57]. However, there is still much work to be done in quantifying the limits of these approaches before they can be applied fully and effectively within a real-world river management context.

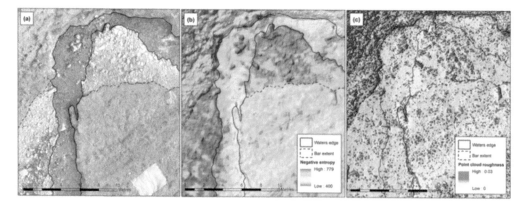

FIGURE 4.12   (a) Orthophoto for a subsection of the Coledale Beck site with (b) negative image entropy and (c) point cloud roughness outputs for the same area [53].

REFERENCES

1. The Guardian. New Zealand river granted same legal rights as human being, https://www.theguardian.com/world/2017/mar/16/new-zealand-river-granted-same-legal-rights-as-human-being (accessed Apr 8, 2017).
2. European Commission Directive 2000/60/EC of the European Parliament and of the Council of 23 October 2000 establishing a framework for Community action in the field of water policy. **2000**, *L327*, 1–72.
3. Leyland, J.; Hackney, C. R.; Darby, S. E.; Parsons, D. R.; Best, J. L.; Nicholas, A. P.; Aalto, R.; Lague, D. Extreme flood-driven fluvial bank erosion and sediment loads: direct process measurements using integrated Mobile Laser Scanning (MLS) and hydro-acoustic techniques. *Earth Surf. Process. Landforms* **2017**, *42*, 334–346.
4. Kriechbaumer, T.; Blackburn, K.; Everard, N.; Rivas Casado, M. Acoustic Doppler Current Profiler measurements near a weir with fish pass: assessing solutions to compass errors, spatial data referencing and spatial flow heterogeneity. *Hydrol. Res.* **2015**, doi:10.2166/nh.2015.095.
5. Pellerin, B. A.; Stauffer, B. A.; Young, D. A.; Sullivan, D. J.; Bricker, S. B.; Walbridge, M. R.; Clyde, G. A.; Shaw, D. M. Emerging tools for continuous nutrient monitoring networks: sensors advancing science and water resources protection. *JAWRA J. Am. Water Resour. Assoc.* **2016**, *52*, 993–1008.
6. Reform. Methods, models, tools to assess the hydromorphology of rivers—part 5 applications, http://www.reformrivers.eu/methods-models-tools-assess-hydromorphology-rivers-part-5-applications (accessed Apr 8, 2017).
7. Environment Agency. http://www.eurisy.org/good-practice-cabinet-office--environment-agency-use-of-copernicus-emergency-service-and-charter-in-the-flooding-of-2014_128 (accessed Apr 8, 2017).
8. Adams, M. S.; Fromm, R.; Lechner, V. High-resolution debris flow volume mapping with unmanned aerial systems (UAS) and photogrammetric techniques. *ISPRS—Int. Arch. Photogramm. Remote Sens. Spat. Inf. Sci.* **2016**, *XLI-B1*, 749–755.
9. Husson, E.; Hagner, O.; Ecke, F. Unmanned aircraft systems help to map aquatic vegetation. *Appl. Veg. Sci.* **2014**, *17*, 567–577.
10. Maddock, I. The importance of physical habitat assessment for evaluating river health. *Freshw. Biol.* **1999**, *41*, 373–391.
11. Maddock, I.; Bird, D. The application of habitat mapping to identify representative PHABSIM sites on the River Tavy, Devon, UK. In *2nd International Symposium on Habitat Hydraulics*; Leclerc, M.; Capra, H.; Valentin, S.; Boudreault, A.; Côté, Y., Ed.; 1996; pp. 203–214.
12. DeBell, L.; Anderson, K.; Brazier, R. E.; King, N.; Jones, L. Water resource management at catchment scales using lightweight UAVs: current capabilities and future perspectives. *J. Unmanned Veh. Syst.* **2016**, *4*, 7–30.
13. Rivas Casado, M.; Ballesteros Gonzalez, R.; Wright, R.; Bellamy, P. Quantifying the effect of aerial imagery resolution in automated hydromorphological river characterisation. *Remote Sens.* **2016**, *8*, 650.
14. Woodget, A. S.; Carbonneau, P. E.; Visser, F.; Maddock, I. P. Quantifying submerged fluvial topography using hyperspatial resolution UAS imagery and structure from motion photogrammetry. *Earth Surf. Process. Landforms* **2015**, *40*, 47–64.
15. Rivas Casado, M.; Gonzalez, R.; Kriechbaumer, T.; Veal, A. Automated identification of river hydromorphological features using UAV high resolution aerial imagery. *Sensors* **2015**, *15*(11), 27969–27989. doi: 10.3390/s151127969.
16. Woodget, A. S.; Visser, F.; Maddock, I. P.; Carbonneau, P. E. The accuracy and reliability of traditional surface flow type mapping: is it time for a new method of characterizing physical river habitat? *River Res. Appl.* **2016**, *32*, 1902–1914.
17. Waterfly on Vimeo https://vimeo.com/107679080 (accessed Apr 8, 2017).
18. Rivas Casado, M. Unmanned Aerial Vehicles (UAVs) for flood extent mapping and damage assessment. Unpublished Presentation, Cranfield University, 15 slides, 2016.
19. Public Finance. Floods demonstrate the value of eyes in the sky http://www.publicfinance.co.uk/opinion/2016/02/floods-demonstrate-value-eyes-sky (accessed May 8, 2017).
20. UAS Sightings Report https://www.faa.gov/uas/resources/uas_sightings_report/ (accessed Apr 8, 2017).
21. FAA Releases Updated UAS Sighting Reports https://www.faa.gov/news/updates/?newsId=85229 (accessed Apr 8, 2017).
22. Ballesteros, R.; Ortega, J. F.; Hernández, D.; Moreno, M. A. Applications of georeferenced high-resolution images obtained with unmanned aerial vehicles. Part I: description of image acquisition and processing. *Precis. Agric.* **2014**, *15*, 579–592.

23. James, M. R.; Robson, S. Mitigating systematic error in topographic models derived from UAV and ground-based image networks. *Earth Surf. Process. Landforms* **2014**, *39*, 1413–1412.

24. Vericat, D.; Brasington, J.; Wheaton, J.; Cowie, M. Accuracy assessment of aerial photographs acquired using lighter-than-air blimps: low-cost tools for mapping river corridors. *River Res. Appl.* **2009**, *25*, 985–1000.

25. James, M. R.; Robson, S.; d'Oleire-Oltmanns, S.; Niethammer, U. Optimising UAV topographic surveys processed with structure-from-motion: ground control quality, quantity and bundle adjustment. *Geomorphology* **2017**, *280*, 51–66.

26. Ribeiro-Gomes, K.; Hernandez-Lopez, D.; Ballesteros, R.; Moreno, M. A. Approximate georeferencing and automatic blurred image detection to reduce the costs of UAV use in environmental and agricultural applications. *Biosyst. Eng.* **2016**, *151*, 308–327.

27. Carbonneau, P. E.; Dietrich, J. T. Cost-effective non-metric photogrammetry from consumer-grade sUAS: implications for direct georeferencing of structure from motion photogrammetry. *Earth Surf. Process. Landforms* **2017**, *42*, 473–486.

28. Unmanned Aircraft Systems https://www.faa.gov/uas/ (accessed May 10, 2017).

29. Department for Transport Consultation on the safe use of drones in the UK www.gov.uk/dft (accessed May 8, 2017).

30. European Aviation Safety Agency, Prototype Commission Regulation on Unmanned Aircraft Operations https://www.easa.europa.eu/easa-and-you/civil-drones-rpas.

31. Kraus, K. *Photogrammetry: geometry from images and laser scans*; 2nd ed.; Walter de Gruyter: Berlin, 2007.

32. Wentworth, C. K. A scale of grade and class terms for clastic sediments. *J. Geol.* **1922**, *30*, 377–392.

33. Wright, A.; Marcus, W. A.; Aspinall, R. Evaluation of multispectral, fine scale digital imagery as a tool for mapping stream morphology. *Geomorphology* **2000**, *33*, 107–120.

34. Reid, M. A.; Thoms, M. C. Mapping stream surface flow types by balloon: an inexpensive high resolution remote sensing solution to rapid assessment of stream habitat heterogeneity? In *Ecohydrology of surface and groundwater dependent systems: concepts, methods and recent developments. Proceedings of the JS.1 at the joint IAHS and IAH convention*; IAHS Publ: Hyderabad, India, 2009, *328*.

35. Hardy, T.B.; Anderson, P.C.; Neale, C.M.U.; Stevens, D. K. Application of multispectral videography for the delineation of riverine depths and mesoscale hydraulic features. In *Effects on human-induced changes on hydrological systems*; Marston, R. A., Hasfurther, V. R., Eds.; Proceedings of the American Water Resources Association: Jackson Hole, WY, 1994; pp. 445–454.

36. Marcus, W. A. Mapping stream microhabitats with high spatial res hyperspectral imagery. *J. Geogr. Syst.* **2002**, *4*, 113–126.

37. Marcus, W. A.; Legleiter, C. J.; Aspinall, R. J.; Boardman, J. W.; Crabtree, R. L. High spatial resolution hyperspectral mapping of in-stream habitats, depths, and woody debris in mountain streams. *Geomorphology* **2003**, *55*, 363–380.

38. Large, A.; Heritage, G. Terrestrial laser scanner based instream habitat quantification using a random field approach. In *Proceedings of the 2007 annual conference of Remote Sensing and Photogrammetry Society*; Mills, J.; Williams, M., Ed.; Remote Sensing and Photogrammetry Society: Newcastle-upon-Tyne, 2007. Publisher Curran Associates Inc., http://www.proceedings.com/03288.html

39. Milan, D. J.; Heritage, G. L.; Large, A. R. G.; Entwistle, N. S. Mapping hydraulic biotopes using terrestrial laser scan data of water surface properties. *Earth Surf. Process. Landforms* **2010**, *35*, 918–931.

40. Hervouet, A.; Dunford, R.; Piégay, H.; Belletti, B.; Trémélo, M.-L. Analysis of post-flood recruitment patterns in braided-channel rivers at multiple scales based on an image series collected by Unmanned Aerial Vehicles, Ultra-light Aerial Vehicles, and satellites. *GIScience Remote Sens.* **2011**, *48*, 50–73.

41. Fonstad, M. A.; Dietrich, J. T.; Courville, B. C.; Jensen, J. L.; Carbonneau, P. E. Topographic structure from motion: a new development in photogrammetric measurement. *Earth Surf. Process. Landforms* **2013**, *38*, 421–430.

42. Spence, C.; Mengistu, S. Deployment of an unmanned aerial system to assist in mapping an intermittent stream. *Hydrol. Process.* **2016**, *30*, 493–500.

43. MacVicar, B. J.; Piégay, H.; Henderson, A.; Comiti, F.; Oberlin, C.; Pecorari, E. Quantifying the temporal dynamics of wood in large rivers: field trials of wood surveying, dating, tracking, and monitoring techniques. *Earth Surf. Process. Landforms* **2009**, *34*, 2031–2046.

44. Michez, A.; Piégay, H.; Jonathan, L.; Claessens, H.; Lejeune, P. Mapping of riparian invasive species with supervised classification of Unmanned Aerial System (UAS) imagery. *Int. J. Appl. Earth Obs. Geoinf.* **2016**, *44*, 88–94.

45. Cheek, B. D.; Grabowski, T. B.; Bean, P. T.; Groeschel, J. R.; Magnelia, S. J. Evaluating habitat associations of a fish assemblage at multiple spatial scales in a minimally disturbed stream using low-cost remote sensing. *Aquat. Conserv. Mar. Freshw. Ecosyst.* **2016**, *26*, 20–34.

46. Woodget, A. S.; Austrums, R.; Maddock, I. P.; Habit, E. Drones and digital photogrammetry: from classifications to continuums for monitoring river habitat and hydromorphology. *Wiley Interdiscip. Rev. Water* **2017**, e1222.

47. Carbonneau, P. E.; Lane, S. N.; Bergeron, N. E. Catchment-scale mapping of surface grain size in gravel bed rivers using airborne digital imagery. *Water Resour. Res.* **2004**, *40*, W07202.

48. Jones, A. F.; Brewer, P. A.; Johnstone, W.; Macklin, M. G. High-resolution interpretative geomorphological mapping of river valley environments using airborne LIDAR data. *Earth Surf. Process. Landforms* **2007**, *32*, 1574–1592.

49. Dietrich, J. T. Bathymetric Structure-from-Motion: extracting shallow stream bathymetry from multi-view stereo photogrammetry. *Earth Surf. Process. Landforms* **2017**, *42*, 355–364.

50. Miřijovský, J.; Langhammer, J. Multitemporal monitoring of the morphodynamics of a mid-mountain stream using UAS photogrammetry. *Remote Sens* **2015**, *7*, 8586–8609.

51. Cook, K. L. An evaluation of the effectiveness of low-cost UAVs and structure from motion for geomorphic change detection. *Geomorphology* **2017**, *278*, 195–208.

52. Tamminga, A.; Hugenholtz, C.; Eaton, B.; Lapointe, M. Hyperspatial remote sensing of channel reach morphology and hydraulic fish habitat using an Unmanned Aerial Vehicle (UAV): a first assessment in the context of river research and management. *River Res. Appl.* **2015**, *31*, 379–391.

53. Woodget, A. S.; Austrums, R. Subaerial gravel size measurement using topographic data derived from a UAV-SfM (SFM) approach. *Earth Surf. Process. Landforms* **2017**, *42*, 1434–1443.

54. Woodget, A. S.; Visser, F.; Maddock, I. P.; Carbonneau, P. E. Quantifying fluvial substrate size using hyperspatial resolution UAS imagery and SfM (SFM)-photogrammetry. In *11th International Symposium on Ecohydraulics*, 7–12 February; Melbourne, Australia. 2016.

55. Tauro, F.; Petroselli, A.; Arcangeletti, E. Assessment of drone-based surface flow observations. *Hydrol. Process.* **2016**, *30*, 1114–1130.

56. Detert, M.; Weitbrecht, V. A vision for contactless hydraulic measurements. In *Session III, Small UAS for Environmental Research Conference*, 28–29 June; University of Worcester, Worcester, 2016.

57. Newson, M. D.; Newson, C. L. Geomorphology, ecology and river channel habitat: mesoscale approaches to basin-scale challenges. *Prog. Phys. Geogr.* **2000**, *24*, 195–217.

58. Dunford, R.; Michel, K.; Gagnage, M.; Gay, H. P.; Tremelo, M.-L. Potential and constraints of Unmanned Aerial Vehicle technology for the characterization of Mediterranean riparian forest. *Int. J. Remote Sens.* **2009**, *30*, 4915–4935.

59. Flynn, K. F.; Chapra, S. C. Remote sensing of submerged aquatic vegetation in a shallow non-turbid river using an unmanned aerial vehicle. *Remote Sens.* **2014**, *6*, 12815–12836.

60. Lague, D.; Brodu, N.; Leroux, J. Accurate 3D comparison of complex topography with terrestrial laser scanner: application to the Rangitikei canyon (N-Z). *ISPRS J. Photogramm. Remote Sens.* **2013**, *82*, 10–26.

61. Vázquez-Tarrío, D.; Borgniet, L.; Liébault, F.; Recking, A. Using UAS optical imagery and SfM (SFM) photogrammetry to characterize the surface grain size of gravel bars in a braided river (Vénéon River, French Alps). *Geomorphology* **2017**, *285*, 94–105.

5 Low-Cost UAVs for Environmental Monitoring, Mapping, and Modelling of the Coastal Zone

David R. Green
UCEMM – University of Aberdeen

Jason J. Hagon
GeoDrone Survey Ltd and UCEMM

Cristina Gómez
INIA-Forest Research Centre (CIFOR)

CONTENTS

COASTAL AND MARINE APPLICATIONS

Unmanned Aerial Vehicles (UAVs) have been used for many different applications around the world in the last 5 years. As their ease of use improves, the number and range of applications has also increased rapidly. In particular, coastal and marine environments requiring high-resolution local data and information or multi-temporal monitoring data and imagery for change detection are well suited to the use of UAVs carrying a range of different sensors including RGB cameras, thermal, and hyperspectral sensors. Klemas (2015) notes that *UAVs (now....) have the capability to effectively fill current observation gaps in environmental remote sensing and provide critical information needed for coastal change research...* (p. 1265).

In recent years, with the widespread availability of small, off-the-shelf, Ready-to-Fly (RTF) UAVs and accompanying photographic and video sensors, and the rapidly evolving Global Positioning System (GPS), navigational, and sensor technology, a number of coastal/marine applications have emerged that take advantage and demonstrate the value of this technology for use by coastal researchers and managers. There are now many different suppliers of UAV platforms or frames: some are ready built, and others are available as kits for construction. These are alongside some very specialist custom frames costing considerably more. Depending on the requirements of a coastal study, there are many examples of small off-the shelf sport cameras e.g. GoPro and DSLR cameras, all of which can be mounted in gimbals that are in use, alongside more expensive sensors, such as thermal cameras, hyperspectral, and light detection and ranging (LIDAR) sensors, and most recently the bathymetric Bathycopter LIDAR by Riegl.

In this chapter, three different examples are used to illustrate the potential these of small airborne platforms and sensors to gather data on various different aspects of the coastal environment. Spanning some 30 years, these provide a good indication of how the aerial platform and sensor technologies have advanced, over a relatively short period of time to the present day, empowering a wide range of people with the potential to acquire and process high-resolution environmental data.

COASTAL UAV APPLICATIONS IN THE LITERATURE

Most of the journal papers on coastal applications of UAVs examined are from the last 5 years, which coincides with the rapidly growing recognition of this technology, its affordability, and ease of use for many different applications ranging from relatively simple photographic and video acquisition to more sophisticated monitoring and small area surveying exercises. Whilst each paper addresses different coastal applications, the large majority of these papers highlight not only the context behind the growth in applications e.g. technology, battery power, GPS accuracy, and survey capability but also the sophistication of the soft-copy photogrammetry software now available to process the imagery, as well as the importance of operational safety, and flight regulations. Additional considerations are the low costs of data acquisition (Ryan, 2012), large-scale mapping, and spatial resolution requirements.

Special UAV environmental considerations are also cited by a number of authors in relation to coastal work. For example, Guillot and Pouget (2015) mention the challenging nature of coastal environments for flying UAVs and the need to take into account parameters such as wind, water, temperature, and climate. Mancini et al. (2013) recorded a number of important observations in relation to the use of UAVs in dune systems which include planning take-off and landing sites to avoid setting the sand in motion by the UAV rotors and avoiding damage to the functioning of the rotors and camera lens. Klemas (2015) observes the value of UAVs to increase operational flexibility and provide greater versatility. In addition, Klemas cites a number of studies that make use of different types of UAV platforms and sensors. Hugenholtz et al. (2012) also note that improvements in the design of flight control systems have transformed these platforms into research-grade tools capable of acquiring high-quality images and geophysical/biological measurements. Klemas (2015) mentions studies that have mapped tidal wetlands, coastal vegetation and algal blooms, sand bar morphology, the locations of rip channels, and the dimensions of surf/swash zones, coastal hazards, fishing surveillance, coastal erosion studies, and combined aerial views from a UAV (drone) with measurements from autonomous underwater vehicles (AUVs) to get an unprecedented look at coastal waters off the coast of southern Portugal. Others have used UAVs in combination with other platforms and sensors to monitor and track marine mammals (e.g. Aniceto et al., 2018).

The role of fixed-wing versus multi-rotor platforms is also discussed in many papers, serving to provide a useful current overview of the status of these platforms and sensors whilst also highlighting the complementary value of these platforms and their sensors for monitoring, mapping, and surveying different aspects of the coastal environment. In addition, reference is still commonly made to other small airborne data acquisition platforms such as tethered kites and balloons of blimps (Ryan, 2012) which also have potential for some applications and can be both cost-effective and easy to use, even if they fall without the UAV domain. As with all such small platforms and systems, however, they also have their limitations.

Some of the applications of UAVs to coastal environments have included monitoring coastal and intertidal habitats such as mangroves, saltmarsh, and seagrass (Ryan, 2012) to gather ortho-photos, as well as the use of video transects for intertidal bathymetric habitat mapping, to assess inaccessible areas; facilitate repeat surveys, extent, and coverage of vegetation; and for monitoring dredging activity and plume extent. Also mentioned are applications that include the generation of 3D point clouds through stereo-photography or LIDAR and routine marine fauna observation (MFO) including cetaceans and turtle nesting activity using fixed-wing aircraft, minor oil-spill contingency tracking, hyperspectral vegetation classification, detailed engineering inspections of pipelines, offshore structures, thermal imaging, and health and safety intervention. Whilst some of the platforms used for environmental applications have been specialised custom examples, the rapid growth in small, low-cost platforms has led to more of these platforms being standard off-the-shelf UAVs e.g. the DJI Phantom has become very popular as an RTF. Inevitably the opportunity to monitor changes to the coastal environment has been one of the applications where UAVs have considerable potential given the ease with which it is now possible to acquire multi-temporal photographic imagery not only to see changes to the coastline in terms of erosion and deposition but also to be able to measure the change e.g. cutback and to determine the rate of coastal erosion. Appeaning Addo et al. (2016) have used a DJI Phantom 3 to monitor and map coastal change due to erosion in relation to coastal protection structures in Ghana along the Volta Delta shoreline. Using a relatively standard and simple platform and sensor configuration, they were able to generate aerial photographs of the protection structures and surroundings, as well the generation of high-resolution ortho-photos and Digital Elevation Models (DEMs) for the detection and analyses of both planimetric and volumetric changes. The practicality of the method for repeated monitoring and survey is also highlighted. Anderson et al. (2015) have utilised UAVs to gather data about spatial ecology and coastal water quality citing key advantages as being that *they can be launched, operated, and the data accessed and analysed within hours* and systems that can be cheaply and easily used by aquaculture businesses (e.g. pelagic fish and shellfish farms) and by coastal environmental managers (e.g. UK Environment Agency (EA)) to easily and repeatedly monitor near-shore water quality.

In another study, Guillot and Pouget (2015) have used UAVs specially adapted for coastal environments (e.g. salt, wind, and moisture) for monitoring and precision mapping of the dunes and dikes of Oleron Island, in France before and after tides and storms. The UAV – known also as a Coastal UAV – is a modified DJI F550 and is capable of flying in high wind conditions and is designed to be resistant to moisture and sand particles. Interestingly, and perhaps sensibly, they also made use of an off-the-shelf rugged 'sports camera' with a standard fisheye lens. Post-processing of the imagery acquired was carried out to remove the fisheye distortion from the imagery and to geo-correct the imagery using the Ground Control Points (GCPs). AgiSoft soft-copy photogrammetric software was used to generate an ortho-mosaic, a Digital Surface Model (DSM), and a 3D PDF (Portable Document File) for input to a Geographic Information System (GIS). A DDVM (Difference of Digital Volumetric Model) was generated for the different dates of imagery to determine change. DSM and DEM files provided the means to generate 2D topographic profiles in the GIS software. Benefits of the research that were identified included the ease with which it is possible to design a low-cost, coastal-specific platform using off-the-shelf technology for use in a relatively inhospitable environment to gather repetitive high-resolution data and imagery from which it is possible to quantify coastal change at a cm resolution from the UAV-derived products. Further projects for coastal monitoring include the study of sand and pebble movement, birds, waste, and seaweed which can provide valuable information to improve ecosystem knowledge in the future.

Mancini et al. (2013) describe how UAVs have the potential to generate data and information to understand coastal processes using low-altitude UAV aerial photographs and Structure from Motion (SfM [SFM]) to generate DSMs comparable to ground surveys of the morphometry of a dune and beach system in Italy. Comparison of ground and aerial surveys using UAVs reveals the considerable potential of the latter to complete accurate surveys of dunes with less cost, time, and manpower requirements. The research used a non-standard UAV with a Digital Single Lens Reflex (DSLR), Real Time Kinematic (RTK) GPS, and autonomous flight control.

Turner et al. (2016) consider UAV technology to be a tried and tested technology for survey-ing work, specifically engineering applications, and to be cost-effective solutions for coastal zone applications. They also note that no more step changes in UAV technology or ease of usability are required. With the aid of RTK GPS, Turner et al. (2016) now consider this technology to be a practi-cal, effective, and routine way to assess post-storm survey tool for coastal monitoring in Australia at spatial and temporal resolutions not previously possible. They also provide a brief overview of some of the recent UAV applications to coastal engineering and management.

Pereira et al. (2009) focus on the benefits of recent evolutions in UAV technology that include with autonomous take-off and landing capabilities for aerial gravimetry, aerial photography, sur-veillance and control of maritime traffic, fishing surveillance, and detection and control of coastal hazards. Such developments include advances in the distance and duration drones can fly for mari-time surveillance in harsh coastal environments.

Goncalves and Renato (2015) have used UAVs imagery and photogrammetry to derive topo-graphic information for coastal areas, whilst Long et al. (2016) also used UAV imagery to monitor the topography of a tidal inlet.

EVOLUTION OF THE TECHNOLOGY

The following three examples provide some insight into the different ways that this technology has been used over time, from an early application of radio-controlled model aircraft, to a very large battery-powered model helicopter and large multi-rotor, and finally a small off-the-shelf multi-rotor drone.

MONITORING AND MAPPING MACROALGAL WEEDMATS IN THE YTHAN ESTUARY, SCOTLAND, UK

As part of a long-term monitoring and mapping project to map macroalgal weedmats in the Ythan Estuary, north of Aberdeen in Scotland, UK (Figure 5.1), using different sources of remote sensing e.g. satellite imagery and aerial photography, part of the work involved the acquisition of large-scale colour and filtered aerial photography from a large-scale model aircraft (Figure 5.2). The Centre for Remote Sensing and Mapping Science (CRSMS) at the University of Aberdeen used

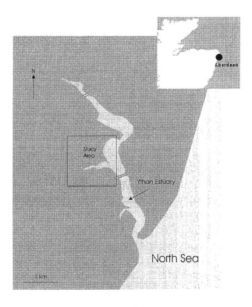

FIGURE 5.1   The Ythan Estuary study area in Scotland, UK.

FIGURE 5.2   Fixed-wing SSA used for aerial overflights of the Ythan Estuary.

FIGURE 5.3   A GCP marker.

the combination of an Small-Scale Aerial (SSA), a GPS, digital image processing software (Erdas Imagine), a GIS (ESRI's ArcView), and 35 mm SLR colour and filtered photography to capture multi-temporal, high-resolution imagery from which to map the extent of macroalgal weedmats in the Ythan Estuary over a number of years (Green and Morton, 1994; Green, 1995).

Prior to the aerial overflights, GCPs were placed in the flight area. Each GCP made use of a white plastic fertiliser bag with a wooden stake (Figure 5.3) the centre location of which was surveyed with the aid of a Trimble GPS. Each overflight was undertaken using a different film/filter combination.

The aerial photographs (Figure 5.4) were printed and then scanned using a desktop scanner. Erdas Imagine was used to mosaic and geo-correct the aerial photographs. The mosaicked imagery was then input to the GIS, and with the aid of the onscreen digitising tools, the Pan and Zoom functionality, and the seven factors of aerial photointerpretation, macroalgal weedmats units were identified and delineated as map layers to create weedmat maps for a number of years. With the aid of GIS, it was then possible to calculate the area of weedmat coverage for each year and to determine the changes in location and areal extent from one year to the next. Combined with other layers of map information e.g. point pollution sources and estuary substrate, it was then possible to establish some correlation between the locations of the weedmats from year to year.

FIGURE 5.4   Aerial photograph of macroalgal weedmats in the Sleek of Tarty, Ythan Estuary.

Spey Bay – Monitoring a Dynamic Coastline with a Model Helicopter and Multi-Rotor UAV

With the assistance of Borich Aircams (www.borichaircams.co.uk), colour aerial photographs and videos were captured using both an Align S690 Hexacopter and an Align S800 Helicopter (Figure 5.5) along the waterfront at Spey Bay in Scotland, UK (Figure 5.6), as part of a funded experimental study to trial two different small-scale UAV platforms carrying Digital SLR cameras as the basis for coastal monitoring and mapping. Aerial overflights were conducted in the summer of 2015 under clear skies with a Panasonic GH4 DSLR camera. The resulting aerial photographs were input to the AgiSoft soft-copy photogrammetry software (www.agisoft.com) to generate photographic mosaics, as well as 3D models of the beach at Spey Bay (Figure 5.7). The project demonstrated the ease with which high-resolution aerial imagery can easily be acquired with small-scale battery-operated UAV platforms and then processed into various different products for subsequent analysis and interpretation.

The project also provided the basis to test the potential of a UAV platform for vertical and oblique aerial photography and video imagery. This data and imagery were used to support a coastal monitoring and mapping application and demonstrated the stability of a large UAV platform and camera system. In addition, there was additional possibility to test new sensors e.g. LIDAR and hyperspectral either on their own or in conjunction with other sensors. Demonstration of this UAV capability provided a number of new aerial acquisition opportunities, with further potential to test the platform and sensors for a number of other marine and coastal tasks.

FIGURE 5.5   Align 800 Helicopter.

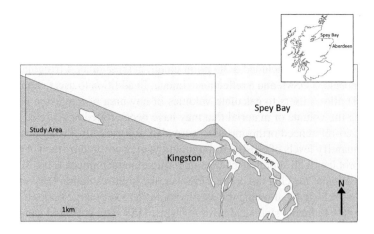

FIGURE 5.6 Spey Bay, Moray, Scotland, UK study area. (Cartography by Jamie Bowie (University of Aberdeen))

FIGURE 5.7 Three-dimensional model of part of Spey Bay beach using AgiSoft.

CHANGE DETECTION AND RATES OF EROSION ON THE NORFOLK COASTLINE USING A MULTI-ROTOR UAV

The poorly consolidated glacial till deposits, separated by silt, clay, and sand, leave sections of the north Norfolk coastline particularly vulnerable to the processes of coastal erosion (Thomalla and Vincent, 2003; Poulton et al., 2006). In recent years, aggressive erosion along the coastline has caused the collapse of several houses and roads in the most vulnerable areas.

Accurately quantifying rates of erosion is important for a variety of coastal management decisions, including creating vulnerability indices and Shoreline Management Plans (SMPs).

In this example, a combination of EA airborne LIDAR data and UAV data was used to identify erosion hotspots along the Norfolk coastline. Airborne LIDAR data was used to identify areas suffering from high levels of erosion along 12 km of coastline, whilst the UAV imagery was used to provide up-to-date rates of erosion and high-resolution 3D models, DSMs, and ortho-mosaics at Happisburgh – one of the worst affected areas.

Airborne LIDAR data, collected by the UK's EA, were obtained for the years 1999, 2009, and 2013. The data were processed using ESRI's ArcGIS software from where a baseline for the cliff toe from 1999 to 2013 was then derived. With the use of the Digital Shoreline Analysis System (DSAS) tool (coast.noaa.gov/digitalcoast/tools/dsas; Thieler et al., 2009), total and average rates of erosion were calculated along 12 km of coastline. The results revealed erosion *hotspot* areas at various locations along the coastline.

In May 2016, a DJI Phantom 3 was flown at the Happisburgh study site to obtain aerial imagery. The aim was to use UAV imagery in combination with the historic LIDAR data to investigate the detection of coastal change in the area. Ninety-five images were collected using Drone Deploy – an autonomous flight planning application – to acquire images of the site. The images were then uploaded from the UAV and imported into Pix4D – a soft-copy photogrammetry software package – where they were processed (Figure 5.8).

The final processed results included a 3D point cloud, a Triangulated Irregular Network (TIN) model, an ortho-mosaic, a DSM, and a reflectance image. In addition to users exporting data for use in software, Pix4D allows users to calculate volumes of any area of interest, a particularly useful tool for identifying the volume of material that may have been eroded or accreted along the coast.

Rectified and geo-referenced ortho-mosaics were also exported for further analysis. The DSAS tool was used to quantify levels of erosion throughout the area covered by the UAV-derived data. A commonly cited problem with shoreline change quantification techniques is the difficulty in identifying the exact location of the cliff top or toe. However, the high-resolution UAV ortho-mosaics can help to address this problem by allowing the user to zoom-in to very large scales to identify the cliff top or toe – ultimately reducing statistical error margins. UAV data collected and processed in this way can provide ortho-mosaics and 3D models with an accuracy of below 1 cm per pixel, something that has traditionally only been possible with very costly LIDAR systems (Figure 5.9).

Figure 5.10 shows how UAV-derived data can be used in conjunction with other data sources – LIDAR and 1:1,000 Ordnance Survey (OS) maps – to accurately quantify spatial change at the coastline over time. The UAV-derived data allows more accurate 3D models and ortho-mosaics to be produced than is otherwise available and at relatively low cost. Furthermore, accurate 3D models can be useful

FIGURE 5.8 Pix4D – the circles indicate where the images were captured and the line shows the UAV flight path.

FIGURE 5.9 A 3D TIN model produced in Pix4D. The location where each photograph was captures is indicated by the spheres and squares.

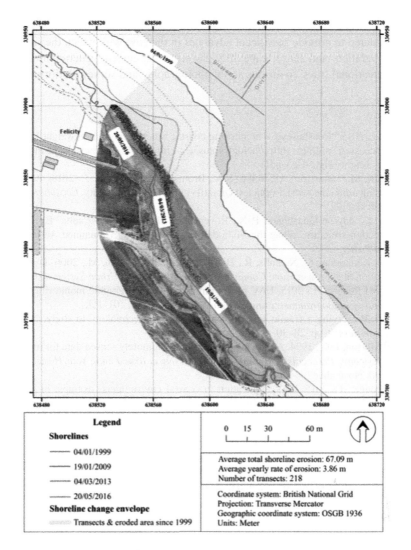

FIGURE 5.10 UAV-derived data used in conjunction with other data sources – LIDAR and 1:1,000 OS maps – to accurately quantify spatial change at the coastline over time.

when explaining complex environmental processes in stakeholder engagement meetings, where audiences may be from a wide variety of non-geographical backgrounds (Brown et al., 2006).

Although it is unlikely that large stretches of the coastline could be mapped using multi-rotor UAVs under current Civil Aviation Authority regulations, it is possible that selective inspection of specific areas using UAVs, used in combination with other data sources, could be advantageous over conventional mapping and monitoring approaches. To conclude, small off-the-shelf UAV platforms can be used to collect high-quality spatial information; the outputs from soft-copy photogrammetric software such as Pix4D can be analysed in industry standard GIS programs to aid in coastal change detection analysis – the results of which can aid in the development of coastal management plans.

SUMMARY AND CONCLUSIONS

The coastal examples used to illustrate this chapter are evidence of the considerable potential these platforms have always had for coastal monitoring and mapping. The technology now available shows that we have now entered a completely different league in terms of what can be achieved using the current UAVs available. In many ways, the technology is still fundamentally the very similar, albeit updated to provide significant advances in the ease with which the small aircraft can be flown, flight duration, and sensor capability elevating the current generation of platforms and sensors from recreational 'toys' to serious aerial photographic and survey platforms.

REFERENCES

Anderson, K., et al., 2015. 'Coastal-eye'—monitoring coastal waters using a lightweight UAV. *Unpublished document* accessed online: (http://nercgw4plus.ac.uk/files/2015/07/Coastal-eye-monitoring-coastal-waters-using-a-lightweight-UAV.pdf)

Aniceto, A.S., Biuw, M., Lindstrøm, U., Solbø, S.A., Broms, F.B. & Carroll, J. (2018). Monitoring marine mammals using unmanned aerial vehicles: Quantifying detection certainty. *Ecosphere*, 9(3). https://doi.org/10.1002/ecs2.2122

Appeaning Addo, K., Jayson-Quashigah, P.-N., Rovere, A., Mann, T., and Caella, E., 2016. Monitoring coastal protection structures along the Volta delta shoreline using Unmanned Aerial Vehicle (UAV). *Unpublished Poster.*

Brown, I., Jude S., Koukoulas, S., Nicholls, R., Dickson, M., and Walkden, M., 2006. Dynamic simulation and visualisation of coastal erosion, *Computers, Environments and Urban Systems*, 30(6). pp. 840–860.

Goncalves, J.A., and Renato, H., 2015. UAV photogrammetry for topographic monitoring of coastal areas. *ISPRS Journal of Photogrammetry and Remote Sensing*. 104. pp. 101–111.

Green, D.R., 1995. Preserving a fragile environment: Integrating technology to study the Ythan Estuary. *Mapping Awareness*. 9. pp. 28–30.

Green, D.R., and Morton, D.C., 1994. Acquiring environmental remotely sensed data for input to geographic information systems. *Proceedings of the AGI'94 Conference: Broadening Your Horizons*, Birmingham, UK, 15th–17th November 1994. pp. 15.3.1–15.3.27.

Guillot, B., and Pouget, F., 2015. UAV application in coastal environment, example of the Oleron Island for Dunes and Dikes survey. *The International Archives of the Photogrammetry, Remote Sensing and Spatial Information Sciences*, XL-3/W3. pp. 321–326, 2015 ISPRS Geospatial Week 2015, 28 Sep–03 Oct 2015, La Grande Motte, France.

Hugenholtz, C.H., Moorman, B.J., Riddell, K., and Whitehead, K., 2012. Small unmanned aircraft systems for remote sensing and earth science research. *Eos, Transactions of the American Geophysical Union.* 93(25). p. 236.

Klemas, V.V., 2015. Coastal and environmental remote sensing from Unmanned Aerial Vehicles: An overview. *Journal of Coastal Research*. 31(5). pp. 1260–1267.

Long, N., Millescamps, A., Benoît, G., and Bertn, X., 2016. Monitoring the topography of a dynamic tidal inlet using UAV imagery. *Remote Sensing*. 8(5). p. 387.

Mancini, F., Dubbini, M., Gattelli, M., Stecchi, F., Fabbri, S., and Gabbianelli, G., 2013. Using Unmanned Aerial Vehicles (UAV) for high-resolution reconstruction of topography: The structure from motion approach on coastal environments. *Remote Sensing*. 5. pp. 6880–6898.

Pereira, E., Bencatel, R., Correia, J., Félix, L., Gonçalves, G., Morgado, J., and Sousa, J., 2009. Unmanned Air Vehicles for coastal and environmental research. *Journal of Coastal Research*. SI 56. pp. 1557–1561. Proceedings of ICS2009.

Poulton, C.V.L., Lee, J.R., Hobbs, P.R.N., Jones, L. and Hall, M., 2006. Preliminary investigation into monitoring coastal erosion using terrestrial laser scanning: Case study at Happisburgh, Norfolk. *Bulletin for Royal Geological Society*. 56. pp. 45–64.

Ryan, D., 2012. Unmanned Aerial Vehicles: A new approach for coastal habitat assessment. Presentation. Worley Parsons Western Operations. OGeo. September 2012. 27 slides.

Thieler, E.R., Himmelstoss, E.A., Zichichi, J.L., and Ergul, A., 2009. Digital Shoreline Analysis System (DSAS) version 4.0—an ArcGIS extension for calculating shoreline change. *U.S. Geological Survey Open-File Report 2008*. p. 1278.

Thomalla, F., and Vincent, E.E., 2003. Beach response to shore-parallel breakwaters at Sea Palling, Norfolk, UK. *Estuarine, Coastal and Shelf Science.* 56(2). pp. 203–212.

Turner, I.L, Harley, M.D., and Drummond, C.D., 2016. UAVs for coastal surveying. *Coastal Engineering*. 114. pp. 19–24.

6 Unmanned Aerial System Applications to Coastal Environments

Francesco Mancini
University of Modena and Reggio Emilia, Italy

Marco Dubbini
University of Bologna, Italy

CONTENTS

INTRODUCTION

Since coastal environments became objects of interest for researchers and professionals, satellite-based and aerial remote sensing have been used for a wide range of applications. These have included mapping of coastal wetlands and rivers mouths, light detection and ranging (LIDAR) bathymetry, mapping of flooded areas, surveillance, oil-spill tracking, land-use/land-cover delineation of coastal areas, investigations of coastal dynamics, and detection of pollutants. The ability to observe large coastal stretches through remotely sensed information and at spatial and temporal resolutions using satellite-based and airborne imaging systems has been a driving factor here. However, investigations of coastal environments at very high spatial and temporal resolutions are still an issue. To date, satellites have offered the advantage of a synoptic view, with optical and multispectral imaging at increasing spatial and spectral resolution and revisiting times. These factors have suited many applications that have focused on coastal monitoring. However, many applications and studies related to coastal geomorphology and vulnerability analysis of coastal environments continue to require additional spatial resolution (Harwin and Lucieer 2012; Klemas 2015). For instance, a hazard that affects a coastal stretch might not be imaged by a satellite sensor due to the orbit design and data acquisition planning. Moreover, airborne surveys over limited areas might come at unsustainable cost.

The recent availability of unmanned aerial systems (UASs) and the joint use of methodologies provided by geomatics engineering, photogrammetry, and computer vision technologies have established new opportunities for environmental monitoring purposes (Colomina and Molina 2014). This is especially the case for traditional survey applications where labour-intensive investigations are required, such as for coastal environments, where logistic difficulties can compromise the quality of the overall product and make the work of the surveyors difficult. The high-resolution and flexible overflights offered by unmanned aerial vehicle (UAV) sensors make these instruments more suitable for a wide range of coastal applications. In particular, all applications that require investigations at desired periods and revisit times of the coastal settings can benefit from the new paradigm offered by UAV surveys. The timely information at the finer spatial resolution that is required in coastline delineation studies and more generally in research into coastal dynamics (e.g., coastal geomorphology, coastal engineering, wetland mapping, vulnerability assessment of coastal stretches exposed to flooding episodes) can be provided by UAVs deployed for such purposes. For example, a UAV survey can be timed to monitor a coastal wetland during low-tide periods to improve the detection of emergent and submerged aquatic vegetation (Klemas 2015; Rafiq 2015).

Among the data recently obtained using UAV surveys for the management of coastal environments, two achievements have to be mentioned. Firstly, the availability of digital surface models (DSMs) and digital elevation models (DEMs), both of which represent the coastal geomorphic features at the centimetre level of accuracy. Secondly, to a lesser extent, the multispectral imaging of vegetation areas. The application to coastal geomorphology falls into the first typology. This requires increasingly accurate topographic information of beach systems, so as to be able to perform reliable simulations of coastal erosion and flooding phenomena, to assess the coastal sediment budget and to investigate morpho-sedimentary changes of the coastal fringe (Delacourt et al. 2009; Pereira et al. 2009; Mancini et al. 2013). For such studies, the availability of a topographic dataset is fundamental, particularly for systems that are characterised by complex morphology. High-resolution topography of coastal environments can also be used as input into hydrodynamics numerical modelling, which covers the gap in the spatial resolution and vertical accuracy of satellite-derived and airborne elevation models, whereby a spatial resolution of >50 cm is not accurate enough for most applications (Delacourt et al. 2009).

The second typology includes applications based on the analysis of multispectral data acquired from sensors on board UAVs, which are specifically designed and miniaturised for light vehicles. Multispectral surveys from UAV platforms are mostly devoted to vegetation mapping at finer spatial resolution (e.g., to a few centimetres), spectral-based land-cover analysis, and construction of vegetation indices. Traditionally, the methodologies used to process multispectral data provided by satellite and aerial imaging are based on classification algorithms that use spectral similarity to group the pixels. Such algorithms can be divided into supervised, unsupervised, and object-based (sometimes used in a combined manner). Their description of the multispectral approach to data analysis goes beyond the scope of this chapter, although the way that they are transferred in the analysis of multispectral data acquired during a proximity survey from UAVs could be a relevant issue.

Thermal imaging of coastal waters from UAVs and their monitoring of pollutant leakage might also represent very interesting applications. Thermal cameras miniaturised for UAVs are now operational, although there are few, if any, studies in the recent scientific literature that depict real-world phenomena that occur in coastal areas.

The objective of this chapter is to acquaint the reader with the potentiality of UAS-based surveys on coastal environments through a discussion of examples available in the very recent literature and through the skills achieved by the authors in this field. After a short introduction on UAV models that can be used to perform aerial surveys and a short description on the available methodologies to process the data acquired by visible and multispectral sensors, the chapter will focus on several UAS applications referred to coastal environments.

Regardless of the specific aerial vehicle used, throughout this chapter we refer to UASs as a complete system that comprises a UAV, a ground control station, and a communication system, with a link for the aerial vehicle (Colomina and Molima 2014).

UAV MODELS: FIXED-WING AIRCRAFT, ROTARY-WING AIRCRAFT, AND OTHER VEHICLES: POTENTIALITIES AND LIMITATIONS FOR COASTAL SURVEYS

Depending on the specific objectives, several UAV configurations have been used for investigations of coastal stretches, coastal lagoons, protective structures, river mouths, and wetlands. Regardless of the specific aims, the deployment of UAVs offers a viable alternative to conventional platforms for acquiring high-resolution remote sensing data from proximity surveys at lower cost, increased operational flexibility, and greater versatility (Klemas 2015). In this section, we provide a short review on unmanned vehicles and the related instrumentation used to date for a wide range of coastal applications and, additionally, some comments upon the potentialities and logistic limitations. For specific applications and investigations on coastal phenomena, the reader should refer to the relevant sections.

FIXED-WING AIRCRAFT

The main applications of fixed-wing UAVs have been in military applications, for damage assessment after natural hazards such as hurricanes, floods, wildfires, volcano eruptions, and seismic events, for the mapping of geologic lineaments, and for industrial plant accidents. To a lesser extent, fixed-wing models have been used for scientific purposes. This is partly due to limitations in their use by national regulations that are restrictive for operations over long ranges. However, several coastal applications based on fixed-wing surveys were summarised in Klemas (2015).

With respect to multi-rotor UAVs, fixed-wind models have increased their capabilities, including their payloads, flight endurance, and travel distance. Due to integrated flight control systems and fully autonomous flight capabilities, fixed-wing models are particularly suited for surveys over large extents and along shores (e.g., for coastline delineation over long ranges). In a similar manner to others UAVs, the acquisition of images from one or multiple sensors is pre-programmed in the flight planning software by uploading of waypoints for data acquisition based on the desired ground sample distance, amount of image overlap, and extent of the surveyed area. Digital cameras and multispectral sensors are often part of the payload. The whole UAS is completed by a ground control station that can communicate with the autopilot on board the aerial vehicle via a telemetry link. During the flight, the UAV can transmit video streaming and images of the coastal landscape in real time to the ground control station. Vousdoukas et al. (2011) operated a UAV with a 1.4-m wingspan (weight, 1.9 kg) to map nearshore sand bar morphology and locations of rip channels, and to investigate the surf zones. Their system provided aerial observations at line-of-sight ranges of up to 10 km after a launch. More details on the technical characteristics of fixed-wing models used for coastal surveys can be found in Pereira et al. (2009), O'Young and Hubbard (2007), and Klemas (2015).

Rotary-Wing Aircraft

Within the rotary-wing models, helicopters and multi-rotors (quad/hexa/octocopters) are the most used for environmental applications and two-dimensional (2D)/3D mapping. One of the major advantages of rotary-wing over fixed-wing aircraft lies in the possibility to hover over a target site, acquire data, images, or video for a selected time and move towards the target for closer inspection at the preferred spatial resolution (Klemas 2015). The need for aircraft with greater manoeuverability and hovering ability has led to a rise in quadcopter use in environmental monitoring and mapping. The UAS regulations for rotary-wing aircraft differ from country to country. However, procedures to authorise UAV flights in nonexclusion areas can be issued by national aviation authorities under similar restriction rules. Vertical take-off and landing UAVs can operate within an aerial work area and under visual line-of-sight conditions (Colomina and Molima 2014). These are now widely used in coastal monitoring and wetland mapping and for flood damage assessment and coastal dynamics studies. Hexacopters and octocopters can carry consumer-grade cameras and fly for 10–20 minutes following a pre-programmed path, with or without stops, and they cost less than US$ 5,000. New systems that can carry multispectral and hyperspectral imagers are being designed, and sensors are being miniaturised to be used on board light UAVs. Due to the ease of use of these models, many of the applications discussed in this chapter have developed for multi-rotor UAV surveys. However, the piloting of multi-rotor UAVs can be problematic under windy conditions (wind speed, >20 km/h) and when there are wind gusts and thermals. Helicopters are less frequently used for coastal investigations, but their performance during a flight can be attractive. Delacourt et al. (2009) developed and tested an unmanned helicopter (DRELIO) to operate under windy conditions of up to 50 km/h and to perform stationary flights from heights up to 300 m through the use of a thermal engine.

Small UAVs are also particularly adapted to survey small areas according to the following needs: no logistic requirements for lifting, high repeatability of a survey, collecting of aerial photographs with high resolution due to the proximity to the subject photographed, positioning and survey of Ground Control Points (GCPs) prior of the surveys, short time needed, or mission planning. All these features allow UAVs to be used quickly, e.g., immediately after a storm.

Blimps, Balloons, and Kites

The main advantages that blimps, balloons, kites, and similar airships have over aircraft are that they can be positioned over specific sites and can monitor environmental changes over long periods. These airships are by themselves naturally low-flying, very slow, and long-endurance platforms (Klemas 2015). As they are steady and vibration free, these platforms are particularly suited to long-term observations and monitoring purposes of a specific area. The main advantage that balloons and kites have over airplanes is that they can be positioned over a site at different altitudes for longer periods of time, with respect to what can be achieved with rotary aircraft (Vierling et al. 2006). In recent years, aerial photography from kites and balloons has been used for scientific surveys, meteorological observations, and military surveillance. Rigid delta kites are more suitable under light to moderate wind conditions, whereas parafoil kites can be flown in moderate to strong winds (Klemas 2015). The size of the kite can be designed depending on the expected wind conditions, and normally the payloads do not exceed 2 kg. Certainly, the usability of kites for coastal survey purposes can be compromised by the absence of winds or the presence of wind gusts. Scoffin (1982) and Rützler (1978) conducted pioneering applications in UAV-based coral reef imaging using a kite and a helium balloon, respectively. On the contrary, tethered balloons can operate under changing conditions, and they provide an excellent platform for imaging coastal processes (e.g., coastal dynamics, sediment transport, flooding). The usefulness of these kinds of UAVs decrease if long distances, complex flights, or imaging of water bodies are planned. This category includes unmanned blimps, which are nowadays more compact, light, and programmable for controlled flights in many environmental applications.

PAYLOADS

Developments in UAV payload design are substantially related to the miniaturisation and adaptation of sensors used by manned aerial surveys. A wide range of sensors, or combinations of sensors, can be included in a payload for UAVs, in addition to a global navigation satellite system (GNSS) for more or less accurate on board positioning, depending on the specific positioning technology adopted.

Such payloads can also include consumer-grade cameras for visible imaging, multispectral imaging systems with variable spectral bands, LIDAR sensors, hyperspectral imaging systems, thermal cameras, chemical analysis tools, sensors for monitoring of key atmospheric parameters, and water samplers (Johnson et al. 2015). The degree of maturity of the design of these sensors for coastal management applications varies significantly. Optical surveys for the imaging of coastal environments and high-resolution topography of coastal features are very common in the recent literature.

Multispectral surveys with cameras specifically designed for UAV platforms are less mature. However, several applications can be found for agricultural purposes. Miniaturised LIDAR systems and some of the other above-mentioned sensors are now available, and applications to the monitoring of coastal processes appear very promising. UAV LIDAR surveys require laser scanning from a moving platform.

To reference the point cloud in a cartographic reference system, the procedure needs rigorous definition of the platform orientation parameters. Thus, the combined use of a navigation system and an inertial mapping unit is required. However, well-documented investigations of coastal areas using LIDAR systems, thermal cameras, and chemical analysis tools are still rare, and their discussion is not an object of this chapter. In the next sections, the potentialities of multispectral and photographic surveys from UAV platforms will be introduced. In the first case, the strategies for multispectral data processing are very similar to those already available from remote satellite and airborne sensing, with a few modifications in consideration of the higher spatial resolution required. In the second case, photography acquired by consumer-grade cameras can be processed following a photogrammetric approach. In most cases, the processing of aerial images is performed by computer vision-based technology, namely the structure-from-motion (SfM [SFM]) approach, and aerial surveys programmed to include SfM (SFM) requirements.

UNMANNED AERIAL SYSTEM-BASED MULTISPECTRAL AND VISIBLE SURVEYS OF COASTAL ENVIRONMENTS

In this section, the methodologies adopted for the processing of multispectral and visible data acquired from UAVs will be introduced, with particular reference to those adopted for documented investigations of coastal environments.

MULTISPECTRAL DATA

In general, the available technologies to process multispectral data in the UAS world were derived from those adopted in satellite and aerial multispectral remote sensing. Certainly, the increased spatial resolution achieved by multispectral sensors designed for UASs poses new questions about the potentialities of proximity surveys. At the time of writing, only a few representative multispectral cameras were available on the UAS mass market. However, in the near future, the photogrammetric and remote sensing community will benefit from the development of new sensors that are specifically designed for UASs, with band allocation and spectral resolution that are more suitable for different applications, considering also very fine (i.e., centimetre level) spatial resolution. The potentialities offered by multispectral UAS-based investigations are very often limited by poor spectral band design. Firstly, miniaturised multispectral cameras use the blue, green, red, and near-infrared (NIR) spectral bands, and sometimes they use a lens filter fitted to the camera. Using a lens filter allows for the targeting of specific bands over the spectrum (Kelcey and Lucieer 2012).

For instance, to improve the details of vegetation features in UAV aerial images, a filter that can absorb the passage of some blue wavelengths can be used to enhance the remaining wavelengths.

With particular reference to the coastal environment, the red and NIR bands are primarily used for identification and monitoring of coastal vegetation through the Normalised Difference Vegetation Index (NDVI). Otherwise, the green and blue bands can be used to study water properties, such as turbidity and eutrophication processes in shallow waters. However, especially for investigations aimed at vegetation detection or vegetation health studies, the representation of the spectral response of an object increases in accuracy when using a dataset with higher spectral resolution. Therefore, more complex spectral sensors that can image an object across several to hundreds of spectral bands are needed by surveyors. Moreover, spectral signatures built from different satellite sensors are well documented, while the more sophisticated development of spectral signatures from UAV aerial images can be the focus for future research.

Many of the applications based on multispectral-based UAV datasets refer to the spatial vegetation distribution and analysis of the degree of vegetation cover of the land, using the NDVI formula. The principle behind the use of the NDVI formula for vegetation studies is related to the properties of the pigments in plant leaves to absorb wavelengths of visible red light and the strong reflection of the NIR wavelengths by leaves. Thus, the NDVI describes the degree of the relative density and health of the vegetation for each pixel in a given image. The size of the pixels in UAS surveys is finer than that achieved during satellite or aerial sensing. Basically, a higher level of spatial detail makes the detection and distinction of terrain covers much easier. Sparse vegetation, such as shrubs, grass, and dune vegetation, results in moderate NDVI values of around 0.2–0.5, while higher values relate to denser and healthier vegetation. Multi-temporal comparisons of NDVI values can provide maps of growing conditions, the appearance of plant diseases due to possible modifications in the spectral responses of the pigments in plant leaves, and changes in the plant phenological cycle. In a coastal wetland or salt marsh environment, the distribution of different land features (e.g., barren soil, submerged areas, sand, vegetation) can be assessed and mapped using the NDVI, with unprecedented spatial detail. These kinds of investigations are well documented for agricultural studies, whereas similar examples for coastal environments are still scarce.

The temporal resolution of sensing is another positive point in favour of multispectral UAS-based surveys, and this goes far beyond that available for satellite and airplane platforms. A UAV can be deployed over an area of interest whenever an updated survey is needed and when the investigated phenomena is of particular interest. Multispectral data can be used by supervised and unsupervised classification algorithms to group pixels on the basis of their spectral similarity. Such methodologies are an established approach in the field of satellite and aerial sensing, although their usefulness when using data from the new generation of multispectral sensors on board UAVs is an open issue.

Visible Imaging

The potentialities offered by UAV photography over the last 10 years have been tightly linked to the availability and widespread use of consumer-grade cameras and computer vision, and traditional photogrammetric-based approaches to processing redundant and unstructured images. Among the range of studies involved in this revolution, coastal monitoring and investigations into the morphodynamic properties of coastal environments are some of the most interesting (Delacourt et al. 2009; Pereira et al. 2009; Vousdoukas et al. 2011; Mancini et al. 2013; Casella et al. 2014; Gonçalves and Henriques 2015; Klemas 2015).

In particular, the majority of investigations aiming at 3D reconstruction of terrains and anthropic features from lightweight (i.e., <5 kg) UAV surveys have integrated an approach based on computer vision technology, namely the SfM (SFM) approach (Snavely et al. 2008). Due to the surprisingly high accuracy and finer spatial resolution achieved by combining UAV surveys and the SfM (SFM) approach, this novel methodology is a rapidly evolving investigative technique for topographic surveys and, more generally, for landscape mapping (Bryson et al. 2013;

Tonkin et al. 2014). The diffusion of UAV-SfM (SFM) methodologies applied to low-altitude aerial images to a wide audience of researchers and professionals has also been related to this methodology being rapid, inexpensive, and highly automated, and to the production of 3D infor mation at unprecedented spatial definition.

In SfM (SFM) methodology, the interior orientation parameters are unknown or derived as approximate values, from the exchangeable image file header of JPEG images collected by a UAV. Approximate values are required in the SfM (SFM) approach, to facilitate the search for conjugate points. A successive refinement of the interior orientation parameters is determined later in the bundle block adjustment, which incorporates the option for camera self-calibration. In the computer vision strategies for image analysis, the search for conjugate points detected in the source photo-graphs is based on point detectors and descriptors of the scale-invariant feature transform (SIFT) type (Lowe 2004). In SfM (SFM) processes, a least squares adjustment is implemented to detect incorrect matches, which are discarded. The whole process produces the so-called 3D sparse point clouds. A successive step, where a multi-view stereo algorithm is performed, produces 3D dense point clouds, and optionally, meshed products and orthophotographs. Moreover, in dense stereo-matching methods, the recent implementation of semi-global matching methods (Hirschmüller 2008) has provided new advances in the generation of 3D point clouds and DEMs. In Figure 6.1, a simplified overview of UAV-based mapping processes with SfM (SFM) methodologies is provided.

The 3D point clouds can be georeferenced to their correct position using GCPs or approximate camera positions obtained by an on board global positioning system (GPS) unit. Accurate position-ing (i.e., centimetre or millimetre level) of GCPs will strengthen the accuracy in the 3D model positioning. To achieve this accuracy in the GCP positions, relative static or network real-time kinematic (RTK) methodologies have to be adopted. However, accurate or coarse positioning of UAVs during flights can reduce the time taken in the initial alignment of the images, in the search for conjugate points, and georeferencing of the sparse/dense point cloud. In most of the software that implement the SfM (SFM) methodology, the identification of GCPs on UAV images is manual. However, the GCP locations are predicted for all of the images, and so the user only needs to refine their position through careful selection of the pixels that correspond to geometric features of artifi-cial targets or natural points (Gonçalves and Henriques 2015).

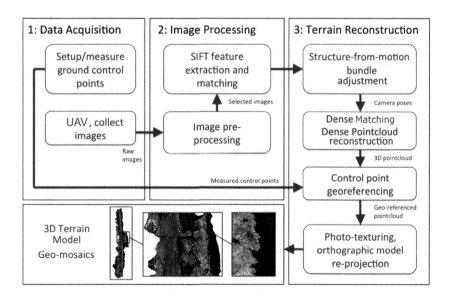

FIGURE 6.1 Overview of UAV-based mapping processes with (SfM [SFM]) methodologies. Three main steps can obtain 3D models: data acquisition, image processing, and terrain reconstruction. (Modified from Bryson et al. 2013.)

When operating on coastal areas in particular, images can be affected by motion blur during wind gusts or by large occlusions of the terrain due to moving features within the scene (e.g., undulating vegetation, presence of water bodies, water from breaking sea waves). Such affected areas can be masked manually to avoid problems in the matching techniques and detection of conjugate points (Bryson et al. 2013). Considering that the SIFT features correspond to distinctive imaged points in the texture of surfaces, the rocky coastal environment is the best candidate for this UAV-SfM (SFM) approach. Nevertheless, various studies have demonstrated the reliability of SIFT-based methodologies also on sandy surfaces (Mancini et al. 2013). More detail on this can be found in the relevant sections.

Several software packages based on the integration between computer vision and traditional photogrammetry have been developed to process UAV images according to the SfM (SFM) philosophy. Among the open-source software, Bundler and CMVS (Furukawa and Ponce 2010), and VisualSfM, Apero, and Mic-Mac (Deseilligny and Clery 2011) can be mentioned. AgiSoft Photoscan (AgiSoft 2015) is a widely used, low-cost, commercial package for very efficient automated processing and surface reconstruction. Another package that can provide very reliable results is Pix4D.

Elevation Models from UAV-SfM (SFM) and Quality Assessment

The validation of reconstructed surfaces is a fundamental issue, as erroneous 3D products can profoundly influence expert decisions based on the interpretation of surface features. In particular, a careful validation assessment of reconstructed surfaces is fundamental for multi-temporal elevation datasets that serve as input data for numerical models, such as the DSMs used in sediment budgeting models (Houser et al. 2008; Wheaton et al. 2010), evolution of morphodynamic features, and high-resolution orthophotographs for coastline monitoring.

A range of recent studies has advanced our understanding of sources of error, accuracy, and precision of SfM (SFM) techniques (Turner et al. 2012; Westoby et al. 2012; Fonstad et al. 2013; Harwin et al. 2015). Bryson et al. (2013) summarised the accuracy that has been achieved in recent studies for the 3D reconstruction of natural landscapes. Position errors of 0.02–0.14 cm were reported for elevation models from images collected at a very low height (15 m) and errors of 0.02–0.06 m for flights programmed at a height of approximately 50 m. In general, all of the mentioned studies used GCPs positioned using a total station survey or survey-grade GPS, with an accuracy from a few millimetres to 1 cm, respectively. Harwin and Lucieer (2012) reported errors of 0.025–0.04 m using the aero-triangulation methodology, from UAV images acquired at 50 m in altitude. Recent studies have faced the accuracy validation of DSMs from SfM (SFM) related to coastal environments. When accurate GCPs were adopted for georeferencing purposes and in the validation process, such studies have reported errors at the level of a few centimetres (Mancini et al. 2013) or below, to 20 cm (Tonkin et al. 2014; Turner et al. 2012; Casella et al. 2014). These results indicate that depending on the UAV elevation and cameras used, spatial resolution of a few centimetres can be obtained from imagery acquired from 50 m altitude, and complex shapes allow for the representation of features on the ground to be reconstructed. In a rare field test with a thermal engine helicopter (DRELIO) and a traditional photogrammetric approach to process the aerial images, Delacourt et al. (2009) produced a DEM of a beach located in French Brittany with an accuracy of 3 cm, after comparison with beach profiles acquired by differential GPS.

However, point clouds from UAV-SfM (SFM) methodology do not densely represent areas of vegetation. This is the case for sand dunes where highly specialised plants find their ideal habitat. In these cases of areas covered by vegetation, and for water bodies or other surfaces with extremely homogeneous texture, the visible attributes needed for algorithms such as SIFT are not provided, and a void appears in the sparse point cloud. During a field experiment at Piscinas di Ingurtosu (a coastal area in western Sardinia, Italy), Vousdoukas et al. (2011) found a number of identified pairs of keypoints between 50 and 1,000, depending on the feature images.

The number, density, spatial distribution, and positioning accuracy of GCPs are also key factors for reliable positioning of UAV-SfM (SFM) products. James and Robson (2012), Clapuyt et al. (2015), and Harwin and Lucieer (2012) demonstrated that the position error of DSMs decreases with the use of an increasing number of GCPs. James and Robson (2012) and Vericat et al. (2009) assessed the effects of GCP distribution on the results. These studies showed that widely dispersed GCPs and their orientation along the perimeter of an area promote decreased reconstruction errors. Goldstein et al. (2015) investigated the GCP requirements for SfM (SFM)-derived topography in a low-slope coastal setting and with the images acquired using a kite-mounted camera. In this study, they determined that beyond 10 GCPs, the error remains constant at 0.04 m. Several experiences carried out by the present authors on coastal areas have suggested that around five accurate GCPs over an area of 2–3 ha are sufficient to achieve a vertical accuracy of 2–3 cm. Similar data were obtained by Harwin et al. (2015) in an accuracy validation procedure under different scenarios placed within a limited portion of an eroding coastal scarp.

GCPs have to be recognised by the user on the collected images. Depending on several parameters (flight elevation, camera resolution, target size), the target geometry will help the user to assign coordinates to the target reference geometry. Some software for automated image processing can print their own GCPs before performing the flight. During the image processing, the pattern will be automatically recognised. Few manual adjustments will then be required. Panels with black, red, or yellow (contrasting with the background) chessboards printed on the upper side are very commonly used.

Mission Planning

Mission planning is perhaps the most crucial aspect for a successful operation. Depending on the expected uses of the data and products, the desired margin of error must be accounted for when designing a mission (Maguire 2014). UAVs can be programmed by users to fly autonomously at fixed altitudes along flight lines, to ensure optimal image overlap for any kind of optical survey. In the applications found relating to coastal areas, UAVs have been deployed at flying heights of 30–50 m above ground level. Visible imaging devoted to 3D reconstruction of surfaces requires a redundant set of images acquired at the selected timing, to facilitate the SfM (SFM) approach to data processing. Images acquired during the take-off and landing operations or images representing features outside of the investigated areas (e.g., the water-edge area) have been filtered out during successive processing. In most applications, images are acquired every second at an average speed of 4 m/s. As already mentioned, the SfM (SFM) approach has resulted in great interest in the processing of images acquired by multi-rotor UAVs, where irregularities that are often experienced in the path followed by a vehicle, varying altitudes, and camera attitudes can be beneficial (Mancini et al. 2013). In the case of bare sand, a minimum number of ten overlapping images should be acquired to allow reliable automated matching procedures (Fonstad et al. 2013). Some precautions have to be highlighted prior to UAV take-off in sandy environments. The take-off and landing operations are crucial because of the sand that is set in motion by the rotors, unless a hard platform is used. Also, the functionality of the rotors and the camera lens can be compromised by finer sand grains raised during these phases. Depending on the features to be images, UAV-borne sensors can acquire data from close range and at multiple viewing angles (i.e., nadir, oblique). The imaging of sub-vertical coastal cliffs will require oblique cameras, whereas sub-horizontal surfaces will be better imaged using a nadir view. The nadir view is commonly used in photogrammetry, but it results in more occlusion, and details can be missed (Harwin and Lucieer, 2012). In the SfM (SFM)-based strategy for surface reconstruction, near-parallel viewing directions and a self-calibrating bundle adjustment approach are used. For image sets with near-parallel viewing angles, the algorithms might not derive radial lens distortion accurately, which will lead to associated 'dome-shaped' DEM deformation. This kind of effect can produce an artefact on the 3D surface and compromise multi-temporal investigations (i.e., the sediments budget computation).

During simulations representative of UAV surveys, James and Robson (2014) highlighted that with near-parallel cameras, and in the absence of control measurements, a domed deformation with magnitude of about 0.2 m over horizontal distances of about 100 m is seen. They observed that the deformation can be reduced if suitable control points can be included within the bundle adjustment and oblique images are collected in addition to the nadir views. Certainly, this requires gimballed camera mounts, which represents a widely used solution on multi-rotor UAVs. Residual systematic vertical errors can remain, depending on the precision of the control measurements. Reference can also be made to Javernick et al. (2014) for further details on deformation, as seen for some UAV reconstructions. If a nadir image set is not required, James and Robson (2014) suggested to fly along overlapping flight lines in opposing directions with an off-nadir installed camera. During field tests that the present authors conducted, an off-nadir angle from 30° to 45° was used for the 3D reconstruction of coastal cliffs.

The above-mentioned study by Harwin et al. (2015) investigated the impact on mapping accuracy of the bundle adjustment and 3D reconstruction process of a coastal cliff surveyed by nadir and oblique cameras. Ranges of typical operating scenarios were tested, and the sensitivity of the process to the calibration procedures and the GCP spatial distribution/accuracy was carefully assessed. Their results indicated that the use of robust self-calibration, additional oblique photography, and a reduced number of accurately surveyed GCPs can improve the final mapping accuracy of natural landforms like coastal cliffs. Thus, the joint use of nadir and oblique cameras in addition to some accurate GPS is a requirement for dual reasons: firstly, to represent the vertical portions of a cliff, and secondly, to incorporate network design elements that promote robust self-calibration.

APPLICATIONS OF UNMANNED AERIAL SYSTEMS TO COASTAL ENVIRONMENTS

HIGH-RESOLUTION COASTAL TOPOGRAPHY BY UNMANNED AERIAL SYSTEM-BASED PHOTOGRAMMETRY

The interest of surveyors in the reconstruction of high-resolution topography of coastal environments through proximity surveys is mainly due to the increasing vulnerability of coastal environments over the last decade. Basically, factors that contribute to the degradation of coastal areas worldwide can be identified in the dynamic action of a rising sea, in winds and storms, and in losses to the coastal sediment budget, with this last related to reductions in the transport of suspended sediments by rivers or through anthropogenic actions. In the Mediterranean basin, several areas of low and sandy beaches have undergone rapid morphological changes, with observable net regression and destabilisation of beach/dune systems (Gonçalves and Henriques 2015). Thus, timely investigations of topographic profiles at increasing spatial resolution and accuracy can detect the evolution from soft topographic profiles, with a high capacity for sea-energy dissipation, to steeper profiles, with partial or complete destruction of foredune systems and reductions to beach stretches. To counterbalance these trends, several hard structures for coastal defence (e.g., walls, groynes, breakwaters) can be placed to contain the effects of the sea. Such structures can induce additional modifications to the nearby geomorphology (that might be more or less desirable). However, the quantitative assessment of such modifications on a regular basis or after severe events requires a timely, accurate, and effective survey methodology to be adopted over large areas. None of the existing technologies can cover this gap at moderate cost.

To monitor sensitive coastal environments and beach morphologies, it is feasible to generate DEMs, DSMs, and orthoimages. Despite the wide range of methods available for the production of high-resolution point clouds, and successive high-quality DSMs, some difficulties are generally experienced when topographic surveys are being carried out on the sandy littoral, based on significant investment in personnel and time, and the lack of benchmarks or other local permanent reference points (Mancini et al. 2013). Permanent reference points that are not affected by coastal processes are fundamental to guarantee the georeferencing of point clouds within studies related to change analysis, where successive surveys have to be compared. All of these difficulties can be limiting factors for ground-based survey methods whenever high spatial resolution and high-quality DSMs are required. The use of UAS-based photogrammetry can cover these necessities, as this can overcome the limitation of other methodologies used in the reconstruction of shore regions or dune morphometry. Among these methodologies, those based on GNSS, total stations, and terrestrial laser scanning (TLS) should to be mentioned. GNSS positioning (which is based on network or traditional RTK) is fast and accurate, although limited in the number of measurable points. TLS can also be very accurate, although this requires long survey sessions due to complex coastal morphologies, large areas to be surveyed, occultation, and significant investment in data processing time. Geomorphic studies based on airborne LIDAR surveys are also relevant to this topic, although despite the exploration of significantly extensive areas using LIDAR, its use is costly and the results do not provide data with comparable spatial and vertical accuracy with respect to TLS, GNSS, and UAV photography. In addition, the costs are not comparable; a TLS survey is two to three orders of magnitude more expensive than a UAV survey. Instead, a traditional topographic survey (e.g., by a total station) would be of similar cost to the UAV system, but the resulting DSM would be much coarser for similar effort in the field. DSMs that are built using aerial or satellite images might also be of some interest in the field of coastal investigation, although the average vertical accuracy (e.g., coarser than 50 cm) does not meet the requirements for most of applications. Certainly, elevation models produced by any of the available photogrammetry-based methodologies (or more correctly, DSMs) are affected by patches with vegetation. These patches should not be included in any validation procedure and so should be filtered out using a reliable filtering process. The availability of contemporary multispectral data at reasonable spatial resolution can help after a preliminary classification procedure. To discuss the feasibility of UAS photogrammetry in the reconstruction of coastal morphologies, the next three sections consider some data obtained during surveys of low sandy beaches, cliffs, and dune systems. For further reading relevant to the use of UASs for surveys of coastal areas, the reader can refer to Table 6.1, where recent studies in the scientific literature are summarised.

Low Sandy Beach Morphometry

Low sandy beaches have high vulnerability towards extreme events, such as coastal inundation, long-term sea-level rises, and deficits in the sediment budget. Depending on the specific hazard that affects low beaches, the following direct and indirect negative effects can be mentioned: temporary coastal inundations, tourism decline, rise in insurance premiums for house owners, and other disruptions of businesses (Sekovski et al. 2015). However, careful assessment of the vulnerability of low sandy beaches to a single or multiple hazards requires detailed knowledge of the coastal topography at significant times. Some studies have recently investigated the potential of UAS surveys for the reconstruction of the surface of sub-horizontal sandy beaches.

Gonçalves and Henriques (2015) generated detailed DSMs for two test sites, as a sand-spit in the mouth of the Douro River (Portugal) and a sandy beach with reduced extent in the vicinity. The equipment consisted of a fixed-wing light UAV (Sensefly Swinglet Cam model), in addition to artificial targets placed on the beach surface. The survey was carried out with survey-grade GPS receivers and RTK technology (for centimetre-level accuracy). The artificial targets used as GCPs and the independent checkpoints consisted of plastic bands or wooden plates with crosses in the middle and of contrasting colours. As a practical recommendation, Gonçalves and Henriques (2015) suggested the use of checkpoints to validate the dataset provided by the analysis and, in particular, to confirm the vertical accuracy of the DEMs and to proceed with orthophotograph generation and mosaicking into a single map. DEMs free from effects due to vegetation are more useful to extract topographic profiles at preferred locations, and to compute volumes, locate the high water line, and delineate the shorelines through an elevation-based approach. However, high-resolution orthophotographs can complement the 3D perception given by the DSMs and can provide useful information on the vegetation cover.

In the area of beach-surface reconstruction, Maguire (2014) investigated an extremely low-lying coastal stretch known as Delray Beach, on the eastern coast of Florida, USA, with elevations at or slightly above the mean sea level. In land, the sandy beach evolved into 50-m-wide dunes. The coastal stretch was affected by coastal flooding episodes during high tides, storms, and hurricanes. To protect the coastal ecosystem from these hazards, several beach nourishment programmes had been performed previously. A 'Y-shaped' hexacopter with an autopilot and a camera (Nex-5R

TABLE 6.1

Recent Studies Related to Surveys of Coastal Environments Using UAVs

Authors (Year)	Journal	Contents
Klemas et al. (2015)	*Journal of Coastal Research*	Overview of coastal and environmental remote sensing from different UAS platform (fixed-wing, rotors, blimps, kites, and balloons)
Gonçalves and Henriques (2015)	*ISPRS Journal of Photogrammetry and Remote Sensing*	Fixed-wing UAV-based photogrammetry for topographic monitoring and DSM generation of sandy coastal areas
Pereira et al. (2009)	*Journal of Coastal Research*	Coastal monitoring with different types of fixed-wing UAS
Bryson et al. (2013)	*PloS One*	Kite-based multispectral imaging system for constructing terrain models of intertidal rocky shores
Harwin and Lucieer (2012)	*Remote Sensing*	Multi-view stereopsis techniques to imagery acquired from a multi-rotor micro-UAV of a natural coastal site
Vousdoukas et al. (2011)	*Journal of Coastal Research*	Semi-automatic technique to process images from fixed-wing small UAV and provide information on the nearshore sand bar morphology, the locations of rip channels, and the dimensions of surf/swash zones
Casella et al. (2014)	*Estuarine, Coastal and Shelf Science*	Study of wave run-up using numerical models and terrain elevation at very high spatial resolution from low-altitude UAV-based photogrammetry
Goldstein et al. (2015)	*PeerJ Preprints* (not peer-reviewed)	Reconstruction of topography in low-slope coastal environments by kite and consumer-grade camera
Harwin et al. (2015)	*Remote Sensing*	Accuracy assessment under several scenarios in the reconstruction of a coastal cliff topography from UAV survey
Mancini et al. (2013)	*Remote Sensing*	Accuracy assessment in the reconstruction of a beach dune system from UAV survey
Delacourt et al. (2009)	*Journal of Coastal Research*	Using an unmanned helicopter for imaging coastal areas
Drummond et al. (2015)	Australasian Coasts & Ports Conference proceedings	UAV applications to coastal engineering, beach erosion assessment, and estuary entrance vegetation mapping
Caprioli et al. (2015)	*The International Archives of the Photogrammetry, Remote Sensing and Spatial Information Sciences*	Survey of coastal cliff by nadiral and oblique imageries for further investigations on rock slope failures

DSLR; Sony) was used as the airframe. This had two motors on each arm, one that acted as a pushing motor and the other as a pulling motor. The multi-temporal surveys allowed elevation change analyses from DEMs created after the repeated UAV surveys. This analysis revealed some contrasting trends of erosion and accretion for distinct sectors of the investigated areas.

For similar purposes, Delacourt et al. (2009) surveyed Porsmillin Beach, in French Brittany, using a helicopter that weighed 11 kg (the DRELIO project) and flew at speeds of up to 70 km/h while carrying an additional payload of 6 kg. At that time, the SfM (SFM) approach to image processing was not in use. The stereo pairs were processed using the traditional photogrammetric approach, and DEMs with a few centimetres of spatial resolution were obtained.

Cliffs

Surveys of coastal cliffs represent very challenging tasks because of the sub-vertical settings and, in many cases, direct contact with the sea. Cliffs that plunge into the sea do not allow the setting up of topographical instrumentations, and for this reason, an aerial methodology is required. However, the aerial nadir view is not very suited for such a task, and coastal cliffs have remained as one of the last challenges for surveyors. UAV surveys with oblique cameras have been exploited very recently for similar purposes, and some applications have also been reported for coastal cliffs.

In recent years, researchers from the University of Tasmania, Australia, proposed some applications related to the use of oblique images for coastal environments. In particular, Harwin and Lucieer (2012) used UAV images to produce a 3D model of a coastal cliff. Point clouds offered a spatial resolution of 1–3 cm and an accuracy of 25–40 mm. Harwin et al. (2015) investigated the joint use of nadir and oblique images to reconstruct a section of an erosion scarp, which were acquired using a multi-rotor UAV. In this study, centimetre-level spatial resolution and positional accuracy were achieved using a reduced number of accurately surveyed GCPs. Genchi et al. (2015) used SfM (SFM) technology to characterise bioerosion patterns (i.e., cavities for roosting and nesting) caused by burrowing parrots on a cliff in Bahía Blanca, Argentina. In this study, several geometric parameters were derived from the fine spatial details of the cliff surface, and then associations between the surface topography and the bioerosion features were established.

Caprioli et al. (2015) proposed the joint use of nadir and oblique images from a hexacopter UAV survey to perform stability analysis along a 500-m-long coastal cliff of the Salento Peninsula (southern Italy). This is a spectacular landscape in the proximity of the towns of San Foca, Torre dell'Orso, and Sant' Andrea, within the municipality of Lecce. The cliff is subjected to widespread instability phenomena due to the steep rocky cliffs that were formed by weathered and fractured carbonate rock through intense erosion processes. In this area, geologists and geotechnical engineers have detected complex mechanisms of rock-slope failure, but very often, detailed 3D information is missing. The UAV system used was a vertical take-off and landing hexacopter that was equipped with a calibrated digital camera (EOS model 550D; Canon) with a 20-mm lens. The camera mount allowed the camera tilt to be controlled by the UAV operator when oblique images were required. The focus was fixed at infinity, and the camera settings were carefully chosen to reduce motion blur when acquiring images at 1 Hz (i.e., one photograph per second). The flight with the oblique camera was performed by manual piloting for two reasons: firstly, to inspect occulted zones like arches and cavities, and secondly, to fly very close to the cliff wall and to take pictures of meaningful details. Figure 6.2 shows the point cloud of the whole area in addition to details of zoomed areas for closer inspection of the point cloud properties. The GCP locations are also visible, as the numbered flags. This processing was performed using the PhotoScan software (AgiSoft 2015 of) that can process nadir and oblique images in a single run. This reconstruction of the ground surface and objects using PhotoScan was a three-step process.

Dune Systems

Coastal dunes have a primary role in the protection of coastal zones and as a natural barrier, while they provide a sediment supply to beaches and protect the inland areas from storm surges. The relationship between dunes and ground saltwater intrusion has also been demonstrated, which has highlighted the importance of dunes for coastal agriculture systems. Nonetheless, the stability of dune systems can be threatened by coastal erosion, which affects numerous beaches and dune systems.

Thus, for reliable modelling of a beach dune system, detailed knowledge of the dune morphometry is necessary (e.g., accurate DSMs). The growth and development of coastal dunes with vegetation relies on the effects of vegetation on sedimentation patterns. However, observing the evolution of a beach dune system requires the reconstruction of the dune morphometry at high resolution and frequency.

Timely topographic and photographic surveys at variable timescales are also fundamental to capture any significant stages in dune evolution. Certainly, the use of UAVs can be a solution for monitoring of the evolution and morphodynamics of coastal dune systems. Goldstein et al. (2015) used a consumer-grade camera attached to a delta kite flying at an altitude of 20 m to photograph the beach and dune of Hog Island (Virginia, USA). They identified this kite aerial photogrammetry as a valid technique for examination of ground surface changes in low-lying coastal environments and highlighted the benefits from synchronous collection of 3D and photographic data. In addition

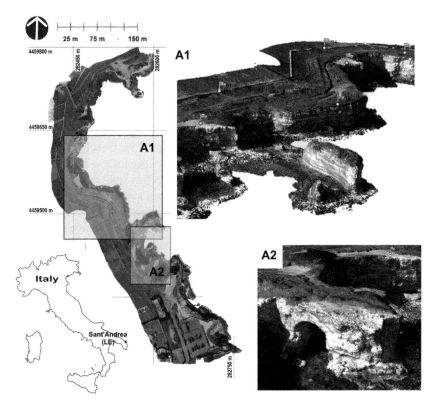

FIGURE 6.2 Textured point clouds that represent a coastal cliff of the Salento Peninsula (Sant' Andrea, southern Italy) after a UAV survey with nadir and oblique cameras. The full extent of the surveyed area is shown on the left, as a nadir view. More detailed features reconstructed through textured point clouds are shown on the right, for the areas coded as A1 (upper right) and A2 (lower right). (Modified from Caprioli et al. 2015.)

to investigations of the growth of coastal foredunes, the restoration of coastal dunes can benefit from the effectiveness of UAV-SfM (SFM) technology.

The morphodynamics of coastal dunes require DSMs that are accurate enough to detect any change in the volume of the primary sand dunes, from the baseline reference DSM. Thus, a careful procedure to assess errors that affect DSMs is a fundamental step. Mancini et al. (2013) reconstructed the topography of a foredune located in Ravenna (Italy) at very high spatial resolution using 550 images that were acquired above an area of about 2 ha using a hexacopter and an SfM (SFM) approach to process the data. This dune system was selected as a test site due to its very complex topographic features, the presence of weathering features, and the relevance of its protective role for the inland areas.

These investigations were aimed at an assessment of the vertical accuracy of the elevation model through comparison with a reference point cloud generated from a TLS survey and using an external set of validation points referenced by a GNSS methodology with centimetre-level accuracy. The accuracy of the TLS- and UAV-based methodologies was comparable, although Mancini et al. (2013) noted that the TLS survey lasted several hours, in comparison with the 7 minutes it took for the UAV to capture the images of the same area. They noted that the assessment of geomorphic features is usually based on multi-temporal analyses of continual surfaces. Thus, an interpolation process was required. Reliable validation of the products from UVA survey needs a preliminary accuracy assessment of the sparse point cloud and then the processing of a dense cloud, with the final interpolation to provide a continuous surface at the desired spatial resolution for further geomorphological investigations (Mancini et al. 2015). The dense matching algorithm produced very distinct features. Areas with coarser sand and sparse dune vegetation showed higher point densities (up to 145 pts/m^2), whereas the completely flat area with finer sand showed reduced density (about 30 pts/m^2). This was due to the difficulties of the features matching algorithm to relate features for homogenous textures. However, this density might be sufficient for most applications.

Monitoring Coastal Dynamics and Geomorphological Processes

The monitoring of coastal areas is a critical task in the reduction of hazard impact and for reduced risk in response to coastal hazards. Surveying of coastal morphological features on a regular basis provides the background for reliable coastal management activities and land-use policies. Effective integrated coastal management programmes must incorporate a realistic picture of the coastal processes at appropriate spatial and temporal scales, and responses based on an understanding of the physical environment obtained through the use of surveys (Delacourt et al. 2009). Very often, remote sensing technologies or aerial surveys provide the basics for coastal monitoring. However, coastal monitoring at finer spatial scales and temporal frequencies has some requirements that cannot be obtained from satellite data or aerial surveys. For instance, the demand for DEMs that represent smaller geomorphic features is increasing (e.g., at decametre resolution and accuracy) to improve the performances of numerical modelling of coastal processes, to analyse changes, and to monitor the effects after restoration projects. Harwin et al. (2015) discussed the use of UAV-SfM (SFM) technology to monitor small changes at temporal resolution that was suited to the requirements for pre-event and post-event mapping at unprecedented spatial resolution. This revolution introduced UAV-SfM (SFM) to aid in the accurate geomorphological mapping, to quantify geomorphological changes, and to provide input into hydrodynamic numerical modelling. Indeed, where the rate of change was below or close to the achievable accuracy of a given survey technique, the quantification of the morphometric change can be problematic (Tonkin et al. 2014).

Among the problems related to coastal dynamics, erosion of a beach ecosystem due to natural or man-made causes can be a significant threat to major cities and to revenues from tourism areas. Erosion and accretion processes can also be altered after intervention of coastal protection by hard structures. To face these problems, coastal monitoring tools need to be cheap, easy to use, and able to accurately map small areas of coastline (Maguire 2014). As already mentioned, LIDAR, GNSS, and aerial photogrammetric surveys are being used by municipalities as methods of erosion monitoring. These survey methods are not within the reach of local municipalities without the availability of highly skilled technicians and expensive equipment to monitor all of the coastal areas under their control. Additionally, monitoring of coastal erosion is cost effective when it is carried out on a large scale, whereas very often the interest of geomorphologists and surveyors is focused on small sections of coastline.

Applications based on UAS surveys to monitor coastal erosion can be found in James and Robson (2012). Harwin and Lucieer (2012) used a multi-rotor UAV survey to reconstruct a detailed DSM for a 100-m-section of the coast in a sheltered estuary in southeast Tasmania, Australia. In this study, they aimed at the determination of the suitability of a UAV technique to provide fine-scale change detection using nadir and oblique photography.

Figure 6.3 shows the data provided from repeated surveys on a coastal stretch in Ravenna (Italy). The UAV flights provided the reconstruction of point clouds prior to and after a severe storm surge episode. The data from the DSM comparisons allowed quantitative assessment of the subtle changes and the erosional processes that occurred over time. To emphasise the changes in beach morphology, six profiles were extrapolated at the chosen locations. The erosion observed for this section of the coastline was subtle and might not have been visible via traditional satellite and aerial change–detection techniques that have less accurate vertical accuracy and spatial resolution. However, the use of UAV-SfM (SFM) surveys to generate the 3D point clouds for coastal monitoring requires careful survey design to ensure sufficient accuracy for the detection of any changes.

Drummond et al. (2015) presented a similar approach for rapid assessment of the impact of coastal storms using a fixed-wing UAV (Sensefly eBee RTK). This was achieved through pre-storm

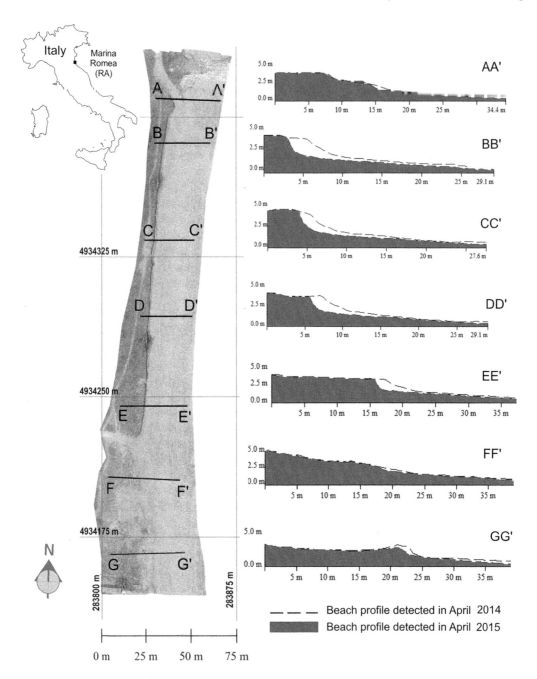

FIGURE 6.3 Assessment of erosive effects through repeated UAV surveys on a coastal stretch in Ravenna (Italy). The comparison of point clouds refers to the years 2014 and 2015, during which time a severe storm surge episode occurred. This has allowed quantitative assessment of subtle changes and erosional processes that occurred over this time. Top: point cloud of the 400-m-long coastal stretch, with the positions of six transects for multi-temporal point cloud comparisons. Bottom: point clouds comparison along transects (white and black, years 2014 and 2015, respectively; unpublished data).

and post-storm surveys, with the aim to quantify the coastal impact of a major storm that occurred for Narrabeen-Collaroy Beach (Sydney, Australia) and that lasted over a 3-day period in April 2015. They performed beach erosion assessment in addition to fine computation of the volume changes between the surveys. Validation of the UAV-derived elevation models was performed through comparisons with in situ RTK-GPS surveys.

In the monitoring of coastal dynamics with UASs, several achievements in other studies can be noted as expansions of the existing mapping methods. Vousdoukas et al. (2011) used small fixed-wing UAVs to monitor the nearshore, including the sand bar morphology, the locations of rip channels, and the dimensions of the surf zones. They inferred that the methodology allows increased spatial cover and a more favourable vantage point, in comparison with ground-based imaging systems (e.g., ARGUS). Pereira et al. (2009) performed video surveillance for coastal hazard detection and in erosion studies using six different types of fixed-wing UAVs (wingspans, 1–6 m) with autonomous take-off and landing capabilities.

To reinforce the effectiveness of coastal monitoring operations from UAV surveys, the repeatability of flights and GCP accuracy/distribution is advised. A single waypoint file provided to the autopilot for successive missions allows the user to pre-programme mission profiles from take-off to landing, and for completely repeatable missions, which ensures that the same area is covered on multiple occasions.

RIVER MOUTH, COASTAL WETLAND, AND INTERTIDAL LANDSCAPE DYNAMICS

River mouths, coastal wetlands, and intertidal landscapes have fundamental roles as the remaining ecosystems for the preservation of littoral biodiversity. Integrated monitoring using the efficient technologies of topographic, chemical, physical, and other environmental parameters is necessary as a basis for any management purposes. For instance, investigations into the vegetation cover of the land and its mapping are a typical application in wetland monitoring (Rafiq 2015). Traditionally, wetland monitoring relates to extensive land surveys that involve ground-based sampling transects, and very often, these environments can be hostile to ground-based methodologies, for many logistic reasons. Topographic surveying of emerged terrains can be extremely time consuming and costly. Currently, the applications of UAS to transitional environments relate to vegetation mapping through multispectral data and the reconstruction of topography through UAV-based photogrammetry.

For instance, miniaturised multispectral and hyperspectral sensors are being developed for incorporation into UAVs, for discrimination of various vegetation species. The task of classification and mapping of coastal vegetation types is currently performed with methodologies taken from satellite and aerial multispectral surveys. Pixel-based supervised/unsupervised classification and vegetation indices are the most common methodologies. For example, a fixed-wing UAV and a helicopter in tandem have been used to map wetlands and to survey estuaries for hazardous algal blooms (Klemas 2015) and to identify long-term changes in swamp hydrological conditions that can affect ecological communities (Lechner et al. 2012).

The combination of visible and multispectral imaging is also an opportunity to map phenomena that occur in coastal wetlands and for river mouths. For instance, a sequence of images of river and coastal plumes at suitable spatial and temporal resolutions can be analysed to study the mixing processes between high-nutrient terrestrial plumes and salt water.

Rafiq (2015) reported on a study focused on land-cover changes for the Huntington Beach Wetlands (southern California, USA) using UAV-acquired multispectral images. Anthropogenic effects have resulted in an increasingly degraded wetland environment and reduction in the tidal flow in the area, which triggered drying up phenomena. Reduction of the extent of the vegetation in this area dramatically reduced the wildlife populations. A methodology was proposed to monitor the extent of the vegetation on a yearly basis through capture and spectral land-cover analysis of aerial UAV images. Both the spectral mapping and the photogrammetric approaches were selected as appropriate methodologies for creating up-to-date land-cover maps. Spectral-based land-cover maps from UAV data were compared with land-cover maps derived from satellite- and airplane-based images.

This comparison highlighted the improvements introduced by multispectral data at very high resolution to these land-cover maps for the representation of small features on the ground. Although the use of UAVs for coastal wetland investigations is not very common, Rafiq (2015) concluded that in the near future, several applications will be possible, including realisation of the vegetation cover of the land.

Kelcey and Lucieer (2012) carried out a study where they tested low-cost consumer-grade multispectral sensors aboard an octocopter UAV platform for vegetation monitoring. They used a UAV platform and installed a miniature camera array that housed six-band multispectral sensors (Tetracam Inc.) and gathered UAV images of salt marsh communities in Ralphs Bay, Australia. In this study, they provided very useful information to gather high-quality reflectance data from the vegetation, including the necessary sensor corrections and radiometric calibration.

Drummond et al. (2015) mapped the distribution of an endangered ecological community that was composed of salt-tolerant plants and was located in a coastal saltmarsh (Shoalhaven River Estuary, 100 km south of Sydney). The UAV was deployed with an NIR camera payload to examine and map the existence and health of the coastal salt marshes and the adjacent dune vegetation. The NIR images were processed to produce an NDVI. Positive NDVI values are representative of dense green vegetation. It is interesting to note that these NDVI values decrease as the leaves become water stressed. NDVI values close to zero represent bare soil or sand, and negative values indicate water surfaces.

The study of intertidal environments by ground-based surveys requires a large amount of logistical effort, and the data are often not recorded in a contiguous manner. In addition, remote sensing allows the acquisition of targeted data, which limits its effectiveness in small-scale environmental science and ecologically focused studies (Bryson et al. 2013). The availability of high-resolution DEMs is another requirement for studies that are focused on the influence of small-scale topographic variability on the distribution of assemblages of plant and animal species. For instance, if elevation data are carefully referred to a geoid model, this allows delineation of shorelines with respect to tides.

Bryson et al. (2013) introduced a kite-based multispectral imaging system for construction of 3D terrain models of intertidal rocky shores, and to build the NDVI indices for identification of intertidal species during imaging, and for the evaluation of algal biomass. The kite was walked in a zig-zag fashion rather than in a straight line along the shoreline to facilitate the collection of images from various perspectives with respect to the terrain and to exploit the capabilities of the SfM (SFM) methodologies. This allowed the division along the shoreline of the upper tidal zone and the mid-tidal zone, both of which were characterised by typical populations, and the relating of this to the fine-scale structure of the rock pools and crevasses. Before the advent of modern UAVs, Guichard et al. (2000) used a helium blimp to take photographic stereo pairs of a section of a rocky intertidal shore using both colour and NIR film.

Applications that use bare soil as the input need the removal of vegetation surfaces from elevation models for further processing. As indicated, the vegetation can be removed after multispectral classification. However, a void remains in the underlying terrain in the reconstructed point clouds. A different approach relies on the use of laser scanners designed for UAVs.

To overcome limitations due to the presence of a canopy, and to better represent the riverbanks of an estuarine area, Mancini et al. (2015) integrated images acquired by a UAV (ground sample distance, 0.03 m) with a TLS survey and an unmanned surface vehicle that collected images of the river from different points of view. The basic idea behind this approach was to map areas that required maintenance due to the large canopy that occluded the identification of the riverbanks. This aspect might be a threat for the security of the river and all of the infrastructure close to the river basin. The unmanned surface vehicle also acted as a docking station to recharge the batteries.

COASTAL ENGINEERING

Among the wide variety of UAS-based applications for coastal environments, those related to coastal engineering are at an early stage, and many contributions are expected in the near future. The provision of a cost-effective solution through UAV surveying for an accurate 3D representation is the main advantage, with the potential use of this methodology for coastal engineering applications. Surveyors are aware of the difficulties that are encountered very often during traditional land-based methods of data collection for coastal sites (e.g., network-based RTK-GPS, TLS, or airborne LIDAR). In particular, for coastal engineering purposes where structural investigations supported by complete 3D surveys are required over large areas, productive methodologies at fine spatial resolution are common requirements. To avoid labour-intensive and costly mapping work, UAV surveying might be a solution if no physical and logistic restrictions are encountered. In this field, the scientific literature remains relatively poor. Drummond et al. (2015) evaluated the structural damage suffered by a rubble-mound breakwater, and volume analyses of the protection were performed after a UAV survey. Traditional methods that are used to determine breakwater conditions following a storm can be time consuming, and they are often ineffective for identifying the individual movement of this protection. The possibility to use a light vehicle to inspect and document a specific structure over time without using water craft provides further advantages. Ground-based operations such as the need to access breakwaters situated on exposed coastlines can also present a hazard to surveyors. Due to the dense point cloud produced by the UAV-SfM (SFM) method during repeated surveys, this allows the detection and tracking of individual breakwater stones following storm damage. This dataset provided the identification of specific sections of the breakwater that required maintenance, as well as analyses of the protection volumes, which allowed for targeted repairs that also minimised expenditure. Certainly, the negative effects of the water in the conjugate point detection have to be removed by masking out these areas. Other very interesting applications in this area will be relevant to beach renourishment projects. To date, the benefits of these interventions are very difficult to track with regularity, because the methods used to monitor and measure the balance between deposited and removed volumes of sand are cost prohibitive.

COASTAL VULNERABILITY AND HAZARD ASSESSMENT

There is an increasing interest in coastal vulnerability assessment, as it represents the preliminary stage of investigation into coastal defence and a tool in the management of the economy of beach-related tourism, which is a significant resource for many communities during the summer season. Very often, the coastal vulnerability and hazard assessment of coastal processes refer to monitoring of the impact of sea storms on coastal areas. Low-lying coasts are particularly prone to erosion and flooding processes. The modelling of these processes can be much improved by the introduction of accurate topographic data as reference surfaces.

Casella et al. (2014) introduced high-resolution DSMs from UAV-SfM (SFM) technology into a wave run-up model and validated the data by comparing the observed and the modelled run-up. They collected aerial photographs (Mikrokopter Okto XL vehicle) after two different swells in a study area in the municipality of Borghetto Santo Spirito (Liguria Region, Italy). Figure 6.4 shows a representation of the studied area, and the data provided by the run-up model.

Since 1944, the study area of this coastal stretch has been the subject of engineering works that have been aimed at its protection from marine erosion. More recently, several interventions of beach nourishment were performed for this shoreline. To obtain a complete beach topography, the point cloud obtained through a UAV survey and the multibeam data were merged to represent the emerged and submerged sectors. Finally, the elevation data were referred to a geoid model and horizontal positions to a cartographic reference system. The use of the elevation referred to a geoid model (above sea level) is fundamental for studies that are focused on coastal vulnerability to storm surges, and the accuracy of the selected geoid model must satisfy the requirements of the model. In Mancini et al. (2015), the DSM of a coastal dune system was referenced to horizontal coordinates in UTM Zone 33N (datum ETRF00), while the vertical values were referred to the mean sea level using the geoid model ITALGEO2005 provided by the Italian Geographic Military Institute. Indeed, the use of a (more or less) accurate GNSS device for GCP positioning produced elevations that referred to an ellipsoidal system. Thus, a geoid model was required.

A report from Casella et al. (2014) highlighted the potential of methodologies that incorporate results from UAV surveys into global information systems to provide reliable information on

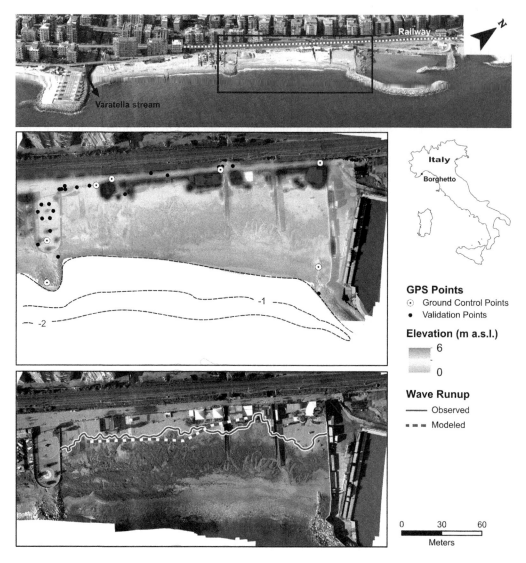

FIGURE 6.4 Wave run-up based on high-resolution beach topography from UAV images acquired at Borghetto Santo Spirito (Liguria Region, Italy). (a) Aerial view, with location of the investigated area. (b) DEM obtained by combination of UAV-based point cloud and multibeam data. (c) Comparison of observed and modelled maximum run-up for an event that occurred in December 2013 over the orthophotograph from UAV images. (Modified from Casella et al. 2014.)

beach topography and geomorphology compared with traditional techniques, without losing accuracy. Researchers and coastal managers might benefit from the use of new elevation datasets in the assessment of the impact of extreme wave events after a storm. However, timely mission planning is fundamental to carry out the aerial survey at significant times. For instance, the net rate of erosion after a sea storm can be more accurately assessed by timely aerial surveys. Again, the extent of flooded areas can be mapped during the event or in the immediate aftermath, before the waters drain out from the flooded areas (Maguire 2014). In addition, the use of high-resolution beach surfaces allows the prediction to some extent of the areas that are more vulnerable to the run-up of extreme swells. Conversely, the deployment of an aircraft at each occurrence with such a short warning period makes the whole mission virtually impossible.

CONCLUSIONS

This chapter has introduced the potential for UASs to effectively fill current observation gaps in coastal environment remote sensing, for applications that include coastal change research, wetland mapping, ecosystem monitoring, vulnerability assessment, natural hazard prediction, and coastal engineering. UAVs can replace many conventional survey methodologies, with considerable gains in spatial resolution, vertical accuracy, and cost of data acquisition. Surveys can be conducted in logistically challenging areas, to investigate unstable or inaccessible terrain, where traditional survey methods are not feasible. Thus, the possibility to survey and monitor dynamic landscapes and morpho-sedimentary processes is now widening.

Some of the weather dependency that affects airborne surveys (e.g., cloud cover) can be circumvented with UAVs. Moreover, the possibility of the reconstruction of the coastal morphology at unprecedented resolution opens the way to micro-topographic investigations and accurate sediment budgeting. However, the benefits of the SfM (SFM) technique might be partially reduced due to the presence of water, reflective surfaces, or very smooth textures on the ground. Careful mission planning is required. Despite the potentiality of UASs for coastal applications, there remain technical and legal hurdles that prevent wider use. Although with some differences across different countries, these difficulties include the strict limits imposed by national regulations that allow UAV flights in visual line of sight at moderate distance from the ground control station and within segregated airspace. Drones would not be allowed to fly at night or over densely populated areas. Fixed-wing flights can cover longer ranges. Thus, for such vehicles, the existing regulations are even more restrictive.

Improvements can be expected from even more accurate direct georeferencing of UAVs; for instance, by integrating low-cost dual frequency GNSS receivers and lighter inertial sensors into the payload. This should provide better altitude determination with very little effort placed on field surveys for GCP collection. Rotary-wing UAVs are susceptible to failure under windy conditions, and their endurance is still limited to 10–20 minutes of flight. For non-tethered UAVs, failure during overflights of coastal areas might result in the loss of the vehicle. These major factors limit their use in applications that are related to coastal environments. In addition, professionals and researchers involved in the use of UAV data still face several difficulties, from the need for metadata and UAV-specific data formats, to the standards for dissemination, standardised calibration of sensors, and effective data validation procedures.

However, the general consensus is that UASs will have increasingly important roles in the investigation of natural resources and land-monitoring needs at reduced costs. In coastal surveys, UAVs will potentially revolutionise the methodologies of data collection, as these can provide timely investigations of logistically challenging areas and inaccessible terrain using an increasing variety of sensors.

ACKNOWLEDGEMENTS

We would like to thank Professor Giovanni Gabbianelli for sharing unpublished data from some repeated UAV surveys over the coastal stretch represented in Figure 6.3.

REFERENCES

AgiSoft. 2015. *AgiSoft PhotoScan User Manual: Professional Edition, Version 1.2.*

Bryson, M., Johnson-Roberson, M., Murphy, R. J., and D. Bongiorno. 2013. Kite aerial photography for low-cost, ultra-high spatial resolution multi-spectral mapping of intertidal landscapes. *PloS One* 8(9): e73550.

Caprioli, M., Trizzino, R., Pagliarulo, R., Scarano, M., Mazzone, F., and A. Scognamiglio. 2015. Management of environmental risks in coastal areas. *ISPRS-International Archives of the Photogrammetry, Remote Sensing and Spatial Information Sciences* 1: 263–268.

Casella, E., Rovere, A., Pedroncini, A., Mucerino, L., Casella, M., Cusati, L. A., Vacchi, M., Ferrari, M., and M. Firpo. 2014. Study of wave runup using numerical models and low-altitude aerial photogrammetry: A tool for coastal management. *Estuarine, Coastal and Shelf Science* 149: 160–167.

Clapuyt, F., Vanacker, V., and K. Van Oost. 2015. Reproducibility of UAV-based earth topography reconstructions based on structure-from-motion algorithms. *Geomorphology* 260: 4–15.

Colomina, I., and P. Molina. 2014. Unmanned aerial systems for photogrammetry and remote sensing:
A review. *ISPRS Journal of Photogrammetry and Remote Sensing* 92: 79–97.

Delacourt, C., Allemand, P., Iaud, M., Grandjean, P., Deschamps, A., Ammann, J., Cuq, V., and S. Suanez.
2009. DRELIO: An unmanned helicopter for imaging coastal areas. *Journal of Coastal Research* S156:
1489–1493.

Deseilligny, M. P., and I. Clery. 2011. Apero, an open source bundle adjustment software for automatic calibration and orientation of set of images. *ISPRS-International Archives of the Photogrammetry, Remote
Sensing and Spatial Information Sciences* 38: 269–276.

Drummond, C. D., Harley, M. D., Turner, I. L., Matheen, A. N. A., and W. C. Glamore. 2015. UAV applications
to coastal engineering. In *Proceedings of Australasian Coasts & Ports Conference*, 15–18 September
2015, Auckland, New Zealand.

Fonstad, M. A., Dietrich, J. T., Courville, B. C., Jensen, J. L., and P. E. Carbonneau. 2013. Topographic
structure from motion: A new development in photogrammetric measurement. *Earth Surface Processes
and Landforms* 38(4): 421–430.

Furukawa, Y., and J. Ponce. 2010. Accurate, dense, and robust multiview stereopsis. *IEEE Transactions on
Pattern Analysis and Machine Intelligence* 32(8): 1362–1376. doi:10.1109/TPAMI.2009.161.

Genchi, S. A., Vitale, A. J., Perillo, G. M., and C. A. Delrieux. 2015. Structure-from-motion approach for
characterization of bioerosion patterns using UAV imagery. *Sensors* 15(2): 3593–3609.

Goldstein, E. B., Oliver, A. R., Moore, L. J., and T. Jass. 2015. Ground control point requirements for structure-
from-motion derived topography in low-slope coastal environments. *PeerJ PrePrints* https://peerj.com/
preprints/1444v1/

Gonçalves, J. A., and R. Henriques. 2015. UAV photogrammetry for topographic monitoring of coastal areas.
ISPRS Journal of Photogrammetry and Remote Sensing 104: 101–111.

Guichard, F., Bourget, E., and J. P. Agnard. 2000. High-resolution remote sensing of intertidal ecosystems:
A low-cost technique to link scale-dependent patterns and processes. *Limnology and Oceanography*
45(2): 328–338.

Harwin, S., and A. Lucieer. 2012. Assessing the accuracy of georeferenced point clouds produced via multi-
view stereopsis from unmanned aerial vehicle (UAV) imagery. *Remote Sensing* 4(6): 1573–1599.

Harwin, S., Lucieer, A., and J. Osborn. 2015. The impact of the Calibration method on the accuracy of
point clouds derived using Unmanned Aerial Vehicle multi-view stereopsis. *Remote Sensing* 7(9):
11933–11953.

Hirschmüller, H. 2008. Stereo processing by semiglobal matching and mutual information. *IEEE Transactions
on Pattern Analysis and Machine Intelligence* 30(2): 328–341.

Houser, C., Hapke, C., and S. Hamilton. 2008. Controls on coastal dune morphology, shoreline erosion and
barrier island response to extreme storms. *Geomorphology* 100(3): 223–240.

James, M. R., and S. Robson. 2012. Straightforward reconstruction of 3D surfaces and topography with a
camera: Accuracy and geoscience application. *Journal of Geophysical Research: Earth Surface* 117:
F03017.

James, M. R., and S. Robson. 2014. Mitigating systematic error in topographic models derived from UAV and
ground-based image networks. *Earth Surface Processes and Landforms* 39(10): 1413–1420.

Javernick, L., Brasington, J., and B. Caruso. 2014. Modeling the topography of shallow braided rivers using
Structure-from-Motion photogrammetry. *Geomorphology* 213: 166–182.

Johnson, R., Smith, K., and K. Wescott. 2015. Unmanned Aircraft System (UAS) Applications to Land and
Natural Resource Management. *Environmental Practice* 17(03): 170–177.

Kelcey, J., and A. Lucieer. 2012. Sensor correction of a 6-band multispectral imaging sensor for UAV remote
sensing. *Remote Sensing* 4(5): 1462–1493.

Klemas, V. V. 2015. Coastal and environmental remote sensing from Unmanned Aerial Vehicles: An overview. *Journal of Coastal Research* 31(5): 1260–1267.

Lechner, A. M., Fletchera, A., Johansen, K., and P. Erskinea. 2012. Characterising upland swamps using
object-based Classification methods and hyper-spatial resolution imagery derived from an Unmanned
Aerial Vehicle. In *Proceedings of XXII ISPRS Congress*, 25 August–1 September, Melbourne (Australia),
pp. 101–106.

Lowe, D. G. 2004. Distinctive image features from scale-invariant keypoints. *International Journal of
Computer Vision* 60(2): 91–110.

Maguire, C. 2014. Using Unmanned Aerial Vehicles and "Structure from Motion" Software to Monitor
Coastal Erosion in Southeast Florida. Open Access Theses. Paper 525.

Mancini, A., Frontoni, E., Zingaretti, P., and S. Longhi. 2015. High-resolution mapping of river and estuary areas by using unmanned aerial and surface platforms. *Proceedings of International Conference on Unmanned Aircraft Systems (ICUAS)*, 9–12 June, Denver, CO pp. 534–542.

Mancini, F., Dubbini, M., Gattelli, M., Stecchi, F., Fabbri, S., and G. Gabbianelli. 2013. Using Unmanned Aerial Vehicles (UAV) for high-resolution reconstruction of topography: The structure from motion approach on coastal environments. *Remote Sensing* 5(12): 6880–6898.

O'Young, S., and P. Hubbard. 2007. RAVEN: A maritime surveillance project using small UAV. *Proceedings of Conference on Emerging Technologies and Factory Automation (ETFA)*, 25–28 September, Patras, pp. 904–907.

Pereira, E., Bencatel, R., Correia, J., Félix, L., Gonçalves, G., Morgado, J., and J. Sousa. 2009. Unmanned Air Vehicles for coastal and environmental research. *Journal of Coastal Research* SI56, 1557–1561.

Rafiq, T. 2015. A temporal and ecological analysis of the Huntington beach wetlands through an Unmanned Aerial System remote sensing perspective. PhD Diss., California State University, Long Beach, CA.

Rützler, K. 1978. Photogrammetry of reef environments by helium balloon. In D.R. Stoddart and R.E. Johannes, eds., *Coral Reefs: Research Methods*. UNESCO, Paris.

Scoffin, T. P. 1982. Reef aerial photography from a kite. *Coral Reefs* 1(1): 67–69.

Sekovski, I., Armaroli, C., Calabrese, L., Mancini, F., Stecchi, F., and L. Perini. 2015. Coupling scenarios of urban growth and flood hazards along the Emilia-Romagna coast (Italy). *Natural Hazards and Earth System Science* 15(10): 2331–2346.

Snavely, N., Seitz, S. M., and R. Szeliski. 2008. Modeling the world from Internet photo collections. *International Journal of Computer Vision* 80(2): 189–210.

Tonkin, T. N., Midgley, N. G., Graham, D. J., and J. C. Labadz. 2014. The potential of small unmanned aircraft systems and structure-from-motion for topographic surveys: A test of emerging integrated approaches at Cwm Idwal, North Wales. *Geomorphology* 226: 35–43.

Turner, D., Lucieer, A., and C. Watson. 2012. An automated technique for generating georectified mosaics from ultra-high resolution unmanned aerial vehicle (UAV) imagery, based on structure from motion (SfM [SFM]) point clouds. *Remote Sensing* 4(5): 1392–1410.

Vericat, D., Brasington, J., Wheaton, J., and M. Cowie. 2009. Accuracy assessment of aerial photographs acquired using lighter-than-air blimps: Low-cost tools for mapping river corridors. *River Research and Applications* 25(8): 985–1000.

Vierling, L. A., Fersdahl, M., Chen, X., Li, Z., and P. Zimmerman. 2006. The Short Wave Aerostat-Mounted Imager (SWAMI): A novel platform for acquiring remotely sensed data from a tethered balloon. *Remote Sensing of Environment* 103(3): 255–264.

Vousdoukas, M. I., Pennucci, G., Holman, R. A., and D. C. Conley. 2011. A semi automatic technique for Rapid Environmental Assessment in the coastal zone using Small Unmanned Aerial Vehicles (SUAV). *Journal of Coastal Research* 64: 1755–1759.

Westoby, M. J., Brasington, J., Glasser, N. F., Hambrey, M. J., and J. M. Reynolds. 2012. 'Structure-from-Motion' photogrammetry: A low-cost, effective tool for geoscience applications. *Geomorphology* 179: 300–314.

Wheaton, J. M., Brasington, J., Darby, S. E., and D. A. Sear. 2010. Accounting for uncertainty in DEMs from repeat topographic surveys: Improved sediment budgets. *Earth Surface Processes and Landforms* 35(2): 136–156.

7 UAV Image Acquisition Using Structure from Motion to Visualise a Coastal Dune System

Andrew Smith
UCEMM – University of Aberdeen

CONTENTS

INTRODUCTION

Coastal zones and the geomorphology of this environment have been of great interest to historians, scientists, and the public for many centuries, mainly due to the large number of people across the world who dwell near to the coast. The knowledge gained in understanding the impacts of a gain or reduction in beach and dune systems, either naturally or by human inducement, is vital as to the role that these fragile regions have in the future. The presence of dune systems must be seriously considered, as many are the first and last defence that rural environments, towns, and cities have against coastal inundation and flooding.

Traditional approaches to coastal monitoring have long been the reserve of airborne methods such as Light Detection and Ranging (LiDAR), Radio Detection and Ranging (RADAR), terrestrial laser scanning (Micheletti et al., 2014, 2015; Fabbri et al., 2017), historic and contemporary aerial photography (Alberico et al., 2017), and satellite imagery (Goméz et al., 2014; Papakonstantinou et al., 2016; McCarthy et al., 2017) to collect remotely sensed data. Many of these methods will generate high-quality point clouds and digital surface models (DSMs), nevertheless there may be significant cost implications, lengthy processing times, georeferencing, and accuracy of vertical elevations (Klemas, 2015). These remote sensing techniques have been important developments for mapping large areas of coastline that are remote, yet they all share drawbacks including lengthy processing times, expensive methods of data capture, and lack the spatial accuracy of a survey closer to the ground (Drummond et al., 2015). Professional standard Unmanned Aerial Vehicles (UAVs) have been around for many years, and applications for their deployment have been growing, mostly developed and utilised by the military, film and television industry, structural and building survey, parcel delivery, surveillance, and more recently for emergency disaster response (Shakhatreh et al., 2018). Successful UAV surveys have been completed on sections of river bank, agricultural applications, coastal salt marshes, estuary vegetation mapping, and coastal dune systems, at low altitudes (<100 m) have been shown to generate very high pixel resolution <5 cm²/pixel (Mancini et al., 2013; Drummond et al., 2015; Marteau et al., 2016; Papakonstantinou et al., 2016; Woodget et al., 2017), which allows for precise mapping compared to traditional remotely sensed imagery, which at very best may range from 10 to 100 m²/pixel (Goméz et al., 2014).

The UAV approach for image capture and analysis within a dynamic coastal environment has only recently been shown as a viable option to aid with mapping and monitoring as the accuracy that can be achieved is significant, compared to traditional methods (Drummond et al., 2015; Moloney et al., 2018). UAV advancements in recent years have included miniaturisation of hardware and software, improved resolution of sensors, battery efficiencies, continued advancements with automated flight, and possibly the most key advancement is the reduction in price (Klemas, 2015; Marteau et al., 2016). Technological advancements within the control systems of UAVs can now be achieved by the existing and improved autonomous flight planning and new technology such as using virtual reality (VR) smart glasses as a control (replacing VR headsets and first-person view). Operation of termed 'off-the-shelf' UAVs (Klemas, 2015; Turner et al., 2016; Moloney et al., 2018) has revolutionised the capacity of individuals and consortiums to enter the survey industry, disaster response, and environmental protection, chiefly through the reduced cost of these UAVs. UAV environmental coastal survey has for many years been conducted by previously mentioned aerial survey techniques which are expensive to employ and are not always guaranteed to achieve clear imagery. Therefore, this type of image acquisition has been largely used by large commercial companies and governments for mapping and monitoring (Moloney et al., 2018).

Structure from Motion SfM (SFM) owes much of its existence to mathematical photogrammetry models developed in the 1950s and 1960s, which were able to demonstrate spatial relationships between images (Micheletti et al., 2015). The traditional SfM (SFM) photogrammetry approach has been around for many years and most notably used in architecture, urban landscapes, engineering, industrial survey, and within medical science and has been heavily influenced by the advancements in computing science (Schenk, 2005). SfM (SFM) applications used within architecture and engineering are less reliant on absolute geographic location (ground control); therefore, unstructured overlapping images, obtained from multiple cameras, from a range of viewpoints, are favourable to generate three-dimensional (3D) visualisations and models (Westoby et al., 2012). SfM (SFM) algorithms within the software are designed to automatically image match common features between each image within the data, positioning and orientating parameters throughout the dataset. Following orientation, production of a high resolution and colour-coded (RGB) point cloud that represents the object is created. Micheletti et al. (2014) suggest that the larger the dataset (increase in the number of images), the easier it is to reject lower quality images, greatly improving the density of the point cloud and refining the accuracy of the model. SfM (SFM) techniques used as a tool for geographical or topographical survey have emerged from the advances in modelling software, computational power, and traditional photogrammetry (Warrick et al., 2017). SfM (SFM) software generates very accurate 3D DSMs and digital elevation models (DEMs) from two-dimensional (2D) imagery which are all powerful visualisation tools. The overriding objective of SfM (SFM) software is to reconstruct real-world 3D geometry utilising 2D photos that depict a static environment and recreating the environment within a 3D VR model. The development of SfM (SFM) within geosciences has provided many opportunities to generate accurate topographic maps and high-resolution and high-quality 3D results from low-cost data acquisition methods such as the UAV (Westoby et al., 2012; Nex and Remondino, 2014; Drummond et al., 2015; Turner et al., 2016; Moloney et al., 2018).

SfM (SFM) does not explicitly require the use of ground control points (GCPs), as many locations have easily identifiable reference points where georeferencing of the image can easily be obtained. The addition of GCP is essential within the coastal margin as fluctuations with the mean low and high water marks constantly change depending on the geographic location. Employing the GCP technique within the SfM (SFM) photogrammetry is essential to orientate the model (defining an absolute geographic location) and must be noted as coastal processes within these environments are subject to constant change from tidal dynamics and weather phenomena (Moloney et al., 2018). Mancini et al. (2013) suggest that the Global Navigation Satellite System (GNSS) or Global Positioning Systems (GPSs) are valuable additions when mapping shoreline locations to accurately define direction (X and Y values), elevation (Z value), and for implementation of GCPs within the environment (Drummond et al., 2015). These spatial datasets can easily be added to a GIS package for analysis.

This project will utilise a multirotor UAV (Quadcopter) to conduct an autonomous overflight of the survey site, collect structured images covering a dune system, and process these within SfM (SFM) software (Pix4D Cloud®, 2018) to create point clouds, DSMs, orthomosaic imagery, and 3D models (Mancini et al., 2013; Nex and Remondino, 2014; Papakonstantinou et al., 2016; Turner et al., 2016). The importance of generating this up-to-date DSM and ground survey (GNSS) is to be able to test the reliability against alternative sourced data such as LiDAR and Ordnance Survey (OS) DSMs, as modelling for potential coastal flood inundations for many areas of coastline use these data types (Warrick et al., 2017). The structured approach to data collection will result in a large dataset (images) covering the whole study site, which hopes to significantly improve the resolution and accuracy of the visualisations (Micheletti et al., 2015). The pre-flight SfM (SFM) application approach to survey facilitates a pre-programmed autonomous flight path and height to be uploaded to the UAV which will result in a structured image capture process ensuring the required 65%–75% overlap of imagery (Pix4D, 2018).

The objectives within this project are to:

1. Identify and define survey area:
   i.   Test effectiveness of mobile mapping application to identify points of interest (POIs)
2. Define GCP locations within the survey area
   i.   Conduct a GNSS survey of GCPs
   ii. Identify two smaller sites within the perimeter to conduct ground truths
3. Conduct a full overflight of the area using an automated UAV platform

4. Process aerial imagery using SfM (SFM) software
   i. Comparison between Pix4D Cloud and Pix4Dmapper to generate DSM, orthomosaic, and 3D models
   ii. Comparison between traditional methods (LiDAR) and SfM (SFM) DSM

SURVEY LOCATION

The area of interest is within the Tentsmuir National Nature Reserve (under the stewardship of Scottish Natural Heritage (SNH)) situated at the north-east tip of Fife, which provides an excellent range of environments; mature dunes, mature slacks, mobile dunes, mobile and fixed ridges, and evidence of severe coastal erosion, all contained within an area of 35 hectares and just over a kilometre in length. This area of coastline is rich in soft sandy beaches and is part of an ancient sand dune system which dates back 6,000 years and extends several kilometres to the west. Extensive geomorphological processes to the dune system both positive and negative have been in evidence within this environment which has shaped the area that we see now (Cunningham, 2014). The reserve generates a large amount of tourism throughout the year including dog walkers, summer sun seekers, educational visits, and outdoor sports enthusiasts, which has undoubtedly led to damage to parts of the reserve. This is certainly a factor as to the speeded-up erosion of sections of path within the site. Three bands can be defined at this location: mature fixed dunes extending many kilometres to the west; dune slacks made up of mature dunes, hummocks, and moist fertile meadow or marsh; and mobile dunes on the seaward side (east) of the reserve that have migrated (west) over many tens of metres over the last 10 years.

METHODOLOGY: PROJECT PLANNING

SITE SELECTION AND MAPPING

Initially, several areas of beach and dunes along the north-east coast had been examined for suitability of each coastal environment, geographic location, access, and feasibility. A visit to Tentsmuir National Nature Reserve (NNR) confirmed the suitability of this stretch of coastline, and following formal authorisation from SNH, a pre-survey walk was conducted to allow familiarisation of the dune system. Although not necessarily the over-arching purpose of this research project, it is hard to ignore the exceptionally dynamic environment that is located here at Tentsmuir Sands and further to the north at Tentsmuir Point that have seen many decades of change. The survey site is located on the north-east edge of Tentsmuir forest which is ~14.3 km² in size and is managed by Forestry Commission Scotland. Access to the site is restricted as there are locked gates at all entrances to stop members of the public from using the unsurfaced roads and to allow ongoing tree felling. Formal authorisation was granted by SNH to gain access to the private roads (8.4 km) that run throughout the forest which is the only way to gain vehicular access to the site.

This project uses a mobile mapping application to help with the initial survey process. ViewRanger™ (VR), created by Augmentra Ltd© (version at time of survey 8.5.46 [3]), is a mobile smart phone application that is designed to help plan routes, navigate safely, and share your walks with other users. The application does not rely on mobile network connectivity (mobile data) to function, rather relying solely on the GPS aspect which allows for use anywhere in the world. VR is primarily used by hillwalkers and lets GPS data (waypoints or POIs) to be collected, stored, and then exported as a GPX file. The VR application has the option to upload or recover pre-collected data, allowing the user to follow pre-determined or past routes. One of the many features of the VR application is that it allows images to be recorded and stored as part of the POI which is useful if time has elapsed between initial survey and post-survey analysis. This application is ideally suited for the purposes of this project, as uploaded GCP positions can be easily located by following the pre-programmed route. VR was used on a Samsung Galaxy SM-G930F android smart phone for two surveys, initially on the familiarisation walk to record POI features and, secondly, to follow and find pre-defined GCP locations. Two smaller sites have been identified to conduct ground truths, one in the south which is mainly untouched by geomorphological processes and made up of mainly mature dunes, slacks, and a small area of beach. The site to the north has been identified as an area where significant erosion and accretion has taken place and is mostly an area of beach, mobile dunes, and dune ridges.

Esri© ArcMap™ 10.5 was used to initially produce a digitised habitat map of the survey site using a basemap of aerial imagery supplied as the template. Everything that was in existence within the area was digitised to produce a base habitat layer: vegetation, sand, trees, a stream, and structures (Figure 7.1). The exported VR GPX file was added to the map (POI and GPS coordinates) which would form the basis of the survey and aid with the choice of favourable GCPs. Using the VR survey which totalled 52 points, 35 ground reference points within the site were identified (Figure 7.1), 6 perimeter points (red triangle markers), 13 inner points (yellow star markers), 8 markers for the green survey site (green circular markers), and 8 for the blue survey site (blue square markers). Each marker added to the database was given an individual identification number, location code, and geolocations utilising both British National Grid and the WGS 1984 coordinate systems for compatibility purposes. The completed GCP map was now added to the VR application again (Figure 7.1, Inset image (a)), which would allow a simple process of following the waypoints and locating the GCPs quickly and efficiently.

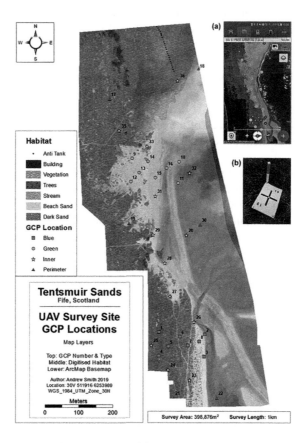

FIGURE 7.1 Digitised map showing the results of the habitat map and the Leica GNSS survey, showing the 35 GCPs distributed throughout the length of the site to be used in the processing within Pix4D to help geolocate imagery. Inset image (a) is a screengrab of the mobile GPS application ViewRanger™ and (b) an example of a GCP in situ and its mooring system.

A series of health and safety assessments were conducted to assess the suitability of operating a UAV within the site and to protect the public from potential harm. Tentsmuir NNR is 7 km to the north of Leuchars Station ATZ and falls out with the high-risk no fly area; however, the Commercial Aviation Authority (CAA) guidelines for flying UAVs (CAA Drone Code, 2018) were followed as part of the research. At the start and conclusion of the survey, contact was made with the air traffic tower (ATC) at Leuchars Station to make them aware of operations within the Tentsmuir site. The purpose of the UAV survey at Tentsmuir was to conduct an autonomous overflight of a specifically selected area that would capture a significant dataset to facilitate the generation of DSMs, contour maps, orthomosaic imagery, and 3D visualisation. Analyses of the data are to investigate the influences on the coastal environment for two key parts of the beach and dune system; the green survey area consisting of mobile dunes and limited vegetation (north of site) and the blue survey area which consists of vegetation-rich mature fixed dunes and slacks (south of site).

GROUND CONTROL POINTS AND GPS LOCATION

GCPs are physical objects that are placed on the surface of the earth that have an absolute geographic location within a coordinate system and are widely used within the survey industries. They are extremely important when surveying within a coastal environment, as there are very few identifiable consistent structures that would appear in multiple maps that aid georectification. In the context of surveying using a UAV, it is important to be able to distinguish the marking on the ground from the photographs, as unidentifiable marks will affect the geolocation of the reference points. Ideally, GCPs should be evenly distributed throughout the survey considering the topography of the area. Thought must be used when considering the method to securely moor it to the ground, considering the environmental factors and weather conditions at this location. This means that they must be sufficiently large and constructed of material that will survive physical processes such as the wind and rain. GCPs used within this project were constructed using a large black cross-shaped template (Figure 7.1, Inset image (b)) created within Microsoft Word and a different number was allocated to each of the 35 template cards. As the UAV uses a high-quality camera, it was enough for them to be printed on A4-sized paper which was laminated to protect it from the elements.

Surveying the dune system and placing the GCPs took place on the afternoon of the first day (28 June 2018), followed by the UAV survey on the following day (29 June 2018). The imported data within the VR application was easy to access and follow, which made it easy to locate, position, and record the absolute GPS at the centre of the GCPs black cross using the Leica Viva GS15 GNSS. The precise placement and recording of these GCPs was necessary for georectification of the images post survey. The Leica GPS system uses a base station which is positioned roughly at the centre of the survey area and at a known GPS location (GCP 28). A rover is connected via Bluetooth which allows the surveyor to walk around the site to each of the control point locations and record directional coordinates for X (latitude), Y (longitude), and Z (elevation) for each GCP. Due to a problem within the Leica GNSS, the coordinates were only recorded in WGS84 coordinate system during the survey which required a transformation later to rectify the correct elevation readings.

Mooring the GCP to the ground was achieved using 30 cm wooden stakes with colour-coded tops, hammered into the ground which would not only hold the cards firmly but also reduce the time spent post UAV survey locating and collecting the GCPs. The laminated cards had a hole punched in each corner of the card, which allowed one side to be attached to the post with a tie wrap and the other to be secured using a tent peg (Figure 7.1, Inset image (b)). Following attachment of GCP cards to the post, the GNSS rover was placed at the centre of the cross to give an accurate reading of the GCP. Following the POIs in the ViewRanger application to locate each individual, GCP locations were impressive and quick, with each displaying an average accuracy error of +/− 40 cm when compared to the Leica GNSS system.

STRUCTURE FROM MOTION (SFM [SFM]): APPLICATIONS

The use of SfM (SFM) technology is key to producing surface visualisations that can recreate the complex dynamics of a coastal environment within a virtual world. Unlike SfM (SFM), traditional photogrammetry required the structured and complex planning of image capture which was very time consuming. The great benefit of SfM (SFM) over traditional photogrammetry is that image direction, sensor type, or whether the images are collected as structured or unstructured is irrelevant. Due to significant advances in technology within the UAV, it is possible to use SfM (SFM) applications to pre-programme autonomous flight, thus saving significant time in the planning stage of survey. The software that has been chosen to process the imagery in this project is Pix4D™, as it has a large armoury of applications designed to simplify the experience. A significant benefit of Pix4D is that it comes with its own tailored autonomous image capture and flight planning application (Pix4Dcapture) that is also free and can be used on a smartphone, tablet, or laptop. As a specific goal of this research is to attempt to utilise free software, is it possible to utilise free (15-day trial subscription) cloud-based image processing (Pix4D Cloud) or is the limited manipulation and interaction with the data a significant problem? The desktop version (Pix4Dmapper) is also available for post-processing should there be any issues with the cloud version.

SURVEY APPLICATION

Pix4Dcapture® is a free autonomous flight application that allows the user to pre-plan and execute strict flight plans that have been specifically designed for individual survey sites, guaranteeing the image overlap is the specified 60% to 80%. The application is flexible, precise, and features automatic upload of data to the cloud (depending on image quantity) for processing. Pix4D supports a range of multirotor and fixed-wing UAV manufacturers such as DJI and Parrot. The system can use a range of different online and offline maps to access and select the location of the survey followed by the type of mission that will be flown. The survey of Tentsmuir will select the Double Grid Mission, which is ideal for generating 3D models but takes considerably longer to complete and will generate a large dataset. The UAV uses GPS to run along each transect of the double grid taking photographs every 2 or 3 seconds that are geotagged and in the JPEG file format. The large quantity of images will allow for low-quality images to be removed from the dataset before processing can begin. Another benefit of using this system is that should the UAV run out of battery midway through flight, the UAV will return to take-off location, and following battery replacement, it will return to the point that it ended at. It is possible to monitor the progress of the survey using the camera view which provides live feed from the camera.

DESKTOP AND CLOUD-BASED PROCESSING

Pix4D© is the leading modern photogrammetry software designed for professional UAV mapping. The Swiss company, originally based in Lausanne, now has four more offices around the world (USA, China, Spain, and Germany). They develop software that uses cutting-edge computer algorithms to transform 2D aerial photographs into 3D maps and models. This software is available on desktop, cloud, and mobile platforms, and completed projects will provide a range of outputs dependant on project type but may include full-colour point cloud, 2D orthomosaic, DSM, contour lines, and 3D textured mesh. A main project aim was to assess the effectiveness and value of the cloud-based processing solution against the desktop version and compare the results between the two platforms.

Pix4D Cloud™ is an image-based cloud processing solution that starts with a free 15-day license and an unlimited upload of imagery-based projects. Performance of the cloud-based platform when uploading large quantities of images can be slower than using the desktop version; however, both applications will produce the same outputs. The cloud-based processing option relies on a good steady Internet speed and a much-reduced requirement for a fast central processing unit (CPU), which is ideally suited to use on a laptop. Due to project requirements, access to a desktop pc is limited which is why the cloud processing option is a convenient solution. A distinct advantage of the cloud-based processing is that multiple projects can be uploaded and processed through the night removing the need to be at a workstation through the day. As with desktop mapper, the 2D and 3D outputs and processing logs are the same.

Pix4Dmapper™ is the desktop application designed for professional UAV mapping which also specialises in automatically converting imagery taken by any camera to produce highly precise 2D maps and 3D models. Desktop mapper produces several high-quality outputs. Algorithms within the software, image match multiple images producing a dense point cloud to which a 3D mesh and 3D textured models can be constructed. High-resolution 2D georeferenced aerial orthomosaic imagery is produced that is an accurate representation of the earth's surface and finally the DSM which is a georeferenced elevation map. Using the DSM within ArcMap, the generation of a contour map for the site is useful to the project. This software was mainly chosen not only for its advanced processing options but also because of the option to test the free-to-use cloud-based image processing version.

UAV Technical and Survey

The survey will utilise a DJI Inspire 2 multirotor quadcopter which is owned and operated by Bristow UAS Ltd. The Inspire 2 is a very versatile and well-equipped UAV, ideally suited to surveying large areas and in a hostile coastal environment. Equipped with a three-axis gimbal for ultra-steady filming, a standard 20 MP Zenmuse X4S HD camera is able to capture and store up to 64 GB of imagery. Image processing system is a CineCore 2.1 which records video up to 6 K in Cinema DNG/RAW formats. It has a dual battery system which prolongs flight time up to 24 minutes depending on payload and can achieve a speed of 58 mph. DJI has revised the autonomous flight within the UAV and developed a new version specifically for the Inspire 2 which provides two directions of obstacle avoidance and sensor redundancy. It also has a return to home function, should it lose contact with the controller (more than 200 m from the controller). This system is fully supported by the Pix4Dcapture application.

Utilising a commercial company to undertake the survey removes the need to obtain authorisations from the CAA for flights and to obtain competency certificates for operating UAVs. On the day of survey, as to comply with the Drone Code and safe flight control procedures, contact will be made to Leuchars Station ATZ (7 km to south) as to notify details of the flight plan followed by a similar notification at the end of survey. A risk assessment carried out on the site prior to the survey indicated that the only significant issue to UAV safety were the trees at the west and north of the site (~25 m) and a former Second World War communication tower (15 m at highest point) which was located at centre of site. Flight height was estimated to be between 50 and 100 m, well above stationary objects and to accommodate the double grid formation of survey. The risk to the public was minimal due to the high flight height, but as a precautionary measure, warning signs were attached to the two main entrance points to warn the public of the survey.

Ground Truth

Collection of ground truthing information allows for the calibration of images and considerably aids with the interpretation of data. It is critical to understand errors that may arise from incomplete data or incorrectly identified vegetation as this may affect the legitimacy of the survey. Truthing visits conducted pre-survey and post-survey are designed to investigate the two small survey areas that were selected for further study. For each of the blue (south) and green (north) sites, a random area within the site will be chosen to analyse and identify the vegetation that are found here. This is ideally done by taking an area of around 2 or 3 m^2 and attempting to identify all vegetation growing in that area (similar to using a quadrat). For both areas, digitised vegetation maps would be constructed using the orthomosaic imagery to help with the identification of all visible vegetation. On day of plant survey, notes are to be taken as to the geographic location, colour, and shape of each plant again to aid identification away from the site. These are then identified later to create a complete vegetation map for both sites. We can use these maps to compare the differences between two areas which are just under 500 m apart within the same coastal environment.

RESULTS

UAV Flight Overview

The UAV survey within Tentsmuir NNR was conducted on the afternoon of 29 June 2018 and used two separate double grids to cover the whole site. The afternoon weather was cold and overcast with a fluctuating wind velocity of 3.88–5.36 m/s and direction from East South East to East. Occasionally cloud and mist would roll in and drop below the UAV survey height of 100 m obscuring some of the images. UAV command using the autonomous function of the drone was excellent as it could easily be programmed using the Pix4Dcapture application to fly specific flight grids and capture images that had the 70%–80% overlap that Pix4D required for processing of the orthomosaic and DSM. As outlined in the methodology, a double grid mission was used twice to cover the entire site at a height of 100 m, flown firstly in the south followed by the north, which resulted in a dataset totalling 1,175 images. As the double grid is time consuming, the UAV returned to the launch site three times during the two surveys to collect new batteries and then return to the sky. Each double grid formation took 1 hour to complete, and immediately following the survey, two videos were recorded of the site to give some context to the area. Post-processing involved working through the entire dataset to remove 252 photographs that were poor quality or displayed cloud or mist. All aspects of the GCP location, stabilisation, and mooring worked very well within this blustery environment. There was only one minor mishap during the survey as GCP 18 was lost to the rising tide (furthest north-east within survey site) by the time the UAV reached that area.

UAV, GNSS, and LiDAR Elevation Errors

The ground survey conducted using the Leica Viva GS15 ground station GNSS system worked extremely well for collecting grid coordinates X and Y though following analyses of the file, the Z values showed flawed readings. Each measurement was inaccurate by over 50 m (range from 54.39 to 59.04 m above sea level) due to an issue within the GNSS that had not been observed during a short practice with the apparatus. Given the problem with the elevation values recorded by the GNSS, it was thought that elevation values could be extracted from the data acquired by the UAV. Upon further investigation, it was found that many UAVs (including DJI) no longer use the more accurate barometer system to measure elevation, and this has been replaced by GPS which has a much higher +/− inaccuracy. As the imagery from the Inspire was processed within Pix4D, the elevation data used to create contour maps in ArcMap showed a variety of negative reading (range between −116.19 and −125.08 m below sea level) values, none of which could be trusted.

As the GNSS, DJI images, and Pix4D-generated DSM had provided inconsistent Z values, it was necessary to obtain DSMs from another source to compare and search for correct Z values. Data was acquired from an OS-generated DSM, which includes spot heights and contour lines (OS Terrain 5, 2018) and LiDAR-generated DSMs for 2012 and 2014 (sourced from the Scottish Spatial Data Infrastructure Portal) which were used for the comparison. It was important to compare the newly acquired data against the GCP locations that have been known not to have changed position or elevation. GCP 5 and 17 are located at the western edge of the north (GCP 17) and south (GCP 5) as both are located within the mature dunes and slacks within the survey area. The results, seen in Table 7.1, show very large differences between the six data sources confirming that the project had no reliable measurements for elevation. Elevation data inaccuracies can now be seen in Figure 7.2, as it shows the elevation discrepancies within the data for three different years covering all the GCPs. A total of 20 GCPs within the survey area are in areas of mature dunes and slacks that have

TABLE 7.1

The Readings below Are an Extract from the Main Database following Acquisition of New Data for Comparison but Importantly Each of the New Sources Display New Values

| | Elevation (z) | | | | | | | |
GCP	OS DSM 2012 (m)	LiDAR DSM 2012 (m)	LiDAR DSM 2014 (m)	Leica GNSS 2018 (m)	DJI UAV (m)	DJI Image (m)	Range (m)	Correct GNSS (m)
5	3.60	3.07	2.92	56.72	−123.71	27.71	180.43	5.35
17	3.03	3.06	3.06	57.07	−119.04	23.69	176.11	5.96

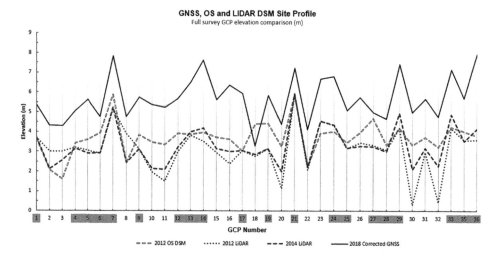

FIGURE 7.2 DSM comparison between four data types: a 2012 OS generated DSM, 2012 and 2014 LiDAR-generated DSMs, and the converted Leica GNSS Z values. Clear differences can be seen throughout all data types, showing the difficulties in sourcing accurate data. GCPs marked in grey demonstrate locations that are within mature dunes and unaffected by changes in elevation.

not seen change by either the tide or storm events (Figure 7.2, GCPs highlighted). The coordinate system set in the Leica was WGS84 and its ellipsoid is known as GRS80, which has an average fit to the geoid around the entire globe which is significantly lower than the British ellipsoid Airy 1830. This system has no mean sea-level height data, so elevation is set above the ellipsoid and thus the potential to give false readings depending on the GNSS, location, and system you are in. A batch transformation to convert the X/Y Leica GNSS data with OS Newlyn (OSGB36 Mean Sea Level Heights) for a defined sea level was required for the UK. The elevation information was derived directly from the GNSS values for Z utilising the coordinate transformer OSTN15 supplied by the OS Net batch transformation. The process returned the correct elevations for the GCPs, which could then be added to the data for processing in Pix4D.

Following transformation of the elevation data, it was obvious that there was a significant difference between the LiDAR DSMs (2012 and 2014) and the corrected GNSS elevations throughout the survey area. For nine or ten of the GCPs on the eastern side of the site, this would be expected as they fall within high tide ranges and seasonal weather events. In stark contrast are the GCPs (21) that fall within the mature areas of the site all along the western edge, which show significant differences in height leading to consideration about how accurate the LiDAR-generated DSMs are. Post-analysis of the LiDAR data suggests that it is also inaccurately averaged over all the GCPs by 2.36m which is a significant elevation, given that these types of datasets are used to predict and model coastal inundation.

STRUCTURE-FROM-MOTION SURVEY COMPARISON

Pix4D Cloud Results

The first process within the cloud platform is to select the images that will be used and upload them into the relevant project. This proved to be a problem for the first project, as 580 images were selected to cover the entire survey area with a data size of 5.1 GB. As each image size was between 7.5 and 10 MB, this relied on access to a rapid and reliable Internet connection to quickly upload files. Due to intermittent and slow network speeds, the first project was started around 10 pm and ran through till around 6 am the following day. The project was terminated after 8 hours as access to the hardware was only available for a period through the evening and night. The images uploaded within the 8-hour bracket totalled 327 and only covered half the site. The terminated project containing 327 images was still able to produce a 2D orthomosaic, DSM, and 3D mesh visualisations with an average ground sampling distance (GSD) of 2.67 cm/ pixel. The completed SfM (SFM) outputs use a GSD to show the quality of the spatial imagery and the relative size of an object identifiable on the ground. The GSD is the measure between two consecutive pixel centres measured on the ground and is related to the altitude that the UAV survey was conducted, the greater the height, the larger the GSD. The lower the value of GSD, the higher the spatial resolution that can be achieved, allowing for smaller objects to be identified on the ground. The GSD of 2.67 cm/pixel refers to one pixel on the ground with the linear measurement of 2.67 cm on the ground or 7.13 cm^2 (2.67×2.67). Completion of the other three projects (Table 7.2) gave similar results to Project 1 with GSDs ranging between 2.64 and 2.71 cm/ pixel. The reduction of imagery chosen for Projects 2–4 rapidly sped the processing within the cloud but ultimately increased the time for pre-processing analysis of suitable images. Following upload and processing of imagery, it was not possible to use the 35 surveyed GCPs to accurately georeference the imagery as this process is automated within the cloud version. The four completed 2D orthomosaics were opened in ArcMap successfully, but the images showed slight misalignment to each other, this meant the images required to be georeferenced as to nudge them into their perfect geographic position. Stitching and georeferencing of the four projects could be achieved within ArcMap, but this showed discolouration between the projects and ultimately warped parts of the map image, producing a less than ideal finished output.

Pix4Dmapper Results

A total of 999 images were selected and uploaded into the software. Following calibration and geolocation of the 999 images, 76 of those images were removed as typically these images were of forest or water located at the peripherals of the survey. The project ran with a slightly reduced 923 images, still more than enough to cover the whole site. Mapper took just 2.5 hours to complete and allowed the attachment and geolocation of the GCPs to be included as part of the

---

TABLE 7.2
Processing Using the Cloud (Projects 1–4) Compared to the Desktop (Project 5) Version and the Relative Resolutions Achieved between the Two Versions

Project No.	Processing Option	No. of Images	Survey Location	GSD (cm/pixel)	GSD (cm²)
1	Pix4D Cloud	327	South	2.65	7.02
2	Pix4D Cloud	82	Middle	2.68	7.18
3	Pix4D Cloud	29	North	2.64	6.96
4	Pix4D Cloud	15	East (sea)	2.71	7.34
5	Pix4Dmapper	999	Whole site	2.66	7.07

process, which generated a georeferenced 2D orthomosaic and DSM image, that once inserted into ArcMap was in the correct geographic position. The GSD for the mapper was 2.66 cm/pixel or 7.07 cm² (2.66×2.66).

RESULTS CONCLUSION

Detailed investigation of total time required to upload Project 1 (580 images) was calculated using the average image upload speed of 1 minute 30 seconds, processing of Project 1 would have taken 14 hours 20 minutes for full upload. Results between the cloud version and mapper suggested the cloud was around ten times slower than mapper, most of this time was lost due to the time taken to upload images, estimating up to 25 hours to complete image upload and processing. Processing the completed imagery was distinctly quicker using the cloud compared to the desktop version as the graph in Figure 7.3 shows. This is because the processing power of the CPU used for cloud processing is greater than the CPU running the mapper desktop version. Results from this test show the unsuitability of using an average powered laptop and with a slow Internet upload speed (~9–11 mbs), if you are considering using the cloud for large project processing.

IMAGE RESOLUTION AND POINT CLOUD

PIX4D CLOUD

At the outset of the project, the choice to use the cloud-based processing solution appeared like a good option with all the complicated technical interaction removed to streamline the experience. Disappointingly for this project, the cloud version failed to work sufficiently enough as the key interaction was removed during processing, uploading, and within the post-processed imagery. Given the issues affecting elevation of the various datasets, not having an option to add GCPs is a significant problem when surveying in a coastal location that needs to be precisely geolocated.

As the cloud does not accommodate adding GCPs, the imagery is geolocated and calibrated automatically meaning that there may be errors when considering the x, y, and z directions. Detailed analysis for three projects: Project 1 (south) had 327 images and had a median of 8,069 matches per 2D image and generated a total of 33,982,755 3D densified points, meaning the average density was 178.06 points/m³; Project 2 (middle) used 82 images and had a median of 11,163 matches per 2D image and generated a total of 10,744,892 3D densified points, meaning the average density was 200.3 points/m³; and Project 3 (north)

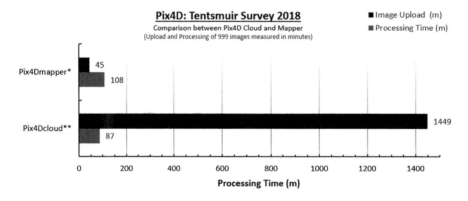

FIGURE 7.3 Comparison between the Pix4D Cloud-based processing tool and Pix4Dmapper. This test was completed to compare processing times for the Tentsmuir survey data: Pix4Dmapper's total time was 2.5 hours (**CPU: Intel® Core™ i7–6900K CPU @ 3.20 GHz, RAM: 69 GB) compared to the Pix4D Cloud which took just over 25 hours (Estimate) to complete (*CPU: Intel® Xeon® Platinum 8124M CPU RAM: 69 GB).

used 29 images and had a median of 6,380 matches per 2D image and generated a total of 5,764,062 densified 3D points, meaning that the average density was 153.91 points per/m³. The report from the four projects shows the average absolute geolocation variance to be 3.55 m for X, 1.58 m for Y, and 0.70 for Z. These values are very high and show how inaccurate this method is and show the great importance of attaching GCPs to the data to give much reduced distance errors.

Pix4Dmapper

From the 35 surveyed GCPs, 21 3D GCPs were calibrated and used as tie points for geolocation, 9 were used as checkpoints to assess the accuracy of the project which is displayed as the root mean square (RMS) value and refers to the error of distance in each direction (x, y, z). The smaller the number, the more accurate the GCPs are. As shown in Table 7.3, the RMS error for the 9 GCP checkpoints chosen is 0.12 m for X, 0.09 m for Y, and 0.06 m for X. The projects 923 images had a median of 3,607.43 matches per 2D image and generated a total of 68,642,311 3D densified points, meaning the average density was 218.79 points/m³. Compared to the cloud-based version, the mapper version shows the importance of manipulating data and adding GCPs to gain significant accuracies with the geolocation of the imagery. The 2D orthomosaic imagery created in Pix4Dmapper and 3D mesh models generated by Pix4D Cloud can be seen in Figure 7.4.

Considering the overall performance of both methods, imagery produced in its finished form is very impressive for the majority of the 2D orthomosaic; nevertheless there are problems within the image of blurring certain areas of the dunes. Further investigation of this blurring would suggest that it is not the fault of the software but a more natural phenomenon. Ground-level study shows that areas within the study that are low to the ground (within 0.30 m) exhibit the detail and fine scale generated by the software. Locations such as within the slacks and on mature dunes are generally sheltered from the wind, but areas that are on tops of dunes and beach areas display blurring caused by the wind blowing through the Marram Grass (as detailed in the case study below).

TABLE 7.3

Geolocation Details Used by Pix4Dmapper of the 9 GCPs Chosen as Checkpoints: All Checkpoints Have Been Calibrated as Accurate

GCP ID No.	Error X (m)	Error Y (m)	Error Z (m)	Projection Error (pixel)
GPS 003	0.0333	−0.1059	0.0315	0.4153
GPS 004	0.1433	0.0156	−0.1026	0.5146
GPS 007	0.1246	−0.0473	−0.0501	0.5839
GPS 011	0.0089	0.0434	0.0340	0.2759
GPS 014	−0.2141	−0.0078	0.0304	0.6888
GPS 029	−0.0106	0.1296	−0.1145	0.5723
GPS 030	−0.2085	0.1969	0.0322	0.8959
GPS 033	0.0375	0.0675	0.0891	0.5489
GPS 035	−0.0926	0.0436	0.0210	0.4892
Mean (m)	−0.0198	0.0373	−0.0032	
Sigma (m)	0.1216	0.0851	0.0654	
RMS error (m)	0.1232	0.0929	0.0655	

Localisation accuracy per GCP and the mean errors in the three coordinate directions. The difference between the computed 3D checkpoint and the original position for x, y and z directions.

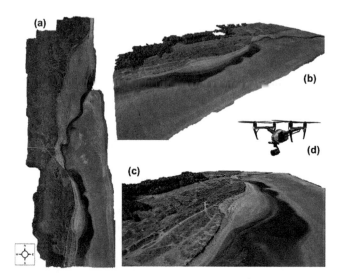

FIGURE 7.4   Pix4D-generated imagery: (a) Pix4Dmapper 2D orthomosaic image of the whole survey site, (b) Pix4D Cloud (b and c) generated 3D textured mesh model, (c) a view from the south end of the survey looking north, and (d) an image of the DJI Inspire 2 used to conduct the survey. (Photo: Andrew Smith.)

GROUND TRUTH

To ground truth the new imagery, the creation of digitised vegetation maps was achieved by using the orthomosaic imagery as a base and digitising (tracing) the outlines of all the plant and vegetation that could be identified from the map. Earlier visits to the site documented the differing vegetation types that could be found within this coastal environment recording information about colour, height, spread, and width of the plant. These could then be populated onto the digital map afterwards within ArcMap. Visualisations within a 3D environment were possible using Esri's ArcScene. The blue survey site is located to the south of the survey area within mature dunes and slacks, measuring 9,089 m² in total and is made up of 75% vegetation and 25% sand. Ground truthing revealed that there were at least 20 species of plants growing in this environment with at least 11 having positive identification from aerial imagery alone prior to ground truthing visits. The green survey site is 9,106 m² and is situated towards the north of the site within an area where considerable erosion and accretion is active. This area contains a more balanced 55% sand and 45% vegetation. Plant species are sparser in this location with only ten plant species identified but only four or five species having positive identification from the aerial images. Results are positive using the imagery to identify vegetation from this survey, nevertheless could be improved by the reduction in height of the UAV and possibly the increased light available from a day freer of cloud and mist.

CASE STUDY

MARRAM GRASS DILEMMA

As a direct result of processing a large block of images (923), the resulting orthomosaic layer displays an impressive level of detail throughout the site; however, on closer inspection, there are problems within the heavily vegetated mature dunes. This problem presents an interesting dilemma which does not appear to have been documented sufficiently within academic literature. This case study will outline the issue identified and a suggested solution.

Within the SfM (SFM) software, there is a process where image matching takes place throughout the dataset multiple times, as highlighted here in the heavily vegetated areas which leads to significant blurring of the imagery. As the drones fly multiple times back and forth (north-south or by west-east) in the double grid format, it will build a point cloud of the survey surface by image matching multiple points within each photograph. For the most part, this is acceptable as the main function of this software, which is to recreate real-life 3D objects with a 3D model. The greater the number of images acquired using the single or double grid, the improved quality of the output is the primary function of SfM (SFM) software. This is ideal for stationary objects such as buildings, highways, monuments, mountains, fields, sculptures, skyscrapers, and car parks. In stark contrast to the aforementioned object types, environmental factors found at coastal locations are harder to control and will have a detrimental effect on surveyed imagery. Wind, rain, fog, and mist are all important factors to consider when considering a UAV survey located at the coast. While these factors may not stop the UAV survey from being conducted within the coastal region, the resulting imagery will suffer from the effects of all environmental factors on the surface vegetation, more specifically on the Marram Grass (*Ammophila arenaria*).

EFFECTS OF WIND-BLOWN VEGETATION

Wind is the major factor that may limit the effectiveness of a survey taking place in a coastal location as these areas endure some degree of wind throughout the year. Many UAVs have max fly thresholds of around 11 m/s, allowing for most operations to take place safely. The wind velocity experienced at Tentsmuir was fluctuating between 3.88 and 5.36 m/s from a direction of ESE to east, which is well below the 11 m/s threshold for the UAV allowing for a successful survey. For most of the site, wind was not considered a significant factor as there are lots of vegetated areas within sunken low-lying sheltered dune slacks on mature dunes and dune ridges. Post-analysis of imagery revealed impressive image resolution of low-lying vegetation within the dune slacks and heavily vegetated mature dunes. Unfortunately, the equivalent could not be said for the taller vegetation on the dune ridges and mobile dunes. The post-processed orthomosaics displayed very sporadic blurred imagery throughout the survey area generally where Marram Grass was located (Figure 7.5 Lower image).

The discovery of blurred imagery within the orthomosaic layer suggested that the Marram Grass was bending over in different directions and angles depending on the direction of the fluctuating wind. Multi-directional images captured by the UAV could have up to 37 images covering the same geographic location from four separate directions (north-south/south-north and east-west/west-east). The orthomosaic layer highlights the problem of multi-directional imagery when recreating models of vegetation especially in areas of high occurrences of Marram Grass. This accentuation of movement made the grass appear to lean in different directions, and more importantly, it lost its clarity and definition needed to identify individual species.

FIGURE 7.5 Comparison of the different quality of orthomosaic imagery. (Lower Image) The blurred Marram Grass orthophoto created using 29 separate images captured through four directions. (Top Image) Orthomosaic created using four specifically selected images from two directions. (Image: Andrew Smith 2018.)

SOLUTION

As a solution to this issue, one orthomosaic layer was created using four carefully chosen images taken from the direction north-south and south-north and couple this with only choosing images where the Marram appeared to be tilted to the same general direction. The resulting image had a greatly improved resolution that allowed for identification of vegetation types and individual Marram Grass plants (Figure 7.5 Top image). There may be an issue here not with quality but with quantity of imagery. As with many surveys, the opportunity to survey an area may only happen once due to many factors such as cost, time, distance, location, access rights, weather, obstructions, and public places. This means that those undertaking surveys will wish to capture as much data as possible as a revisit may not be possible. Batch uploads using a photo editing suite were used to improve the quality and sharpness of images, a useful tool as it was able to rescue several images. While this process has been greatly improved, the concept of this solution may not be suitable for all data types as concentrated post-survey image processing would be required to remove blurred images. The issues faced within this project highlight a lack of knowledge of using this platform to survey a coastal environment that is subject to environmental constraints such as wind.

CONCLUSION

The concept of conducting an aerial survey on the face of it seemed to be a relatively simple task – choose a suitable location, hire or purchase a UAV, utilise autonomous flight to collect imagery, and then process it into something that resembles true life. This is certainly not the case for complex and dynamic locations that can have many different factors that may affect your results. A coastal location will always pose difficulties in collecting aerial imagery; access to remote locations, local environmental factors (wind, rain, cloud, or mist), difficulties with collecting accurate elevation (Z values), and complex issues with SfM (SFM) software will always need to be a consideration.

Initial research suggested in the planning stage of the project that a flight height of 100 m above sea level would be too high to generate high-resolution imagery. This was not the case as the average resolution of 2.66 cm/pixel (7.07 cm^2) was generated throughout the site, identifying vegetation and even directional footsteps on the beach was possible. The standard 20 MP sensor carried by the Inspire 2 was equally impressive, generating imagery far more than what was initially theorised. Consideration must be given for the less-than-perfect weather experienced on survey day, with periods of reduced visibility, the cold gusty fluctuating wind, and thick clouds hampering final survey imagery. The increased flying height also allowed for a larger area to be surveyed in a shorter time. In addition to the discovery and solution of the Marram Grass problem, the 100 m flying height has shown itself to be acceptable when using a 20 MP sensor in less-than-ideal weather conditions.

The focus from the outset was to attempt to utilise cloud-based image processing technology that produces the same outputs as the desktop software version. Initially these 2D and 3D maps and models were very visually pleasing. This was short lived as problems with geolocation, GCP interaction, accurate contour representations, resolution, and elevation caused endless difficulties. Following extensive research, many UAVs no longer carry accurate barometer systems (including DJI) to measure ground elevation, coastal locations are notoriously difficult to calculate elevational data as accurate mean sea level fluctuates greatly. The option to utilise the cloud-based process had limited features to modify data and critically does not allow you to attach the GCPs to the data to geolocate the model of the site. Furthermore, applying the Leica Viva GNSS ground station survey (GPS) should have recorded relatively accurate GPS elevation data (Z values) but did not.

Regardless of difficulties experienced with the collection of accurate elevation data by the Leica GNSS ground survey, the subsea-level elevation values generated by the UAV manufacturers and the inaccuracies with external sourced LiDAR DSMs, elevation values were eventually calculated using the original GNSS ground survey data following an ellipsoid conversion. Accuracies within the GNSS survey proved to be very accurate following successful interaction of GCPs within the

Pix4Dmapper project: 0.12 m for **X**, 0.09 m for **Y**, and 0.06 m for **Z**, with an average reading of 0.09 m in the X, Y, and Z directions. In comparison, geolocation within Pix4D Cloud was far less accurate, only determining absolute geolocation variance to be 3.55 m for **X**, 4.58 m for **Y**, and 0.70 for **Z** showing an average of 2.94 m in the X, Y, and Z directions mainly as no GCPs could be added to improve the accuracy of data within the cloud.

The use of the UAV platform greatly speeds up the process of survey, collection of a large dataset, and processing using SfM (SFM) technology. Without doubt, this project has utilised cutting-edge technology to capture imagery, record very accurate spatial data, and generate high-resolution imagery and 3D visualisations to help identify fragile habitat within this coastal environment. There are tremendous benefits to exploiting UAV-SfM (SFM) compared with expensive time-consuming traditional methods such as LiDAR. LiDAR elevation data has for many years been successful in improving identification of coastal areas that are at risk from sea-level rise or inundation, nevertheless clearly from the analysis in this project, there may be some issues with the accuracy of this elevation data. The UAV-SfM (SFM) approach to survey is a rapid method to collect, process, and create impressive real-world geometry within the 3D VR environment. The UAV-SfM (SFM) approach to data collection right through to completed visualisation has notably changed from weeks or months to a matter of days or even hours. This advancement has the power to revolutionise the coastal survey industry and, more importantly from a coastal monitoring perspective, to speed up the mapping and modelling of fragile environments that can change significantly in a matter of hours.

ACKNOWLEDGEMENTS

The author would like to thank Tom Cunningham, Reserve Manager for Tentsmuir NNR and Scottish Natural Heritage for the opportunity to conduct field work within the reserve in 2018. Many thanks to Dr. David R. Green for his continued support, Dr. Dmitri Mauquoy for his help in identifying vegetation, and my very supportive family.

REFERENCES

Alberico, F., Cavuoto, G., Di Fiore, V., Punzo, M., Tarallo, D., Pelosi, D., Ferraro, L., and Marsella, E., 2017. Historical maps and satellite image as tools for shoreline variations and territorial changes assessment: The case study of Volturno Coastal Plain (Southern Italy). *Journal of Coastal Conservation.* doi:10.1007/s11852-017-0573-x

Augmentra Ltd. 2018. Mobile Mapping Application ViewRanger™. Version 8.5.46[3]. Software used on Samsung Galaxy SM-G930F. [online] Available from: http://www.viewranger.com

CAA Drone Code. 2018. The Drone Code. *Guidelines on conducting flights with an Unmanned Air Vehicle.* [online] Available from: https://dronesafe.uk/wp-content/uploads/2018/06/Dronecode_2018-07-30.pdf [Accessed 20 Dec 2018]

Cunningham, T., 2014. *The Story of Tentsmuir National Nature Reserve.* 2nd Edition. Scottish Natural Heritage, Inverness. [online] Available from: https://www.nature.scot/sites/default/files/2018-02/Tentsmuir%20NNR%20-%20The%20Story%20of.pdf [Accessed 20 Dec 2018]

Drummond, C.D., Harley, M.D., Turner, I.L., Matheen, A.N.A., and Glamore W.C., 2015. UAV applications to coastal engineering. *Australasian Coasts & Ports Conference 2015*, Auckland, New Zealand. [online] Available from: https://www.researchgate.net/publication/282606382_UAV_Applications_to_Coastal_Engineering

Fabbri, S., Giambastiani, B.M.S., Sistilli, F., Scarelli, F., and Gabianelli, G., 2017. Geomorphological analysis and classification of foredune ridges based on Terrestrial Laser Scanning (TLS) technology. *Geomorphology* 295: 436–451. doi:10.1016/j.geomorph.2017.08.003

Goméz, C., Wulder, M.A., Dawson, A.G., Ritchie, W., and Green, D.R., 2014. Shoreline change and coastal vulnerability characterization with Landsat imagery: A case study in the Outer Hebrides, Scotland. *Scottish Geographical Journal* 130(4): 279–299. doi:10.1080/14702541.2014.923579

Klemas, V., 2015. Coastal and environmental remote sensing from Unmanned Aerial Vehicles: An overview. *Journal of Coastal Research* 31(5): 1260–1267. doi:10.2112/JCOASTRES-D-15-00005.1

Mancini, F., Dubbini, M., Gattelli, M., Steechi, F., Fabbri, S., and Gabbianelli, G., 2013. Using Unmanned Aerial Vehicles (UAV) for high-resolution reconstruction of topography: The structure from motion approach on coastal environments. *Remote Sensing* 5(12): 6880–6898. doi:10.3390/rs5126880

Marteau, B., Vericat, D., Gibbins, C., Batalla, R.J., Green, D.R. 2016. Application of Structure-from-Motion photogrammetry to river restoration. *Earth Surface Processes and Landforms* 42: 503–515. doi: 10.1002/esp.4086

McCarthy, M.J., Colna, K.E., El-Mezayen, M.M., Laureano-Rosario, A.E., Méndez-Lázaro P., Otis D.B., Toro-Farmer, G., Vega-Rodriguez, M., and Muller-Karger, F.E., 2017. Satellite remote sensing for coastal management: A review of successful applications. *Environmental Management* 60: 323–339. doi:10.1007/s00267-017-0880-x

Micheletti, N., Chandler, J.H., and Lane, S.N., 2014. Investigating the geomorphological potential of freely available and accessible structure-from-motion photogrammetry using a smartphone. *Earth Surface Processes and Landforms*. doi:10.1002/esp.3648

Micheletti, N, Chandler J.H., and Lane S.N., 2015. Structure from motion (SfM [SFM]) photogrammetry. *British Society for Photogrammetry*. In: Geomorphological Techniques. Chap 2(2.2). 1–12.

Moloney, J.G., Hilton, M.J., Sirguey P., and Simons-Smith, T., 2018. Coastal Dune surveying using a low-cost Remotely Piloted Aerial System (RPAS). *Journal of Coastal Research* 34(5): 1244–1255. doi:10.2112/JCOASTRES-D-17-00076.1

Nex, F., and Remondino, F., 2014. UAV for 3D mapping applications: a review. *Applied Geomatics* 6: 1–15. doi: 10.1007/s12518-013-0120-x.

OS Terrain 5 [XYZ geospatial data] Scale 1:1000. Tile:no52nw. Ordnance Survey (GB). EDINA Digimap Ordnance Survey Service [online] Available from: http://digimap.edina.ac.uk Download: 2018-06-15 12:10:13.394

Papakonstantinou, A., Topouzelis, K., and Pavlogeorgatos, G., 2016. Coastlines zones identification and 3D coastal mapping using UAV spatial data. *International Journal of Geo-Information* 5(75): 1–14. doi:10.3390/ijgi5060075

Pix4D., 2018. *Pix4Dmapper Pro version 4.2.27*. Pix4Denterprise version 4.2.27. [online] Available from: http://www.pix4d.com/ [Accessed June 2018].

Schenk, T., 2005. *Introduction to photogrammetry*. Department of Civil and Environmental Engineering and Geodetic Science. The Ohio State University, Columbus. [online] Available from: http://www.mat.uc.pt/~gil/downloads/IntroPhoto.pdf [Accessed 18 Dec 2018]

Shakhatreh, H., Sawalmeh, A., Al-Fuqaha, A., Dou, Z., Almaita, E., Khalil, I., Othman, N.S., Khreishah, A., and Guizani, M., 2018. Unmanned Aerial Vehicles: A survey on civil applications and key research challenges. *arXiv* 1–49. arXiv:1805.00881v1 [cs.RO]

Turner, I.L., Harley, M.D., and Drummond, C.D., 2016. UAVs for coastal surveying. *Coastal Engineering* 114: 19–24. doi:10.1016/j.coastaleng.2016.03.011

Scottish Spatial Data Infrastructure Metadata Portal, 2018. [online] Available from: http://www.spatialdata.gov.scot/geonetwork/srv/eng/catalog.search#/home

Warrick, R., Ritchie, A.C., Adekman, G., Adelman, K., and Limber, P.W., 2017. New techniques to measure cliff change from historical oblique aerial photographs and structure-from-motion photogrammetry. *Journal of Coastal Research* 33(1): 39–55. doi:10.2112/JCOASTRES-D-16-00095.1

Westoby, M.J., Brasington, J., Glasser, N.F., Hambrey, M.J., and Reynolds, J.M., 2012. Structure-from-Motion Photogrammetry: A low-cost, effective tool for geoscience applications. *Geomorphology* 179: 300–314. doi:10.1016/j.geomorph.2012.08.021

Woodget, A.S., Fyffe, C., and Carbonneau, P.E., 2017. From manned to unmanned aircraft: Adapting airborne particle size mapping methodologies to the characteristics of UAS and SfM (SFM). *Earth Surface Processes and Landforms* 43: 857–870. doi:10.1002/esp.4285

8 Monitoring, Mapping, and Modelling Saltmarsh with UAVs

David R. Green
UCEMM – University of Aberdeen

Dmitri Mauquoy
University of Aberdeen

Jason J. Hagon
GeoDrone Survey Ltd and UCEMM – University of Aberdeen

CONTENTS

INTRODUCTION

Coastal and estuarine saltmarsh has long been recognised as having key physical, ecological, and recreational value, acting as sediment and nutrient traps and as natural coastal protection structures functioning as protective buffers between the land and the sea. Additionally, they are important areas for shelter, feeding, and breeding of diverse forms of wildlife (Doody, 2007).

Recent work suggests that saltmarshes are particularly good for demonstrating how the coast can change in response to environmental influences, including relative sea-level rise. The pace of current Scottish relative sea-level rise (Rennie and Hansom, 2011) may lead to inundation of coastal saltmarsh, with the potential for rapid change and loss, and so it is critical to be able to monitor the response of saltmarsh to sea-level rise, map saltmarsh topography, and model rates of marsh elevation change on a real-time basis. However, changes in saltmarsh vegetation differ according to local circumstances and can be very difficult to map from the land; and even if there is enough survey time to walk vegetation community boundaries using GPS, some areas are so soft and muddy that they cannot be surveyed safely on foot.

This is where aerial imagery comes into its own. Not only does it provide an overview of the saltmarsh that cannot be obtained from the ground, but with geo-rectified photography, it becomes possible (with more than one survey) to map both area and volumetric changes accurately. In contrast to conventional vegetation mapping, aerial imagery is non-selective in its data capture, and multispectral and hyperspectral imagery adds even more information about these changes for scientists who are attempting to interpret coastal change from a variety of perspectives.

The aims and objectives of this chapter are to demonstrate the practical potential of using unmanned aerial vehicle (UAV)-based remote sensing platforms and sensors to monitor, map, and model coastal and estuarine saltmarsh.

REMOTE SENSING AND SALTMARSH

The Saltmarsh Survey of Scotland (SSS) (Haynes, 2016) identified key areas relevant to the study of saltmarsh including the value of remote sensing (past, present, and future) to provide valuable insight into sediment movement, its interrelationship with saltmarsh vegetation, and changes to saltmarsh communities; the need to collect time-series high-resolution topographic data (e.g. LIDAR) to provide real-time accurate information about the location and rates of saltmarsh erosion and accretion; the need to investigate the decline in the extent of saltmarsh pioneer communities; and the potential for saltmarsh surveys to inform management practices and policy in Scotland and improve the conservation and management of saltmarsh.

Saltmarshes can be difficult to access, and ground monitoring and inventory is problematic and time-consuming. Collection of data from remote sensing platforms therefore has considerable potential. Numerous studies of saltmarshes have been undertaken using both airborne and satellite remote sensing platforms and sensors (Klemas, 2013), including studies using multispectral, thermal, Compact Airborne Spectrographic Imager (CASI) hyperspectral (Kumar and Sinha, 2014), and LIDAR sensors (Rosso et al., 2006). Many different types of satellite imagery and data including Landsat TM (Hobbs and Shennan, 1986); ROSIS, CASI (Wang et al., 2007; Silvestri et al., 2003; Sadro et al., 2007), MIVIS, IKONOS, QuickBird (Belluco et al., 2006), AVIRIS, and MODIS (Mishra and Gosh, 2013) have all been used in various studies to monitor and map saltmarsh environments and characteristics.

Classification of remotely sensed data has been shown to provide a means to help accurately monitor the spatial structure and evolution of saltmarsh vegetation over time; for accurate vegetation mapping and quantitative characterisation of the spatial distribution of saltmarsh plants; for understanding saltmarsh soil elevation and its dynamics; monitoring biophysical characteristics, photosynthetic capacity, nitrogen content, blue carbon storage, and the physiological status of saltmarsh vegetation; to infer overall condition and productivity; and allow for the development of effective management strategies in high-priority areas. In addition, other studies have shown the value of remote sensing to characterise saltmarsh topography at a scale relevant to ecological processes; for evaluating changes in saltmarsh morphometry and community structure over long time scales; to quantify and map the dynamics of saltmarshes, specifically to wetland topography and vegetation structure; and to integrate plant community mapping and local tidal hydrodynamics with fine-scale topographic analysis.

Remote sensing has been shown to offer considerable potential for the long-term study, monitoring and mapping of coastal saltmarsh at different spatial, temporal, and radiometric resolutions to provide information that is not always accessible through field observations. There are now opportunities to acquire, process, and integrate high-resolution datasets derived from miniaturised remote sensing platforms (UAVs or drones) with the aid of image processing software, soft-copy photogrammetry, and Geographic Information System (GIS) into multi-dimensional geo-visual representations of saltmarsh for analysis, interpretation, and communication.

Effective management of our coasts in a time of rapid climate change and uncertainty requires the availability of multi-temporal, high-resolution spatial data. Coastal saltmarshes are a vital part of a functioning coast and provide a significant coastal protection function (SNH, 2016). The research reported here seeks to take advantage of an emerging new area of remote sensing technology to acquire high-resolution aerial photography and imagery from small-scale UAV platforms and sensors as a basis for (i) providing new data and information about the physical and ecological characteristics of coastal saltmarsh, (ii) mapping saltmarsh vegetation communities, (iii) identifying areas of erosion and accretion, (iv) providing a high-resolution topographic model of the saltmarsh, and (v) utilising geo-visualisation tools and techniques to communicate key results to the public as a novel way to raise awareness and educate about saltmarsh as living natural coastal protection structures.

UAVs

UAVs can provide very high-resolution maps and models of changes in the saltmarsh environment in virtually real time. A number of recent studies have highlighted the potential of small drone platforms to acquire high-resolution aerial photography and the use of Structure from Motion (SfM [SFM]) to construct 3D models of the surface (Green and Hagon, 2017). Such studies have focused on the role of remote sensing and information extraction. UAV sensors can: (i) offer comprehensive (i.e. non-selective) high-resolution, spatio-temporal data capture over inaccessible areas of saltmarsh; (ii) create 3D models of saltmarsh micro-topography, soil, and vegetation; and (iii) enhance knowledge and understanding of saltmarsh functioning and the spatial extent of vegetation composition and habitat change over time that affects ecosystem functioning. Such data can (i) inform coastal policy for improved and sustainable coastal management and (ii) raise awareness and educate the public about the importance of saltmarsh, the benefits of natural coastal protection, and adaptation management.

RESEARCH

The research reported here is currently exploratory in nature and sets out to test the viability of using UAV platforms and sensors to monitor and map saltmarsh. More specifically, the questions we wish to address are as follows:

- How practical are UAV remote sensing platforms and miniaturised sensors for collecting remotely sensed data and imagery of coastal saltmarsh?
- How easy is it to acquire high-resolution data and information from UAV remote sensing?
- Is the spatial and temporal resolution of UAV data and imagery useful for the monitoring and mapping of coastal saltmarsh?
- Can a high-resolution 3D terrain and surface model of a saltmarsh be acquired via SfM (SFM) to inform saltmarsh functioning (such as interrelationships between vegetation patterns, distribution, and micro-topography) and better understand saltmarsh dynamics (such as sediment erosion and accretion processes)?
- Can the integration of remote sensing, soft-copy photogrammetry, GIS, and geo-visualisation tools and techniques be used to help communicate detailed information about saltmarsh morphology and ecology to a wider audience to aid in raising awareness and to educate?

The primary purpose of the work is to develop a robust methodology for the practical use of low-cost UAV-based platforms and sensors to acquire aerial data and imagery. The study is being conducted at a number of selected saltmarsh sites around the Scottish coastline based upon the SSS and following consultation with saltmarsh experts to optimise the identification of different types of saltmarsh *hotspots* where information is needed.

UAV overflights of the selected saltmarsh sites are being conducted at regular intervals using a small commercial, off-the-shelf, drone carrying a number of small colour cameras and a number of other miniaturised sensors e.g. a Normalised Difference Vegetation Index (NDVI) camera and a thermal camera to acquire colour stereo-photography and imagery. The flights are undertaken at a low altitude e.g. <50 m in order to maximise the spatial resolution of the photography and imagery.

Ground Control Points (GCPs) are surveyed in at each site using a ground-based Real-Time Kinematic Global Positioning System (RTK GPS) unit or a small RTK GPS unit mounted on the UAV platform to ensure accuracy of the subsequent geo/ortho-correction of the imagery and mosaics. Pix4D soft-copy photogrammetric SfM (SFM) software is used to generate high-resolution DTMs and DSMs and ortho-mosaics of the selected saltmarsh sites. Single date imagery, Digital Terrain Models (DTMs), and DEMs of Difference (DOD) will be interpreted to create vegetation maps and establish where sediment has been eroded, where there is no change, and where accretion has taken place. On-screen digitising, image interpretation, and digital image processing (e.g. classification techniques) will be used to prepare maps of the vegetation types and communities in the saltmarsh. Reference documentation and co-incident ground-truthing (where access is possible) is used to ground-truth the saltmarsh vegetation. Pix4D soft-copy photogrammetric software will be used to develop 3D models of the saltmarsh micro-topography (a DSM and DTM) and will be used to (i) provide insight into the saltmarsh surface characteristics, (ii) identify surface geomorphological features of the saltmarsh, (iii) identify areas of erosion and accretion within the saltmarsh, and (iv) correlate vegetation patterns and distribution with the geomorphology and topography of the saltmarsh. The imagery and 3D model will form the basis for developing geo-visualisations of each saltmarsh site. Integrating the vegetation map and topographic surfaces within a GIS will allow us to develop novel ways to communicate information about saltmarsh functioning to the general public in order to raise awareness about the nature and value of saltmarsh as a key element in coastal protection and management.

NOVEL APPROACH

This is a novel, practical, and integrated data collection, processing, analysis, modelling, and communication approach to the acquisition of high-resolution spatial and temporal information about coastal saltmarsh. At its core is the goal to develop an inherently low-cost and practical solution to the acquisition of information about the functioning of coastal and estuarine saltmarshes, their utility and management, together with communication with the public (Figures 8.1–8.7).

Ultimately, outputs from the UAV-acquired aerial data will provide input to a GIS and offer the potential to explore new ways to raise awareness and engage stakeholders in understanding the importance of safeguarding the roles and functions of saltmarshes via coastal management. The use of geo-visualisation tools and techniques to communicate information will provide a novel way to promote greater awareness and understanding of saltmarshes and engage the public in understanding the ecosystem services provided. The use of 3D models and virtual reality (VR) tools

FIGURE 8.1   (a and b) Dornoch saltmarsh and some of the team in the field (Photograph: David R. Green).

FIGURE 8.2   Aerial data collection equipment (Photograph: David R. Green).

FIGURE 8.3   The DJI Inspire 1 UAV aerial platform (Photograph: David R. Green).

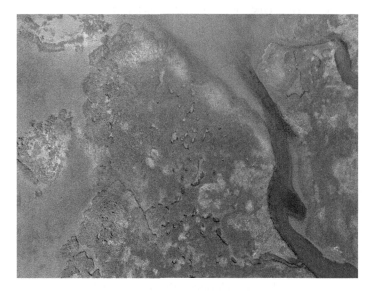

FIGURE 8.4   A UAV aerial photograph of the saltmarsh (Photograph: Jason J. Hagon).

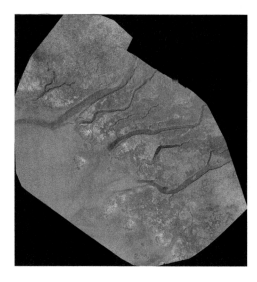

FIGURE 8.5   An UAV aerial mosaic of the saltmarsh (Photograph: Jason J. Hagon).

FIGURE 8.6   A 3D model of part of the saltmarsh (Photograph: Jason J. Hagon).

and technologies will be used to communicate information to coastal managers, in planning and informing decision-making, and to drive scenario discussion with policy and community stakeholders (e.g. the in-combination impacts of sea-level rise, flood risk management, and coastal reclamation and their effect on saltmarshes and management responses).

SUMMARY AND CONCLUSIONS

Saltmarshes are recognised to be a key feature of our coast, offering not only important habitat for wildlife and vegetation but also a major part of our current and future climate change mitigation and adaptation responses at the coast, helping to safeguard our homes and activities. They can be e.g. low-cost, carbon-storing, self-repairing, coastal flood protection features that have the ability to keep pace with sea-level rise. Low-cost UAVs now provide a practical means to provide high-resolution, small-area multi-temporal aerial surveys of coastal and estuarine saltmarsh with the potential to collect data more cheaply and quickly and to present it in novel ways that can significantly enhance our understanding of their characteristics, form, and function over both space and time. Such information will be very relevant to the scientist and policy maker in managing this

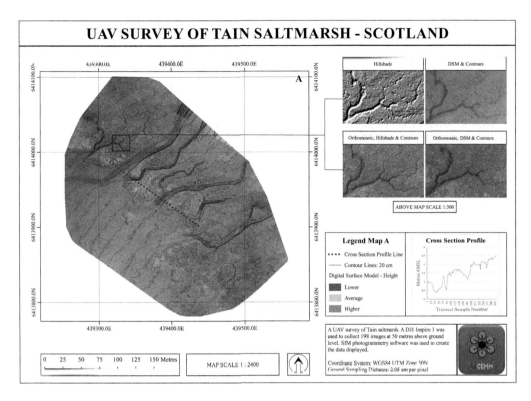

FIGURE 8.7   GIS map of the saltmarsh (GIS Cartography: Jason J. Hagon).

resource and also in an educational and awareness capacity for the public to perceive and better understand the importance of managing this area of the coast.

ACKNOWLEDGEMENTS

- John Cleave and Harvey Mann – Buzzflyer Ltd. For technical expertise and information and for supplying the DJI Inspire 1 UAV to AICSM/UCEMM, Department of Geography and Environment, School of Geosciences, University of Aberdeen.
- MASTS – Coastal Processes and Dynamics Forum for equipment grant support.

REFERENCES

Belluco, E., Camuffo, M., Ferrari, S., Modenese, L., Silvestri, S., Marani, A., and Marani, M., 2006. Mapping salt-marsh vegetation by multispectral and hyperspectral remote sensing. *Remote Sensing of Environment*. Vol. 105: 54–67.

Doody, J.P., 2007. *Saltmarsh Conservation, Management and Restoration*. Springer, Berlin. p. 219.

Green, D.R., and Hagon, J.J., 2017. Coastal data collection: Applying geospatial technologies to coastal studies. *Chapter 7: Marine and Coastal Resource Management: Principles and Practice*. Green, D.R., and Payne, J., 2017. (Eds), Routledge (Earthscan Oceans Series), Abingdon. p. 312.

Haynes, T.A., 2016. *Scottish Saltmarsh Survey National Report*. Scottish Natural Heritage Commissioned Report No. 786. p. 195. (www.snh.org.uk/pdfs/publications/commissioned_reports/786.pdf)

Hobbs, A.J., and Shennan, I., 1986. Remote sensing of saltmarsh reclamation in the Wash, England. *Journal of Coastal Research*. Vol. 2:181–198.

Klemas, V., 2013. Using remote sensing to select and monitor wetland restoration sites: An overview. *Journal of Coastal Research*. Vol. 29(4):958–970.

Kumar, L., and Sinha, P., 2014. Mapping salt-marsh land-cover vegetation using high-spatial and hyperspectral satellite data to assist wetland inventory. *GIScience & Remote Sensing*. Vol. 51(5):483–497.

Mishra, D., and Gosh, S., 2013. Moderate resolution satellite data for mapping salt marshes. *SPIE Newsroom*. p. 4. doi: 10.1117/2.1201304.004800

Rennie, A.F., and Hansom, J., 2011. Sea level trend reversal: Land uplift outpaced by sea level rise on Scotland's Coast. *Geomorphology*. Vol. 125(1):193–202.

Rosso, P.H., Ustin, S.L., and Hastings, A., 2006. Use of LIDAR to study changes associated with Spartina invasion in San Francisco Bay marshes. *Remote Sensing of Environment*. Vol. 100 (3):295–306.

Sadro, S., Gastil-Buhl, M., and Melack, J., 2007. Characterizing patterns of plant distribution in a southern California salt marsh using remotely sensed topographic and hyperspectral data and local tidal fluctuations. *Remote Sensing of Environment*. Vol. 110(2):226–239.

Silvestri, S., Marani, M., and Marani, A., 2003. Hyperspectral remote sensing of salt marsh vegetation, morphology and soil topography. *Physics and Chemistry of the Earth, Parts A/B/C*. Vol. 28(1–3):15–25.

SNH, 2016. *Scottish Saltmarsh Survey National Report*. Commissioned Report No. 786. Scottish Natural Heritage, Inverness. p. 195.

Wang, C., Menenti, M., Stoll, M-P., Belluco, E., and Marani, M., 2007. Mapping mixed vegetation communities in salt marshes using Airborne Spectral Data. *Remote Sensing of Environment*. Vol. 107 (4):559–570.

9 Autonomous UAV-Based Insect Monitoring

Johannes Fahrentrapp
ZHAW Zurich University of Applied Sciences

Peter Roosjen and Lammert Kooistra
Wageningen University & Research

David R. Green
UCEMM – University of Aberdeen

Billy J. Gregory
DroneLite

CONTENTS

INTRODUCTION

Drosophila suzukii Matsumura, the spotted wing drosophila (SWD), has become a serious pest in Europe attacking many soft-skinned crops such as several berry species and grapevines since its spread in 2008 to Spain and Italy. An efficient and accurate monitoring system to identify the presence of *D. suzukii* in crops and their surroundings is essential for the prevention of damage to economically valuable fruit crops. Existing methods for monitoring *D. suzukii* are costly, time and labour intensive, prone to errors, and typically conducted at a low spatial resolution. To overcome current monitoring limitations, we are investigating and developing a novel system consisting of traps that are monitored by means of cameras from Unmanned Aerial Vehicles (UAVs) and an image processing pipeline that automatically identifies and counts the number of *D. suzukii* per trap location.

To this end, we are currently collecting high-resolution RGB imagery of *D. suzukii* flies in sticky traps taken from both a static position (tripod) and a UAV, which are then used as input to train deep learning object detection models. Preliminary results show that a large part of the *D. suzukii* flies that are caught in the sticky traps can be correctly identified by the trained deep learning models.

In the future, an autonomously flying UAV platform will be programmed to capture imagery of the traps under field conditions. The collected imagery will be transferred directly to cloud-based storage for subsequent processing and analysis to identify the presence and count of *D. suzukii* in near real time. This data will subsequently be used as input to a decision support system (DSS) to provide valuable information for farmers.

DROSOPHILA SUZUKII AS AN EXAMPLE OF AN INSECT PEST

During its global spread in the last years (Asplen et al., 2015), *D. suzukii* has become a serious pest all over Europe (Cini et al., 2012, 2014) attacking many unwounded soft-skinned crops such as several berry species, cherry, and grapevines as well as many non-crop plants (Kenis et al., 2016). All these fruits have a high nutritional value due to high vitamin content. *D. suzukii* are present wherever berries ripen in forests, hedges, gardens, and backyards as well as in commercially used plantations. *D. suzukii* place their eggs in (almost) ripe fruits, and within days, feeding larvae bring the fruits to collapse. Plant protection products (PPPs) are often not applicable due to proximity to harvest dates and their specific retention times within the fruit. With trap detection of *D. suzukii* in the 'zero tolerance' berry crops, five to eight insecticide applications may be needed to prevent crop loss (Van Timmeren and Isaacs, 2013).

Monitoring of *D. suzukii* is essential for all existing control strategies, including (i) mass trapping, (ii) pest exclusion by net covers, (iii) PPP applications, (iv) early harvest, and (v) sanitation. However, the monitoring of *D. suzukii* is challenging since (i) the insect is rather small (2–4 mm), (ii) there are several similar looking Drosophila species (Bächli, 2016), (iii) lures are less attractive than ripe fruits, and (iv) current strategies such as using liquid-baited traps are too laborious. Today, monitoring is conducted by placing cup-style traps containing a liquid bait, such as a wine/vinegar mixture, in the field. After a certain period, typically around a week, the cup traps are collected, and their contents are manually analysed by looking at the caught insects one by one under a stereo microscope. Hereafter, the cup traps are refilled with fresh liquid bait and placed again in the field. This is a very time-consuming and labour-intensive process. As such, it is neither used as a continuous and area-wide monitoring method in crops and for diverse and habitats that are difficult to access nor is an automatic integration with a Geographic Information System (GIS)-based DSS possible.

In order to develop an alternative approach, we defined the following project aims: (i) selection of the most efficient photographable trap and a UAV-mounted camera system data collection setup for each crop, (ii) development of a fully automated *D. suzukii* identification and counting algorithm on trap images, and (iii) a novel DSS-implementable threshold definition. The project outcomes defined in this way are in line with the fruit producers' needs including an easy-to-use, non-laborious monitoring system providing an early indication of the presence of the pest.

TRAPPING *D. SUZUKII*

Male *D. suzukii* can be identified and discriminated from other Drosophilids by the naked eye due to an obvious black spot on both of their wings (hence the name: spotted wing drosophila). The most notable feature that separates female *D. suzukii* from other Drosophila species is that female *D. suzukii* have a lignified ovipositor. However, this can only be seen under a stereo microscope. Therefore, the male individuals are the targets for our image-based monitoring because they are more likely to be correctly detected by image analysis. The ratio of males to females found in field trails was reported to be in the range of 0.7–1.2 (Landolt et al., 2012; Lee et al., 2012; Jüstrich, 2014; Burrack et al., 2015). Current field-derived data indicate mainly 40%–60% presence of one sex (Fahrentrapp, personal communication). However, the ratio may vary by trap, crop, and crop status (Escudero-Colomar et al., 2016) and, therefore, has to be determined for each variable. At the 9th International Conference on Integrated Fruit Production (September 2016, Thessaloniki, Greece), during dedicated sessions on *D. suzukii*, many trap-testing trials were presented. All experiments compared the efficiency of different cup-style traps. These traps have the major drawback of hand counting of the caught insects in the liquid, frequent change of the bait, and loss of bait attractiveness over time (Fahrentrapp et al. unpublished). Therefore, we suggest using alternative traps such as sticky traps or an in-house developed prototype in attractive colours and shapes and only if needed baited with evaporating lure capsules (e.g. Pherocon SWD High Specific Lure) or as a bait–glue mixture.

The flat surface of the sticky trap can be monitored by UAV-based cameras. Commonly, yellow, white, red, orange, and blue sticky traps are available attracting different insect species. In-depth analysis of the visual capability including colour-choice experiments with the close relative of *D. suzukii*, *Drosophila melanogaster* suggested a preference in R1–R6 deficient mutants of *D. melanogaster* for blue and UV, thus indicating that yellow sticky traps may not attract *D. suzukii* (Heisenberg and Buchner, 1977; Salcedo et al., 1999; Briscoe and Chittka, 2001; Yang et al., 2004; Yamaguchi et al., 2008; Yamaguchi et al., 2010). The possibilities to increase trap attractiveness by colour and shape were also demonstrated for the cherry fruit fly *Rhagoletis cerasi* and for *D. melanogaster*, respectively (Basoalto et al., 2013; Daniel et al., 2014). Current research demonstrates red and black as most attractive colours for *D. suzukii* (Rice et al., 2016). Red spheres and panels baited with a Scentry lure were found more attractive than cup traps with the same lure (Kirkpatrick et al., 2018).

To increase the level of trap attractiveness, odour-based attractants and trap position in crops may also be considered for optimisation (Walsh et al., 2011; Yu et al., 2013). Preliminary experiments suggested a putative repellent activity of used glues on the traps (Fahrentrapp et al., unpublished). A small number of reports are available on trials with sticky traps (e.g. Rice et al., 2016; Kirkpatrick et al., 2018), with only one report using baited sticky traps (Iglesias et al., 2014). Unfortunately, knowledge about the mechanisms of the olfactory sensory machinery of Drosophilids in general is not directly transferable to *D. suzukii* since, in contrast, the latter species is attracted by ripe fruits rather than by rotten ones. Isopentyl acetate (IPA) and furaneol methylether were identified as attracting *D. suzukii* (Keesey et al., 2015; Ramasamy et al., 2016) and hence could be used to increase trap attractiveness.

TRAPPING *D. SUZUKII* – WORK IN PROGRESS

Attracting high numbers of *D. suzukii* is possible with cup traps baited with liquid lure. However, these traps are not suitable for image-based monitoring. For aerial monitoring of traps, a basic requirement is a planar photographable surface with the target object. Sticky traps could be a solution. However, Drosophilids are mainly guided by odour that in turn is hard to include on sticky traps. We included a curling, inaccessible odour to sticky traps leading to 10× more target insects on the sticky traps compared to unbaited sticky cards. However, compared to cup traps containing the same lure, the baited sticky traps were 25× less attractive (Figure 9.1; Fahrentrapp et al,

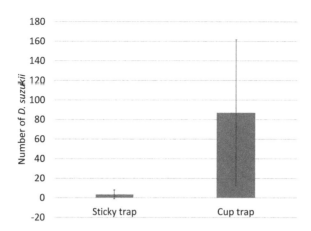

FIGURE 9.1 Trap efficiency comparing the number of caught *D. suzukii* on red sticky traps, with inaccessible attractant, and cup traps filled with lure. As attractant, a commercial wine vinegar mixture was used (Riga Lockmittel, Riga, Kesswil, Switzerland). The error bars indicate the standard deviation.

unpublished); visual cues also contribute to attractiveness. For instance, cup traps with black lids were significantly preferred over ones with red lids (Cahenzli et al., 2018). Rice et al. (2016) demonstrated red and black to be significantly more attractive as six other colours using coloured sticky cards. In field trials using red, orange, and blue sticky cards, we found a dependency of fruit crop species on colour preference in *D. suzukii* during the period with ripe fruits (Table 9.1; Fahrentrapp et al. unpublished).

The glue used on sticky traps is mostly Tangle-Trap® or Tropical-Tangle-Trap® (Tanglefoot®, The Scotts Company LLC, Marysville, Ohio, USA) (e.g. Rice et al., 2016; Kirkpatrick et al., 2018). As there were no alternatives, we used the same glue, although this was not found to be optimal because *D. suzukii* were able to walk on the glue and sometimes to escape the trap (see video on Twitter.com: @AAPMPROJECT). This may contribute additionally to low catch numbers on sticky traps compared to cup traps. To overcome this problem, we developed a trap prototype which allows attraction of *D. suzukii* with liquid lure that is smelling but inaccessible (Figure 9.2). The fly enters the trap prototype through small holes and is trapped behind a plastic window, through which the insect can be photographed. The design of the prototype makes it less likely for *D. suzukii* to escape. Preliminary results under artificial conditions indicate a higher catch rate as recorded for sticky traps (Fahrentrapp et al. unpublished).

COMPUTER VISION TO IDENTIFY INSECTS ON TRAPS

An image classification model takes an image and predicts what the object inside the image is, that is, it predicts its class. This is by itself already a complicated task, and when multiple objects are present in the image, it will become even more difficult for a model to make a correct prediction. Moreover, after classification, the location of the object or objects in the image is still unknown. When multiple objects of the same class are present in an image, for example, in a photo of a sticky trap in which multiple *D. suzukii* flies are caught, a classifier on its own will not be able to count how many flies are present. This is an issue that needs to deal with both predicting where objects are in the image and predicting which class each of these objects belongs to. The combination of both the prediction of the location and the class of the object in digital images or videos is called object detection. For the case of detecting and counting *D. suzukii* flies in images of sticky traps, these issues also need to be resolved: both the location and the class of each caught *D. suzukii* individual in the image need to be predicted, after which they can be counted.

TABLE 9.1
Colour Preferences in Different Crops with Ripe Fruits

Crop	Species Name	General Colour Preference/p-value	Colour Combination	Specific Colour Preference/p-value	Colour Combination	Specific Colour Preference/p-value	Observations/N
Plum	*Prunus domestica*	0.002	Red over orange	0.001	NA	NA	216
Strawberry	*Fragaria x ananassa* L.	0.033	Red over orange	0.012	NA	NA	114
Grapevine	*Vitis vinifera*	0.0014	Red over blue	0.005	Red over orange	0.001	150
Raspberry	*Rubus idaeus*	0.101	NA	NA	NA	NA	264
Wild blackberry	*Rubus fruticosus* aggr.	0.514	NA	NA	NA	NA	39
Cultivated blackberry	*R. fruticosus* aggr.	0.079	NA	NA	NA	NA	234
Blueberry	*Vaccinium myrtilius*	0.001	Red over orange	<0.001	NA	NA	174
Hardy kiwi	*Actinidia arguta*	0.001	Red over blue	0.001	Red over orange	0.001	155

NA, not applicable.

FIGURE 9.2 Prototype of baited trap providing a planar, photographable surface. The basin with net (D) is filled with liquid lure and placed below the 'window' of the trap body (C). Flies can enter through the four small holes on top of the trap body. The flies move forward to the attractant behind the lamella of the inner part of the back of the trap (A and B). Through the small slot on the bottom of the lamella, the fly enters the photographable 'showroom' behind the window. A and B, inner sideview from back part of the trap; C, sideview through 'window' of trap; D, basin with net for liquid lure.

To build an accurate vision-based detection and counting system of trapped *D. suzukii* flies, all instances of them in an image need to be found and classified. For a computer to learn to recognize objects, features of the object of interest need to be extracted, which can then be taught to a classification algorithm. In recent years, deep learning for object detection has rapidly gained increased popularity. Convolutional neural networks (CNNs) are one of a number of deep learning models and the most preferred for object detection in image data (Yamashita et al., 2018). CNNs are a class of artificial neural networks that are specialised for processing data that has a grid-like structure, as is the case with image data (Goodfellow et al., 2016). As opposed to traditional object detection using machine learning where features need to be hand-crafted, CNN models have the ability to learn the features of an object from the data itself without manually defining them, making them independent of prior knowledge and human effort (Agarwal et al., 2018). Deep learning strategies, in combination with CNNs, have been the dominant method for complicated image classification tasks (Russakovsky et al., 2015), and due to the complexity of identifying *D. suzukii* flies from images of sticky traps, an approach that uses deep learning and CNNs for their detection is ideal.

Recently, the detection of insects in images using deep learning methods has gained increased attention in the scientific literature. For example, Zhong et al. (2018) developed a vision-based counting and recognition system for six species of flying insects. In their approach, a digital camera was used to take images from a fixed distance of yellow sticky traps which were pasted on a metal back plate. In the images collected by the camera, detection and counting of the insects were performed with the You Only Look Once (YOLO) (Redmon et al., 2016) algorithm and support vector machines (SVMs) (Cortes and Vapnik, 1995). They were able to obtain an average counting and classification accuracy of 92.50% and 90.18%, respectively, for bees, flies, mosquitos, chafers, moths, and fruit flies. Sun et al. (2018) applied a deep learning method for the detection of different bark beetle species in artificially prepared cup traps based on images taken with a digital camera. In their setup, the camera, and an LED to ensure stable and consistent illumination conditions, were placed at a fixed distance above the cups, and images were collected of beetles in these cups.

By training a modified version of the RetinaNet algorithm (Lin et al., 2017), they were able to detect beetles with an average precision of 74.6%.

COMPUTER VISION TO IDENTIFY INSECTS ON TRAPS – WORK IN PROGRESS

To create a robust classification model, it is important to train the model using imagery in which the aforementioned challenges are also present. A deep learning model will take these effects into account during the training phase of the model, and as a result, it can provide more robust classification results. For the detection of *D. suzukii*, we therefore constructed a training database which not only contains images taken under optimal conditions (e.g. taken perpendicular to the photographable surface at fixed and optimal distance, under controlled illumination conditions), but also with images taken under sub-optimal conditions. Currently, we have a training database that consists of just over a hundred images of sticky traps, containing nearly 4,800 annotations of *D. suzukii* individuals with an equal ratio of males and females. The images of the traps were taken with a Sony DSC-RX100 IV compact camera under varying illumination conditions: sunny, clouded, and partially clouded. A part of the images was taken from a tripod perpendicular to the photographable surface and from an optimal distance. However, roughly 20% were taken under an angle and at different distances. Currently, we are training different deep learning detection and classification models using this database for the detection of *D. suzukii* on sticky traps. Due to the strong resemblance of female *D. suzukii* flies with other Drosophila species, we are focusing on the detection of the male flies, which have distinctive black dots on their wings that can be observed when photographs are taken perpendicular to the sticky trap surface and the caught insects are viewed from the top.

By training Alexnet (Krizhevsky et al., 2012), one of the pioneering deep learning classification architectures, using 70% of the labelled male *D. suzukii* flies for training and 20% for validation, we were able to achieve a classification accuracy of 79.95% when feeding the network the remaining 10% of the images to classify. During an experiment, we mounted the same camera as was used for the collection of the training data under a RotorKonzept (Abtsteinach, Germany) RKM 4X UAV, and captured images of sticky traps in which *D. suzukii* and other insects were caught during a manual flight. As a first step, we applied the Selective Search (Uijlings et al., 2013) algorithm, an algorithm that generates region proposals that could contain the initial locations of the flies on the traps, to these images. Hereafter, each of these locations was classified by the Alexnet classifier. Figure 9.3 shows an example image of this exercise. In this figure, all detections of male DS that were made with a probability >0.5 are displayed (blue rectangles). Of the eleven male DS that were labelled in this image by an expert (green rectangles), six were correctly classified (true positives, green circles) by applying this strategy. Four male DS were not found (false negatives) and 13 false detections (false positives) were made, translating into a precision of 0.35 and a recall of 0.63. This indicates that there is potential for UAV-based detection of insects as small as DS flies (Figure 9.3). Other, more state-of-the-art deep learning models such as Faster R-CNN (Ren et al., 2015), Region-based Fully Convolutional Network (R-FCN) (Dai et al., 2016), Single Shot Multibox Detector (SSD) (Liu et al., 2016), and YOLO (Redmon et al., 2016) need to be explored for the detection of *D. suzukii* flies. These architectures can predict both the location and class of the flies in images, and due to their rapid processing speeds, these models also provide the opportunity for (close to) real-time monitoring of sticky traps.

UAV-BASED IMAGE ACQUISITION

UAV PLATFORMS

There are now many different commercially available UAV platforms and sensors with a growing trend towards smaller and more compact platforms with higher resolution sensors and greater functionality. Several well-known UAV platforms were therefore considered for this research.

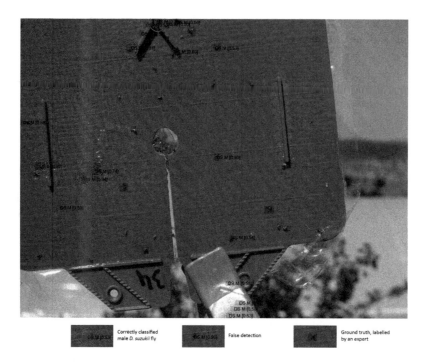

FIGURE 9.3 Example of detection result for identification of male *D. suzukii* flies on a sticky trap, photographed using a camera mounted under a UAV. The values between the squared brackets indicate the class probability.

As many drone cameras available at the outset had relatively modest resolution cameras, attention early on focused on the larger UAV platforms that had the capability to carry a larger camera payload such as digital single-lens reflex (DSLR) cameras with the potential for better resolution cameras. Whilst this would provide the capability to acquire high-resolution images – providing a stronger basis for being able to detect the feature of interest – there are a number of reasons why such a choice would not be ideal; larger platforms and sensors are less compact, more expensive, and unlikely to offer an affordable solution for the farmer; a larger drone is also less likely to be as manoeuvrable in practice, more difficult to fly in confined spaces, and in many cases will not offer autonomous flight capability; larger platforms have larger rotors and would therefore be more difficult to position accurately; many of the larger drones are now also relatively old technology and in the longer term will be out of date. By comparison, the smaller UAVs are more modern, more compact, easier, and more accurate to position (with the aid of prop guards and cages), affordable, easier to fly in confined areas, have autonomous flight capability, and increasingly good miniaturised camera technology.

AERIAL IMAGE ACQUISITION

The use of UAVs for aerial image acquisition is well documented in the literature (Gómez and Green, 2015) for a wide range of environmental applications and increasingly as one element of Precision Agriculture (PA) and Precision Viticulture (PV). As a very rapidly evolving area of technological research small, off-the-shelf UAV platforms are becoming increasingly sophisticated with the addition of RTK (Real-Time Kinematic) and PPK (post-processed kinematic) Global Positioning Systems (GPS) (https://www.swiftnav.com/) for improved locational positioning accuracy. In addition, the inclusion of object avoidance systems and navigational aids in the form of autopilot systems for take-off, landing, and flight have improved significantly. Ready-to-Fly (RTF) UAVs are improving the ease with which it is possible to operate these platforms. Alongside, the platform developments have also been the miniaturisation of sensors and cameras to acquire a greater range of aerial data and imagery, sophisticated soft copy photogrammetry, and digital image processing (DIP) software for image processing, information extraction, and model building (Gómez and Green, 2015). Low-cost, high-resolution cameras have also provided the means to acquire very high-resolution multispectral and hyperspectral digital imagery with high resolving power to aid in feature recognition and extraction. State-of-the-art flight planning software supports opportunities for autonomous flights and complicated flight paths. Different modes of autonomous flight may also mean that in the near future it will be possible to utilise these – aside from the photographic and recreational uses now so popular – for professional data acquisition tasks in the field. There are now many free and low-cost APPs for undertaking autonomous flight (e.g. DJI Go4, Litchi), all of which offer increasingly greater control over the entire drone flight and photographic mission, including the flying height, the mode of flight, the positioning of the camera, and acquisition of the imagery amongst others. With the growing realisation and acceptance of UAVs and the related technologies for scientific research, and practical airborne data and image acquisition systems, has come the need for flight and operation regulations to ensure safe operation of such platforms, aided by guidelines and examples of best practice (Gómez and Green, 2015).

Aerial image acquisition faces two major challenges: (i) varying light conditions such as sun, clouds, shadow, and reflections, and (ii) the lack of a fixed position (angle and distance) of the camera towards the photographable surface. To complicate things further, the size of *D. suzukii* is only 2–4 mm; thus, sharp and focused images with a very high resolution (submillimetre) are needed to be able to detect them.

Images that are typically used in training detection systems are taken under controlled illumination conditions, from a fixed distance, and perpendicular to the surface being photographed. Hand-held image collection allows for properly positioning a camera towards a trap to optimally photograph it. For UAV-based imagery, captured under field conditions, this is not the case. Illumination conditions will vary depending on the presence of clouds, the position of the sun, the orientation of the photographable surface, and the orientation of the camera towards this surface. Moreover, due to positioning inaccuracies that affect UAVs, the distance between the camera and the photographable surface and the angle towards it will be quite variable. This will result in a far from optimal image quality both for training of a classification model and in an operational context for the imagery to do the actual detection and counting on.

UAV-BASED IMAGE ACQUISITION – WORK IN PROGRESS

MANUAL AND AUTONOMOUS FLIGHTS

As part of this research, careful consideration was given to the UAV platform that would be best suited to the tasks required, bearing in mind the desire to develop a practical, low-cost, affordable, and easy-to-fly system that could be developed into a practical solution for the commercial fruit grower.

Choice lies with either a custom-built platform or the adoption and adaptation of an existing commercially available platform. Typically custom-built platforms are usually a specialist one-off, and whilst they can be adapted and fine-tuned to a task, perhaps even boasting a larger payload capacity; they are generally more expensive; less likely to have autonomous flight capability; require more work to maintain and fly; and may have little long-term support and development.

Off-the-shelf drones, by comparison, are generally well developed; continually evolving; have good technical and development support; and whilst possibly not able to carry larger payloads (e.g. high-resolution DSLR cameras) can perform the flight and camera tasks much more easily. Being smaller, they can also be flown more easily; are more manoeuverable in confined spaces; and with additional navigational functionality and controls can be flown by almost anyone. Additional benefits include autonomous flight capability, an array of proximity sensors, FPV (first-person view) glasses that permit visual monitoring of the drone position, and flight cages/prop guards that can help to protect the drone from unwanted collisions and at the same time allow the camera to be positioned closer to the target.

COMMERCIAL UAVs

Four well-known commercially available off-the-shelf drone makers were considered: DJI, Parrot, Yuneec, and 3DR. Several different size UAV platforms offered by these companies were initially considered for this research project ranging from the larger to the smaller platforms (e.g. from the DJI S900 multi-rotor (hexacopter) to the DJI Spark (quadcopter)). Despite the capability of the larger platform to lift and carry larger high-resolution cameras (>20 MP and higher DSLR), large drone platforms, such as the S900, were ruled out from further testing for a number of reasons, including being too costly; requiring specialist camera gimbals; too large; lacking autonomous flight capability; requiring special flight operator qualifications; needing high maintenance; and having a potential lack of manoeuvrability and safe operation in confined spaces, particularly in orchards and vineyards, especially when attempting to navigate from one fruit-fly trap to another.

Several small UAV platforms are also available which are low cost, provide autonomous flight capabilities, are RTF, and affordable. Their main limitation – in most cases – is the lack of a high-resolution on-board camera. The smaller platforms considered were the DJI Spark, DJI Phantom 2/3/4, DJI Mavic/Mavic Pro 2 (and Zoom), and DJI Inspire 1 (all quadcopters); the Parrot Bebop 2 (quadcopte4); Yuneec Typhoon H (hexacopter); 3DR Iris +, and 3DR Solo. A number of these were, however, ruled out on the following grounds: Bebop 2 (14 MP fisheye lens); Yuneec (12 MP camera and untested autonomous flight APP); and 3DR (unsupported technical and development in the long term). Trialling of DJI Phantom 4 platform with a PPK GPS unit on-board – to try to improve the flight and photographic positioning accuracy of the waypoints in autonomous flight mode for positioning the on-board camera to acquire the highest resolution photographs of the flytraps – was discontinued on the grounds of the cost of the PPK (GPS positioning corrections after the image capture) unit.

The smaller UAVs available are more modern, more compact, easier and more accurate to position (with the aid of prop guards and cages), affordable, easier to fly in confined areas, have autonomous flight capability, and increasingly good miniaturised camera technology. Low-cost RTF solutions are gradually becoming available now facilitating their operation by a wider range of people without the need for specialists. A good example, which is currently state of the art, is the Parrot Disco AG with a Parrot Sequoia sensor on-board and the Parrot Bluegrass multi-rotor, both designed with the end user (farmer/fruit grower) in mind. These are a finely tuned platform and sensor combination catering to the demand for simple operational solutions that are affordable and easy to use with a focus on the acquisition of information for decision-making.

SENSORS AND CAMERAS

In the past, many small UAV platforms have carried GoPro cameras or latterly MapIR cameras, which have camera resolutions of 11 or 12 MP and in many cases mounted on fixed camera supports with no gimbal.

More recently, UAVs come with their own built-in camera systems, and although many remain around the 12 MP spatial resolution, with some being 14 MP or better (DJI Mavic 2 Pro with its 20 MP Hasselblad camera), they are all mounted in a gimbal. The Hasselblad camera appears to be the highest resolution miniaturised camera currently available for *off-the-shelf* UAV platforms. In the past year, DJI have also added a 14 MP camera with a zoom capability on the Mavic 2. The zoom camera now provides up to 6X (software) zoom, with the potential to capture better images suitable for this project.

The miniaturisation of other sensors and cameras has yielded multispectral and hyperspectral digital imagery to aid in feature recognition and extraction. Whilst such cameras have potential, many are unfortunately lower resolution than the standard RGB cameras and are currently very costly. For example, the Parrot Sequoia and MicaSense RedEdge are around £3,500–£4,500 in the UK and are not suitable as the basis to develop a cost-effective solution for fruit growers. To this end, the project continued to make use of standard RGB cameras.

As many drones available at the outset of the project had only relatively modest resolution cameras, attention early on focused on the larger UAV platforms that had the capability to carry a larger camera payload such as DSLR cameras with the potential for higher resolution images (e.g. the use of a Panasonic GH4 24 MP camera). Whilst this would provide the capability to acquire the high-resolution images required – providing a higher chance of being able to detect the features of interest – there are a number of reasons why such a choice would not have been ideal in the longer term; for example, larger platforms and sensors are less compact, more expensive, and unlikely to offer an affordable solution for the commercial fruit grower; a larger drone is also less likely to be as manoeuverable in practice, more difficult to fly in confined spaces, and in many cases will not offer autonomous flight capability; larger platforms also have larger rotors and would therefore be more difficult to position accurately; and many of the larger drones are now also relatively old technology and in the longer term will soon be out of date, and unsuitable as a basis for a long-term solution.

Compact multi-rotor quadcopters were therefore deemed to be the most practical platforms for this project and any future developments.

SOFTWARE

State-of-the-art flight planning software also now supports opportunities for autonomous flights and complicated flight paths (e.g. Pix4D, DJIGo4, DroneDeploy, Litchi). These provide complete control over the flight planning, autonomous flights, flight modes, camera settings, and gimbal position. The apps also provide control for the drone height and positioning.

PLATFORM SELECTION

Based on the above observations and considerations, several different UAV platforms were examined for further use. These were the DJI S900, the Inspire 1, the Parrot Bebop 2, DJI Spark, and the DJI Mavic 1, Mavic 2 Pro, and Mavic Zoom. Initially the S900 was considered as a possibility but ruled out as being too large. The Inspire 1 has potential, but the platform and camera are now quite dated. The Parrot Bebop 2 is small, with a built-in forward-facing camera, but this is of limited use as it has a fisheye lens. The DJI Spark has potential, but the camera is not suited to the task, and there is only an untested autonomous flight app available. Of the Mavic platforms and sensors, the Mavic 2 Pro and Mavic 2 Zoom would seem to have the most potential for the tasks. They are compact, have the best drone cameras available, tried and tested autonomous flight software, and good GPS positioning.

FLIGHT TESTS

Early flight tests undertaken with the UAV platforms and sensors involved manually controlled flights of a number of both larger and smaller examples primarily with the aim of testing their ability to fly the data acquisition task, namely flying to a location with a sticky fruit-fly trap, positioning of the camera, and capturing the image.

Trials were made to establish the ease with which it is possible to achieve this – using off-the-shelf commercial UAV or drone technology to fly an area autonomously using known waypoints in a pre-planned flight plan. This was conducted with a standard DJI Inspire 1 UAV platform carrying the standard RGB camera and using the DJI Go app and Pix4D app to fly a circuit of mock-ups

FIGURE 9.4   Experimental waypoint trials with the DJI Inspire 1.

of some coloured flytraps (A4 laminated coloured sheets (based on flytrap research using colour codes), with simulated fly shapes and patterns) (Figure 9.4).

Knowing the GPS location of waypoints (the locations of the flytraps), an autonomous flight planner can be programmed to allow the drone to (i) take off automatically; (ii) fly to the pre-defined location (e.g. the flytrap); (iii) descend in altitude to the height of the flytrap in the orchard/vineyard; (iv) take a single photograph (with the on-board camera (also automated)) of the flytrap (positioned sufficiently close to the trap to allow for a high-resolution photograph); (v) ascend to a nominal flight altitude; (vi) fly to the next flytrap; (vii) repeat steps (iii) to (vii); and finally (viii) return to the take-off point and land autonomously.

Latterly, flight tests included both manual and autonomous flights, UAV and camera positioning, camera resolution, and image acquisition with different cameras to see if it was possible to position different platform/sensor combinations close enough to the target flytraps to obtain an image of a sufficiently high resolution to resolve the fruit flies of interest.

The tests suggest that this is achievable, which makes the long-term task of monitoring possible with minimum input from the drone operator, something that is essential if the potential of monitoring for SWD is to be operationally practical for the fruit grower.

CAMERA RESOLUTION TESTS

Flight tests with the UAV platforms and available sensors were flown manually. These were conducted out in the field (at Knock Farm near Huntly, Moray (courtesy of Roger Polson), and on a farm near Findhorn, Moray (courtesy of Ed White), and in a large UAV flying cage at the Commercial UAV Show in London ExCel Centre. This involved flying the UAV as close as possible to the sticky flytrap mounted on a wooden post/wall. Flytraps had to be fixed in position to avoid movement generated by prop wash (each rotor on the UAV generates movement of air surrounding the prop) from the UAVs.

Tests revealed that most cameras could not resolve the fruit fly. The best result was provided by the 20 MP cameras on the DJI Mavic 2 Pro and the Zoom (Figure 9.5).

In addition, some hand-held tests were undertaken with cameras (hand-held DSLR camera and hand-held drone camera) in order to (i) provide images to test the feature recognition algorithms; and (ii) to test the drone cameras in lieu of suitable outside flying conditions, together with some flights of a small drone (DJI Spark and DJI Mavic) inside a building and a large flying cage. These were undertaken to provide supplementary data for the project (Figure 9.6).

Camera tests were undertaken inside a building by placing a sticky flytrap upright against a wall and then photographing the trap by holding the drone and camera as close to the flytrap as possible (approximately 4 inches) (both lit and unlit). Photographs were taken both with the 20 MP and 14 MP Zoom cameras to determine if the resolution was adequate for resolving the fruit fly.

FIGURE 9.5   Sticky fruit-fly trap (red) (Mavic Zoom camera) (hand-held).

FIGURE 9.6   Sticky fruit-fly trap (red) (Mavic 2 Pro camera).

The results of these flight tests revealed that in theory, these platforms and sensors can be used to acquire the required imagery providing that they can be flown close enough to the flytrap and positioned accurately relative to the feature of interest. The advantage of the zoom camera is that the drone does not need to fly as close to the target.

AUTONOMOUS UAV FLIGHT, PLATFORM, AND CAMERA POSITIONING

A key component of this research has been to provide an autonomous solution as the outcome of the research. To that end, a few additional flight tests were carried out to test the UAV platforms for autonomous flight and automated image data capture.

It should be noted that within the timeframe of this project work, the rapid evolution of drone and sensor technology has subsequently provided the opportunity to test a number of other more up-to-date solutions that would – if suitable – and bearing in mind the desire to provide a practical RTF solution – provide a smaller, lower-cost platform with the capability to fly closer to the fruit-fly trap with a smaller and lower resolution camera, autonomously, with better X, Y, and Z positioning.

Additional developments such as high-resolution FPV glasses (e.g. Epson Movario) and prop guards and ball flying cages also provide potential to position the platform and camera safely closer to the target.

FURTHER FLIGHT TESTS

Early tests with the Parrot Bebop 2 showed that the FreeFlight APP autonomous flight planning software can easily be configured to position the UAV and the camera. In practice, however, whilst the X and Y coordinates are quite accurate, repeat flights using the same flight plan could not be relied upon to position the UAV and camera reliably regarding the height (Z) of the UAV (Figure 9.7).

To this end, a few other small commercially available DJI platforms were selected to trial further. The DJI Mavic 1 (12 MP) and Mavic 2 Pro, carrying either a Hasselblad (20 MP) or Zoom (12 MP) camera, were selected to test the potential of currently available low-cost drones for the task. The advantages of these drones are that they are relatively affordable, RTF platforms, carrying state-of-the art digital cameras that can be flown autonomously and accurately positioned sufficiently close to the fruit-fly traps to acquire the imagery needed to extract information about the presence/absence of the fruit-fly (Figure 9.8).

Autonomous flight tests using the DJI Go app revealed that repeat flights using the same app flight plan retains the X, Y, and Z positions and hence the sensor. There is some variability from flight to flight but not as much with the Mavic UAV as with the Bebop 2.

LIMITATIONS AND CONSTRAINTS

Whilst in theory the entire task is practically achievable, that is autonomous flight and image capture, in practice, however, several factors can and do affect the success of the image data acquisition task currently.

These are identified as (i) *flying and environmental conditions* and (ii) *accurate platform and camera positioning*.

FIGURE 9.7   Testing the Bebop2 (Findhorn) (photos courtesy of author).

FIGURE 9.8   Testing the Mavic 2 Pro (courtesy of Roger Polson Knock Farm, Huntly).

FLYING AND ENVIRONMENTAL CONDITIONS

The ideal conditions for flying and image capture *in situ* are quite important to allow high-resolution imagery to be acquired in practice. Successful flights need to be conducted when there is little or no wind. All UAVs (like real aircraft) are subject to the effects of wind speed and direction and can easily be blown around, which is an important consideration when it comes to flying in a confined space, for example, orchard or vineyard, and in particular when trying to position the UAV and camera perpendicular to the flytrap so as to be able to capture the *ideal* image. Even with additional manual positioning of the drone and camera, it can be still be difficult under such conditions.

The main concern is of course colliding with the flytrap, but this could be assisted with the aid of either small prop guards for the UAV rotors or the use of a flying cage to surround the UAV. The use of an FPV screen and FPV glasses can also help to better judge the position of the UAV. Capturing a suitable image, for example, one that is perpendicular to the target and focused, is also difficult under the best of conditions, especially if the UAV platform must be positioned as closely as possible to the target. The use of a Zoom camera can help with this as it means the lower resolution camera can be flown further away from the target of interest. Aside from autonomous flight, one other option to facilitate image capture is to fly the drone autonomously from trap to trap but position the drone and camera manually at each location, capture the image, and then resume autonomous flight.

In addition, the lighting conditions of the target are also important, as the surface of the sticky flytraps under certain combinations of the *sensor-sun-target* geometry means that they are very reflective. To counter this, the image acquisition should coincide with times when the illumination geometry minimises reflectivity. This may mean also that the flytraps need to be carefully positioned throughout the fruit crop to ensure that the image acquired is not compromised. It also means that all fruit-fly traps must be positioned the same throughout the canopy, that is, facing the same direction to the camera, and images acquired within a short period of time and ideally under the same conditions. The flytraps must also be fixed in position to minimise movement, for example, in wind, or as a result of prop wash from the drone.

To summarise, ideally, the UAV flights should take place when the illumination is clear and bright; the altitude of the sun is high (to minimise shadows and reflections); there is little or no wind; the target (the flytrap) is fixed and preferably in a position within an unobstructed view; and the camera should be positioned perpendicular to the flytrap. In practice, however, some or all these conditions are unlikely to be met. If this is the case, it will be necessary to identify the importance of these factors and to identify the possible impact on the images captured.

PLATFORM AND CAMERA POSITIONING

Accurate positioning of the drone and camera is crucial to ensure that suitable images of the flytraps are captured. This requires accurate positioning in the X and Y positions and the Z (height). With a single GPS receiver on-board a UAV, accurate positioning on its own may be inadequate and inaccurate for the purpose, which could be made worse by other factors such as the wind strength and direction.

Several autonomous flight tests were therefore undertaken to test both the horizontal and the vertical positioning of the drone and camera at each location. This involved setting the waypoints for the drone flight path, by walk-flying the drone to each flytrap, and recording each waypoint in the APP. This can be done for all the waypoints in each location and saved as the autonomous flight path (Figure 9.9).

Initially the UAV was flown to each waypoint and positioned at a height above the flytrap to avoid the risk of collision during each test. Each flight test was then repeated several times using the same flight-path to determine the repeatability of each autonomous flight which would be needed in practice.

Once tested, the autonomous flight path was augmented with a height adjustment to position the drone camera to acquire the image at the level of the flytrap, aided by the addition of prop guards (or a UAV cage) to help position the drone safely without the risk of collision, and the use of an automated camera trigger function to capture an image of the flytrap at each waypoint location.

For some of the smaller and cheaper drones trialled (e.g. the Bebop 2) – in GPS mode – it was found to be quite stable – even in high wind – and held its X and Y horizontal positions well. By contrast, the Z position was found to be quite variable. In autonomous flight, it was not possible; as demonstrated by repeat flights using waypoints to ensure the same vertical position, even when carefully set in the autonomous flight APP. By contrast, the DJI Mavic 2 Pro performed much better, and repeat autonomous flights of the platform appear to be far less variable. Although further tests need to be carried out, it would suggest that repeat autonomous flights of the UAV might be acceptable in the horizontal positioning, whilst the vertical position may be out quite considerably leading to no photograph of the flytrap target being captured.

To try to ascertain the accuracy of the X and Y positioning of the DJI Mavic, some simple positioning experiments were undertaken. A diagonal cross (24 inches×24 inches) was sprayed onto level ground with biodegradable white marking paint. The Mavic drone was placed in the centre of

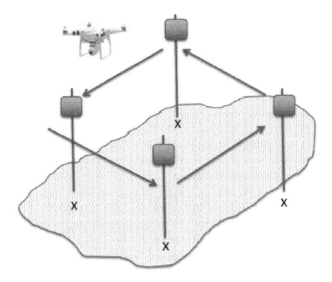

FIGURE 9.9   Autonomous flight path using app waypoints (courtesy of author).

the cross. Auto take-off was activated and the position on take-off and landing noted, measured, and recorded. Return to the take-off position used the *Return to Home* function. This test was repeated five times and tabulated visually. The DJI Mavic 2 Pro performed much better than a DJI Phantom 4 in this respect, both in terms of positioning (X, Y, and Z) and stability. At worst, the Mavic 2 Pro – even with a 3 MP wind – was only about 10 cm inches out from its take-off position (Figure 9.10). By contrast, the Phantom 4 was up to 2 m out.

Current UAV models now also carry vision sensors to aid in positioning (e.g. in relation to the *Return to Home* function); this may help to reduce the error in the X and Y positioning when flying from location to location or the waypoints for each flytrap location.

As the effects of, for example, wind were already considered, only optimum conditions would be used in practice for flying.

Early tests with UAV platforms and sensors involved manually controlled flights of a number of larger and smaller examples primarily with the aim of testing different cameras to see if it was possible to position different platform/sensor combinations close enough to the target flytraps to obtain an image of a sufficiently high resolution to resolve the fruit flies of interest. In addition, some hand-held camera tests were undertaken, together with some flights of the drones inside a building. Results of the initial trials suggested the need for a high-resolution camera (e.g. at least 20 MP) to provide suitable images to allow for detection.

To this end, several small commercially available DJI platforms were selected to trial. The DJI Mavic 1 (12 MP) and Mavic 2, carrying either a Hasselblad (20 MP) or Zoom (12 MP), were selected to test the potential of currently available low-cost drones. The advantages of these drones are that they are relatively affordable, RTF platforms, carrying state-of-the art digital cameras, that can be flown autonomously, and accurately positioned and sufficiently close to the fruit-fly traps to acquire the imagery needed to extract information about the presence/absence of the fruit fly.

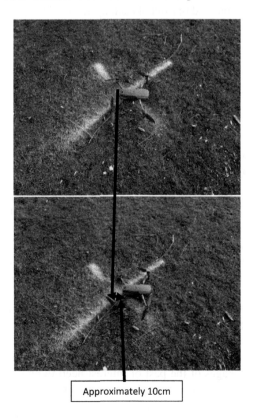

Approximately 10cm

FIGURE 9.10   An X and Y positioning test: top photo before take-off and bottom photo after landing.

Flight tests reveal that, in theory, these platforms and sensors can be used to acquire the required imagery (Figure 9.4). A few factors, however, affect the success of the image data acquisition. Ideally, the flights should take place where the illumination is clear and bright; the altitude of the sun is high (to minimise shadows and reflections); there is still wind conditions; the target (the flytrap) is fixed and preferably in a position within an unobstructed view; and the camera should be positioned perpendicular to the flytrap. In practice, however, some or all these conditions are unlikely to be met. If this is the case, it will be necessary to identify the importance of these factors and to tabulate the possible impact on data collection.

Furthermore, accurate positioning of the drone and camera is crucial to ensure that suitable images of the flytraps are captured. This requires accurate positioning in the X and Y dimensions (spatial coordinates) and the Z (height). With a single GPS receiver on-board the drone, accurate positioning on its own could be inadequate and inaccurate for the purpose, which could be made worse by other factors such as the wind strength and direction. However, current drones now also carry light sensors to aid in positioning, for example, in relation to the *Return to Home* function which may help.

To ascertain the accuracy of the X and Y positioning of the DJI Mavic, some positioning experiments were undertaken. A white cross was sprayed onto level ground with marking paint. The Mavic drone was placed in the centre of the cross. Auto take-off was activated and the position on take-off and landing noted and measured. Return to the take-off position used the *Return to Home* function. This test was repeated five times and tabulated visually. The Mavic 2 Pro performed much better than the Phantom 4 in this respect, both in terms of positioning and stability.

A few autonomous flights were then undertaken to test the vertical positioning of the drone and camera. This involved setting the waypoints for the drone, by walk-flying the drone to the flytrap and recording the waypoint. This can be done for each location and saved as the autonomous flight path. A white cross was spray painted in front of each flytrap to aid in positioning. Initially the drone was flown to each way point but at a height above the flytrap to avoid the risk of collision. Each flight test was repeated five times to determine the repeatability of each autonomous flight which would be needed in practice. Once tested, the autonomous flight path was augmented with a height adjustment to position the drone camera, the addition of prop guards (or a cage) to help position the drone safely without the risk of collision, and a camera trigger function to capture an image of the flytrap at each waypoint location.

TRANSFERABILITY AND ALTERNATIVE IMAGE-BASED MONITORING STRATEGIES

- **Hand-held solutions:** the image analysis technique with the emphasis of pest detection and monitoring is transferrable to any kind of image. For example, Nieuwenhuizen et al. (2018) used images of yellow sticky traps taken both with a hand-held smartphone and a Scoutbox (AgroCares, https://www.agrocares.com/en/products/scoutbox/) to train and validate a deep neural network for the detection of three different insects. They achieved a weighted classification accuracy of 87.4% for whitefly, *Macrolophus*, and *Nesidiocoris*.
- **UAVs:** the concept of UAV-based monitoring of sticky traps for the detection of *D. suzukii* flies as we described in this chapter can relatively easily be transferred to monitoring other pests as well. From the object detection perspective, in case detection of another type of pest needs to be implemented, the greatest efforts will be in the collection of a large database of annotated images of the new pest. To achieve a robust detection algorithm, at least several thousands of images of the pest of interest need to be collected to properly train a model. In addition, the performance of different detection algorithms for the specific pest needs to be tested. We have demonstrated the feasibility of detecting objects as small as *D. suzukii* flies (2–4 mm) in imagery collected by a UAV. For larger insects, detection

might even be easier, since it will be a lot less difficult to capture high-quality images of them, and likely, it is also easier for deep learning models to learn specific features to recognise larger sized pest species. Even though collecting image data is a relatively simple task, it still requires a large amount of manual labour to take photographs in different locations or at multiple times. By using an autonomous UAV that carries a camera to capture images of traps that are placed at strategic locations, this manual labour would be significantly reduced. The concept of monitoring animals with UAVs in combination with object detection also received attention in scientific literature recently. For example, Kellenberger et al. (2018) demonstrated the detection of animals in wildlife reserves with UAV imagery and CNNs. Rivas et al. (2018) trained a CNN for the detection of cattle in images collected with a UAV. Similar to the case of detecting *D. suzukii* flies, Bouroubi et al. (2018) demonstrated the potential of UAVs for the detection of potato beetles using images collected with a DJI Phantom 3 in combination with CaffeNet (Jia et al., 2014), which is also part of the CNN family.

- **Ground-based vehicles:** another possible solution would be to use cameras mounted on an autonomous ground-based platform that would allow a camera on an extendable arm to be positioned in front of the sticky flytrap to collect photographs. Autonomous ground-based vehicles have already been used to collect canopy photographs in a vineyard, for example, Green (2013) and other researchers.

CONCLUSION AND PERSPECTIVE

For this technology to be successfully adopted, it needs to be affordable, easy to operate by a farmer, and the image processing required to be offered as a cloud-based information service, whereby the farmer uses the UAV to capture the required imagery, to upload the data/imagery to a cloud-based service, and to receive back information in real time that allows for planning and decision-making, thereby providing a low-cost early warning system for fruit-fly infestation.

UAV platforms and miniaturised sensor technologies are now increasingly affordable opportunities to develop such a solution with better on-board GPS and advanced autonomous flight capability that can perform the required tasks to facilitate the capture of remotely sensed data at the scale and resolution needed.

In theory this is possible, aided by improved camera technology, longer battery life, cloud-based processing, autonomous flight apps, improved on-board sensors, object avoidance sensors, and built-in safety functions that provide for safe and easy operation. GPS positioning is also improving, and as can be seen from the tests described above, the X, Y, and Z positioning of the low-cost UAVs is rapidly improving.

Whilst there are some limitations to the current technology available, UAV platforms and sensors are evolving very quickly, and it is likely that the ability to position the platform more accurately will improve considerably in the future. Already the autonomous flight apps are providing more control over the drone and sensors, and with advances in image processing, it should be possible to fine-tune the image acquisition.

Results of the initial flight trials and image acquisition tests revealed that:

a. All UAVs have the capability to be flown manually, and in some cases autonomously, and in theory can complete an image acquisition task, albeit with a number of limitations which include size, cost, practicality, ease of flying, maintenance, camera resolution, autonomous flight capability, and positioning accuracy

b. Although the larger drones can carry a large, high-resolution camera, best suited to the image acquisition required, they are not a practical solution for several reasons, including size, cost, practicality, maintenance, and ease of flying

c. There is a definite need for the highest resolution camera to provide suitable images to allow for detection and identification of SWD

d. Some smaller platforms now offer a practical combination of the requirements to complete the task. Whilst not yet operationally perfect, the small platform and miniaturised camera combination of the Mavic 2 Pro and/or the Mavic 2 Zoom can easily be flown autonomously and positioned to automatically capture photographs of the flytraps. Future developments in drone technology, through improved GPS and autonomous flight APPs, together with higher resolution cameras, are rapidly improving this capability.

The limitations of the current drones can, to a large extent, be helped by flights being timetabled for optimum lighting and flying conditions to maximise the acquisition of the best possible image that has the best contrast and resolution, as well as using aids to flying, and manual UAV/sensor positioning.

At the same time, apps are being improved, and additional hardware makes it possible to fly the drones more safely which allows the camera to be positioned closer to the target, for example, in windy conditions. This can be aided with the use of FPV glasses and camera zoom functions, which in turn allow for more accurate positioning.

The problem that remains is that of accurate height positioning of the camera in relation to the target, namely the fruit-fly trap relying on the single GPS unit of the drone for navigation and positioning.

The UAV, sensor, and technology currently available can indeed be used to acquire images autonomously. However, there are some limitations at present relating to the drone positioning capability and the on-board camera technology. Further work is required to address this. One possible solution now being investigated may well make use of the UAV on-board vision sensors to land the drone on a small shelf next to the sticky flytrap using image/pattern recognition on the landing surface (Figure 9.11); a precision landing system.

State-of-the-art examples of the capability of both fixed-wing and multi-rotor examples for PA and PV can already be seen in the form of two products from Parrot, namely the Disco Pro AG (fixed-wing) (https://www.parrot.com/business-solutions-us/parrot-professional/parrot-disco-pro-ag) and the Bluegrass (multi-rotor) (https://www.parrot.com/business-solutions-us/parrot-professional/parrot-bluegrass), both of which are relatively recent products using the Parrot Sequoia multispectral camera, a platform that can be flown by a farmer, the data collected processed in

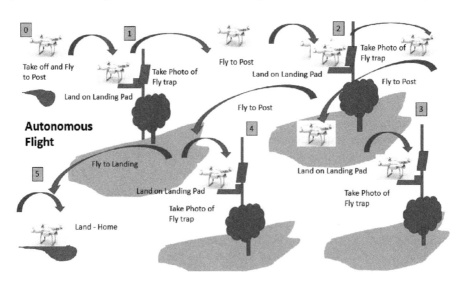

FIGURE 9.11 A precision drone landing system.

the cloud, and the information of interest being returned in the form of a product for planning and decision-making. This type of approach is already finding favour, and more recent developments such as crop-scouting are beginning to emerge, for example, DroneAg's Skippy Scout (https.// droneag.farm/skippy-scout/). Furthermore, the potential of real-time monitoring services based on remote sensing, for example, satellite imagery and aerial photography, is already being developed, for example, the work of D-CAT to provide online real-time information to farmers (https://www.thefusionplatform.com/).

The UAV solution proposed in this chapter is a novel application of drone imaging technology, which together with the use of artificial intelligence to extract information for real-time decision-making offers another example of the potential role of drones in Precision Horticulture.

ACKNOWLEDGEMENTS

This work was supported by ERA NET Coordinated Integrated Pest Management in Europe, C-IPM, project "Automated Airborne Pest Management AAPM of *Drosophila suzukii* in Crops and Natural Habitats" and the Swiss Federal Office of Agriculture grant: 627000782. In the UK, the project was supported by DEFRA. The contribution of Wageningen University & Research was supported by the Netherlands Organisation for Scientific Research (NWO) (project number ALW.FACCE.7). We would also like to thank the farmers for providing access to their fields for the experiments and students for collecting data in the field.

REFERENCES

Agarwal, S., Terrail, J.O.D., and Jurie, F. (2018). Recent advances in object detection in the age of deep convolutional neural networks. *CoRR* abs/1809.03193.

Asplen, M.K., Anfora, G., Biondi, A., Choi, D.-S., Chu, D., Daane, K.M., et al. (2015). Invasion biology of spotted wing Drosophila (*Drosophila suzukii*): A global perspective and future priorities. *Journal of Pest Science* 88(3), 469–494. doi:10.1007/s10340-015-0681-z.

Bächli, G. (2016). TaxoDros: The database on taxonomy of Drosophilidae, Database, compiled by Gerhard Bächli, and hosted at https://www.taxodros.uzh.ch/.

Basoalto, E., Hilton, R., and Knight, A. (2013). Factors affecting the efficacy of a vinegar trap for *Drosophila suzikii* (Diptera; Drosophilidae). *Journal of Applied Entomology* 137(8), 561–570. doi:10.1111/jen.12053.

Bouroubi, Y., Bugnet, P., Nguyen-Xuan, T., Gosselin, C., Bélec, C., Longchamps, L., et al. (2018). Pest detection on UAV imagery using a deep convolutional neural network, In: *14th International Conference on Precision Agriculture*, Montreal, Canada. unpaginated, online.

Briscoe, A.D., and Chittka, L. (2001). The evolution of color vision in insects. *Annual Review of Entomology* 46(1), 471–510.

Burrack, H.J., Asplen, M., Bahder, L., Collins, J., Drummond, F.A., Guédot, C., et al. (2015). Multistate comparison of attractants for monitoring *Drosophila suzukii* (Diptera: Drosophilidae) in blueberries and caneberries. *Environmental Entomology* 44(3), 704–712.

Cahenzli, F., Buhlmann, I., Daniel, C., and Fahrentrapp, J. (2018). The distance between forests and crops affects the abundance of *Drosophila suzukii* during fruit ripening, but not during harvest. *Environmental Entomology*. doi: 10.1093/ee/nvy116.

Cini, A., Anfora, G., Escudero-Colomar, L.A., Grassi, A., Santosuosso, U., Seljak, G., et al. (2014). Tracking the invasion of the alien fruit pest *Drosophila suzukii* in Europe. *Journal of Pest Science* 87(4), 559–566. doi: 10.1007/s10340-014-0617-rz.

Cini, A., Ioriatti, C., and Anfora, G. (2012). A review of the invasion of *Drosophila suzukii* in Europe and a draft research agenda for integrated pest management. *Bulletin of Insectology* 65(1), 149–160.

Cortes, C., and Vapnik, V. (1995). Support-vector networks. *Machine Learning* 20(3), 273–297. doi: 10.1023/A:1022627411411.

Dai, J., Li, Y., He, K., and Sun, J. (2016). R-FCN: Object detection via region-based fully convolutional networks, Advances in Neural Information Processing Systems, 30th Conference on Neural Information Processing Systems (NIPS 2016), Barcelona, Spain, 5 December 2016 through 10 December 2016, 379–387.

Daniel, C., Mathis, S., and Feichtinger, G. (2014). A new visual trap for *Rhagoletis cerasi* (L.) (Diptera:Tephritidae). *Insects* 5(3), 564–576.

Escudero-Colomar, L., Arnó, J., Gabarra, R., and Riudavets, J. (2016). Studies carried out on *Drosophila suzukii* in Catalonia (Spain) and its management in cherries, In: *9th International Conference on Integrated Fruit Production*, (Thessaloniki, Greece: IOBC-WPRS).

Gómez, C., and Green, D.R. (2015). *Small-Scale Airborne Platforms for Oil and Gas Pipeline Monitoring and Mapping*. (Aberdeen: REDWING/AICSM –UCEMM - University of Aberdeen Report).

Goodfellow, I., Bengio, Y., and Courville, A. (2016). *Deep Learning*. (Cambridge, MA: MIT Press).

Green, D. R. (2013). Ground and Airborne Platforms for Data Collection in the UK Vineyard. Presentation to Geography of Wine – Old World. AAG Annual Meeting. Los Angeles. 12th April 2013.

Heisenberg, M., and Buchner, E. (1977). The role of Retinula cell types in visual behavior of *Drosophila melanogaster*. *Journal of Comparative Physiology* 117(2), 127–162.

Iglesias, L.E., Nyoike, T.W., and Liburd, O.E. (2014). Effect of trap design, bait type, and age on captures of *Drosophila suzukii* (Diptera: Drosophilidae) in berry crops. *Journal of Economic Entomology* 107(4), 1508–1518. doi: 10.1603/ec13538.

Jia, Y., Shelhamer, E., Donahue, J., Karayev, S., Long, J., Girshick, R., et al. (2014). Caffe: Convolutional architecture for fast feature embedding, MM 2014 – Proceedings of the 2014 ACM Conference on Multimedia, Orlando, 3 November 2014 through 7 November 2014,, 675–678.

Jüstrich, H. (2014). Kirschessigfliegen in Bündner Rebbergen. Schweiz. Z. Obst-Weinbau (18).

Keesey, I.W., Knaden, M., and Hansson, B.S. (2015). Olfactory specialization in *Drosophila suzukii* supports an ecological shift in host preference from rotten to fresh fruit. *Journal of Chemical Ecology* 41(2), 121–128. doi: 10.1007/s10886-015-0544-3.

Kellenberger, B., Marcos, D., and Tuia, D. (2018). Detecting mammals in UAV images: Best practices to address a substantially imbalanced dataset with deep learning. *Remote Sensing of Environment* 216, 139–153. doi: 10.1016/j.rse.2018.06.028.

Kenis, M., Tonina, L., Eschen, R., van der Sluis, B., Sancassani, M., Mori, N., et al. (2016). Non-crop plants used as hosts by *Drosophila suzukii* in Europe. *Journal of Pest Science* 89(3), 1–14. doi: 10.1007/s10340-016-0755-6.

Kirkpatrick, D., Gut, L., and Miller, J. (2018). Development of a novel dry, sticky trap design incorporating visual cues for *Drosophila suzukii* (Diptera: Drosophilidae). *Journal of Economic Entomology* 111(4), 1775–1779.

Krizhevsky, A., Sutskever, I., and Hinton, G.E. (2012). Imagenet classification with deep convolutional neural networks, *Advances in Neural Information Processing Systems*, Vol. 2, 26th Annual Conference on Neural Information Processing Systems 2012, NIPS 2012, Lake Tahoe, NV, 3 December 2012 through 6 December 2012, 1097–1105.

Landolt, P., Adams, T., and Rogg, H. (2012). Trapping spotted wing drosophila, *Drosophila suzukii* (Matsumura) (Diptera: Drosophilidae), with combinations of vinegar and wine, and acetic acid and ethanol. *Journal of Applied Entomology* 136(1–2), 148–154.

Lee, J.C., Burrack, H.J., Barrantes, L.D., Beers, E.H., Dreves, A.J., Hamby, K.A., et al. (2012). Evaluationof monitoring traps for *Drosophila suzukii* (Diptera: Drosophilidae) in North America. *Journal of Economic Entomology* 105(4), 1350–1357.

Lin, T.Y., Goyal, P., Girshick, R., He, K., and Dollar, P. (2017). Focal loss for dense object detection, Proceedings of the IEEE International Conference on Computer Vision, 22 December 2017, 16th IEEE International Conference on Computer Vision, ICCV 2017, Venice Convention Center, Venice, 22 October 2017 through 29 October 2017, Article number 8237586, 2999–3007.

Liu, W., Anguelov, D., Erhan, D., Szegedy, C., Reed, S., Fu, C.Y., et al. (2016). SSD: Single shot multibox detector, In: *Lecture Notes in Computer Science* (including subseries Lecture Notes in Artificial Intelligence and Lecture Notes in Bioinformatics), 9905 LNCS, 14th European Conference on Computer Vision, ECCV 2016, Amsterdam, 11 October 2016 through 14 October 2016, 21–37.

Nieuwenhuizen, A.T., Hemming, J., and Suh, H.K. (2018). Detection and classification of insects on stick-traps in a tomato crop using Faster R-CNN, The Netherlands Conference on Computer Vision, Eindhoven, 26 September 2018 through 27 September 2018, Contribution in proceedings. https://library.wur.nl/WebQuery/wurpubs/542509.

Ramasamy, S., Ometto, L., Crava, C.M., Revadi, S., Kaur, R., Horner, D.S., et al. (2016). The evolution of olfactory gene families in Drosophila and the genomic basis of chemical-ecological adaptation in *Drosophila suzukii*. *Genome Biology and Evolution* 8(8), 2297–2311.

Redmon, J., Divvala, S., Girshick, R., and Farhadi, A. (2016). You only look once: Unified, real-time object detection, In: *Proceedings of the IEEE Computer Society Conference on Computer Vision and Pattern Recognition*, 29th IEEE Conference on Computer Vision and Pattern Recognition, CVPR 2016, Las Vegas Article number 7780460, 26 June 2016 through 1 July 2016, 779–788.

Ren, S., He, K., Girshick, R., and Sun, J. (2015). Faster R-CNN: Towards real-time object detection with region proposal networks, In: *Advances in Neural Information Processing Systems*, edited by C. Cortes and N.D. Lawrence and D.D. Lee and M. Sugiyama and R. Garnett 91–99.

Rice, K.B., Short, B.D., Jones, S.K., and Leskey, T.C. (2016). Behavioral responses of *Drosophila suzukii* (Diptera: Drosophilidae) to visual stimuli under laboratory, semifield, and field conditions. *Environmental Entomology* 45(6), 1480–1488. doi: 10.1093/ee/nvw123.

Rivas, A., Chamoso, P., González-Briones, A., and Corchado, J.M. (2018). Detection of Cattle Using Drones and Convolutional Neural Networks. *Sensors* (Basel, Switzerland) 18(7), 2048. doi: 10.3390/s18072048.

Russakovsky, O., Deng, J., Su, H., Krause, J., Satheesh, S., Ma, S., et al. (2015). Imagenet large scale visual recognition challenge. *International Journal of Computer Vision* 115(3), 211–252. doi: 10.1007/s11263-015-0816-y.

Salcedo, E., Huber, A., Henrich, S., Chadwell, L.V., Chou, W.-H., Paulsen, R., et al. (1999). Blue- and green-absorbing visual pigments of Drosophila: ectopic expression and physiological characterization of the R8 photoreceptor cell-specific Rh5 and Rh6 rhodopsins. *The Journal of Neuroscience* 19(24), 10716–10726.

Sun, Y., Liu, X., Yuan, M., Ren, L., Wang, J., and Chen, Z. (2018). Automatic in-trap pest detection using deep learning for pheromone-based *Dendroctonus valens* monitoring. *Biosystems Engineering* 176, 140–150. doi: 10.1016/j.biosystemseng.2018.10.012.

Uijlings, J.R.R., Van De Sande, K.E.A., Gevers, T., and Smeulders, A.W.M. (2013). Selective search for object recognition. *International Journal of Computer Vision* 104(2), 154–171. doi: 10.1007/s11263-013-0620-5.

Van Timmeren, S., and Isaacs, R. (2013). Control of spotted wing drosophila, *Drosophila suzukii*, by specific insecticides and by conventional and organic crop protection programs. *Crop Protection* 54, 126–133.

Walsh, D.B., Bolda, M.P., Goodhue, R.E., Dreves, A.J., Lee, J., Bruck, D.J., et al. (2011). *Drosophila suzukii* (Diptera: Drosophilidae): Invasive pest of ripening soft fruit expanding its geographic range and damage potential. *Journal of Integrated Pest Management* 2(1), G1–G7. doi: 10.1603/IPM10010.

Yamaguchi, S., Desplan, C., and Heisenberg, M. (2010). Contribution of photoreceptor subtypes tospectral wavelength preference in Drosophila. *Proceedings of the National Academy of Sciences* 107(12), 5634–5639.

Yamaguchi, S., Wolf, R., Desplan, C., and Heisenberg, M. (2008). Motion vision is independent of color in Drosophila. *Proceedings of the National Academy of Sciences of the United States of America* 105(12), 4910–4915. doi: 10.1073/pnas.0711484105.

Yamashita, R., Nishio, M., Do, R.K.G., and Togashi, K. (2018). Convolutional neural networks: An overview and application in radiology. *Insights into Imaging* 9(4), 611–629. doi: 10.1007/s13244-018-0639-9.

Yang, Y., Zhang, Y.-P., Qian, Y.-H., and Zeng, Q.-T. (2004). Phylogenetic relationships of *Drosophila melanogaster* species group deduced from spacer regions of histone gene H2A-H2B. *Molecular Phylogenetics and Evolution* 30(2), 336–343.

Yu, D., Zalom, F., and Hamby, K. (2013). Host status and fruit odor response of *Drosophila suzukii* (Diptera: Drosophilidae) to figs and mulberries. *Journal of Economic Entomology* 106(4), 1932–1937.

Zhong, Y.H., Gao, J.Y., Lei, Q.L., and Zhou, Y. (2018). A vision-based counting and recognition system for flying insects in intelligent agriculture. *Sensors* 18(5). doi: 10.3390/s18051489.

10 UAV Imagery to Monitor, Map, and Model a Vineyard Canopy to Aid in the Application of Precision Viticulture to Small-Area Vineyards

Luca Zanchetta
UCEMM – University of Aberdeen

David R. Green
UCEMM – University of Aberdeen

CONTENTS

INTRODUCTION

Unmanned Aerial Vehicles (UAVs) have become a popular low-cost and very convenient small-scale remote sensing platform for aerial monitoring, image acquisition, and more recently survey applications. In Precision Agriculture (PA), many studies have already proved the potential of small, low-cost, off-the-shelf aerial platforms for small-area studies providing valuable information on crops and crop status for farm managers. With developments in the technology, specifically platforms, sensors, and apps, there is now the capability to capture imagery and convert into information in near real time. An example of this is the Drone AG Skippy Scout app (https://droneag.farm/skippy-scout/) for use with UAV platforms such as the DJI Mavic Pro.

Applications of this technology have also been demonstrated to yield useful information in viti-culture for vineyard managers and small vineyards in both the UK and Switzerland (e.g. Green, 2012a, b; Green and Szymanowski, 2012a, b; Fahrentrapp et al., 2015). A novel application of drone technology for the monitoring of soft fruit crops is described in Chapter 9 in this volume.

As the drone and related technologies continue to evolve, there are many more applications now to exploit the potential of this technology in applied roles.

PRECISION VITICULTURE

Precision Viticulture (PV) is based on the existence of spatial and temporal variability within vine-yards and is made possible by using four important geospatial technologies: (i) Global Positioning Systems (GPS); (ii) remote sensing, digital image processing (DIP) and softcopy photogrammetry; (iii) Geographical Information Systems (GIS); and (iv) digital mapping.

PV can be implemented as a cyclical process comprising three steps:

1. **Observation and data collection:** this step is performed through either proximal sensing (physical contact) and/or remote sensing (at a distance)
2. **Data interpretation and evaluation:** the raw data acquired needs to be processed and analysed requiring the use of GIS software
3. **Implementation or modification of the vineyard management plans according to the information acquired:** the information acquired and analysed allows the vineyard man-ager to plan and make decisions.

SMALL AIRBORNE PLATFORMS AND SENSORS

Though not often taken seriously as a platform for capturing remotely sensed imagery for environ-mental applications in the past, large-scale model airborne platforms, e.g. fixed-wing (aircraft) and rotary wing (helicopters), have, on a number of occasions, been considered as a low-cost alternative to capturing imagery flown by larger-scale platforms (see http://www.spaceref.com/news/viewpr.html?pid=12357).

With recent technological developments, such platforms now provide a powerful basis to gather data and information that can be tied into a vineyard GIS database and in situ data collection on the ground. There are now a wide range of small-scale platforms available to anyone considering the acquisition of low-cost aerial remotely sensed imagery. These include kites, model aircraft, balloons, helicopters, UAVs, and microlights (Johnson et al., 2003a, b; Johnson and Herwitz, 2004; Berni et al., 2009).

Far from being 'toys', many of these airborne platforms can now be equipped with traditional but lightweight single-lens reflex (SLR) and digital cameras as well as video cameras for the collec-tion of panchromatic, true colour, and colour infrared photography and video footage. Additionally, multispectral imagery can easily be captured through repeated overflights of the vineyard with the aid of camera filters or more recently using on-board multispectral and hyperspectral camera systems.

Some of the so-called model platforms are also large enough to carry multiple payloads and can even make use of wireless transmission for real-time data capture (http://www.uav-applications.org/projects/vineyards_1.html). Johnson et al. (2003a, b), Johnson and Herwitz (2004), and Berni et al. (2009) have all used UAVs and lightweight sensors to collect multispectral and thermal imagery of vineyards.

Whilst the photography and imagery acquired can then be manually interpreted using the tech-niques of aerial photo interpretation, DIP software – much of it now low cost – can also be used to geo-correct and mosaic the photographic prints or images for onscreen interpretation and the mapping of thematic information for input into a GIS. DIP software has considerable potential for the small-scale vineyard, allowing the vineyard manager to overfly a vineyard on a regular basis for monitoring and mapping.

These smaller platforms offer a relatively low cost and flexible means for the frequent capture of airborne imagery to both monitor and map a small vineyard whilst the larger ones can be used for larger area coverage.

The lower cost of PCs and accompanying hardware, e.g. storage media, scanners, printers, and software, also provides an opportunity for the small vineyard to be able to capture, store, process, and map the data on a regular basis in-house without the need to hire-in consultants.

AN EXAMPLE

In the example reported here, two off-the-shelf Ready-to-Fly (RTF) UAVs, carrying two types of low-cost commercial miniaturised cameras, were used to acquire imagery of a small vineyard in Italy, at a number of different spectral wavelengths, to demonstrate their potential role in the provision of easy-to-acquire information of use to the vineyard manager.

This study focused on analysing intra-vineyard variability to provide information about differential management practices. The primary intention of this was to increase awareness about how such low-cost technologies can be of benefit to the vineyard manager and potentially lead to a change in some viticultural practices that may in the longer term reduce costs and/or improve the quality of the wine and increase profits.

STUDY AREA

This study is of a small area devoted to the production of wine grapes that are the basis of *Prosecco di Valdobbiadene*, a sparkling wine named after the area where the vines are grown, namely Prosecco di Conegliano e Valdobbiadene DOCG, which lies in the hilly northern part of Treviso province (north-eastern Italy).

Prosecco is the 44th Italian wine to obtain the Denomination of Origin from the European Union (EU) (Figures 10.1 and 10.2).

The study area lies in the village of Santo Stefano di Valdobbiadene (Province of Treviso) at a Latitude of 45.89° N and a Longitude of 12.02° E and covers around 1,950 m^2. The vineyard surveyed was 8 ha in total and scattered over several different locations with vine plot sizes ranging from 3 to 0.02 ha.

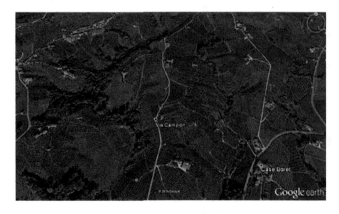

FIGURE 10.1 Map showing the location of the area studied.

FIGURE 10.2   The vineyard.

THE VINEYARD

The vineyard selected for study is characterised by a mixture of deep soils having a high clay content and some shallower soils comprised of sand and marlstone. The presence of different soils associated with areas of varying slope and aspect (gentler in Conegliano and steeper in Valdobbiadene) has led to the identification of several micro-zones within the vineyard.

Furthermore, the location near the Alps and the coast provides a stable, mild climate, with a yearly mean temperature of 12.3°C and a constant breeze that dries the grapes after the rain. Annual rainfall is around 1,250 mm with heavy showers in the summertime. The altitude ranges from 100 to 500 m above sea level (asl) with a good daily thermal range, and the hills, stretching from east to west, allow the southern facing slopes to be planted with vines. Management practices are traditional manual labour, and there is currently no digital record keeping.

STUDY EQUIPMENT

Details of the UAV and sensor equipment used for the study are shown in Table 10.1. Both the platforms and sensors selected are low-cost, off-the-shelf products.

The GoPro Hero 3+ camera was used to capture standard colour (RGB) imagery, whilst the MapIR NDVI (Normalised Difference Vegetation Index) camera – sensitive to both red (650 nm) and infrared (850 nm) wavelengths – was used to capture images containing information about the health and status of the vines.

A total of 1263 RGB images and 202 Red-NIR images were acquired from overflights of the two aerial platforms. As the DJI Phantom 2 does not store telemetry data on-board, a small IgotU

TABLE 10.1

UAV and Sensor Equipment Used in the Study

No.	UAV	Gimbal	Camera	Comments
1	DJI Phantom 2	Zenmuse H3-3D	GOPro Hero 3+ Silver Edition	IGotU geotagger was attached to the UAV
2	3D Robotics IRIS +	Tarot 2D	Mapir Survey 2 – NDVI model	NA

FIGURE 10.3   UAV Platforms and Sensors (a) DJI Phantom 2 and (b) 3D Robotics Iris+ (c) GoPro Hero 3 (d) MapIR.

GPS logger (http://www.i-gotu.com) (see Table 10.1) was attached to one of the landing legs on the Phantom 2 in order to record the spatial coordinates of the imagery. After each flight, the log file recorded was then downloaded and used to geo-tag the images (Figure 10.3 a-d).

In order to geo-reference the imagery collected from the UAVs, four Ground Control Point (GCP) markers were placed across the vineyard area studied. These markers consisted of three yellow, non-reflecting pieces of plastic placed on top of vine trellis posts with an additional GCP being the corner of a building (Figure 10.4).

Image pre-processing assigned a coordinate to each image, as neither the GoPro nor the MapIR cameras are GPS enabled, and therefore do not assign a spatial location to the photographs. The geo-tagged images were then sorted to allow for the choice of the best images for processing.

FIGURE 10.4   A plastic wrap used as a GCP.

The images were subsequently geo-coded using two different software packages:

1. The images collected with GOPRO camera were synchronised initially with TripPC software provided by IgotU GPS logger.
2. The images acquired with MapIR camera were geo-tagged using Mission Planner, free software provided with the 3DR Iris+ that uses a three-step process to pair the images with the flight GPS coordinates.

BUILDING A 3D MODEL OF A VINEYARD USING STRUCTURE FROM MOTION (SFM [SFM])

Structure from Motion (SfM [SFM]) is a softcopy photogrammetric technique in which a few RGB images taken from different positions are overlapped in order to re-create a 3D model of physical objects. The points created using this technique have 3D(imensional) information (X, Y, and Z coordinates) to create a 'point cloud' to which colour information is then added (Mathews and Jensen, 2013).

Geo-tagged images were input to the AgiSoft Photoscan softcopy photogrammetry software (www.agisoft.com) for processing into a point cloud. Subsequently, a Digital Surface Model (DSM) and a Digital Terrain Model (DTM) were rasterised from the dense point cloud.

The last step was to construct a high-resolution ortho-mosaic from the source photos and the DST. The DTM and DSM were also used to derive a DDM (Digital Difference Model) representing only the vineyard canopy using Equation 10.1 (Figures 10.5–10.7):

$$DDM = DSM - DTM \qquad (10.1)$$

INFORMATION LAYERS

TOPOGRAPHIC WETNESS INDEX (TWI)

A Topographic Wetness Index (TWI) map was also computed using the Topographic Index tools (http://www.arcgis.com/home/item.html?id=b13b3b40fa3c43d4a23a1a09c5fe96b9) in ESRI's ArcGIS software.

FIGURE 10.5 DTM of the area studied.

FIGURE 10.6 DSM of the area studied.

Orthoimage of the area studied

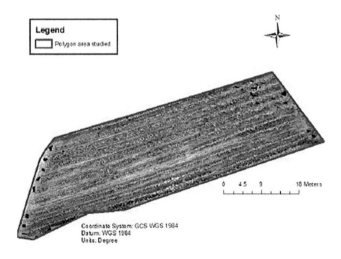

FIGURE 10.7 Ortho-image of the study area from RGB images.

(*Continued*)

FIGURE 10.7 (CONTINUED)   Ortho-image of the study area from RGB images.

This index was calculated according to Equation 10.2:

$$TWI = Ln(FA + 1)/\left(\tan\left(\left(\left(S\right)3.141593\right)/180\right)\right) \qquad (10.2)$$

where
   FA is the flow accumulation
   and
   S is the slope degree.

Knowledge of the TWI is very important in viticulture as it is highly correlated with soil depth, organic matter content, phosphorous content, and percentage of silt (Figure 10.8).

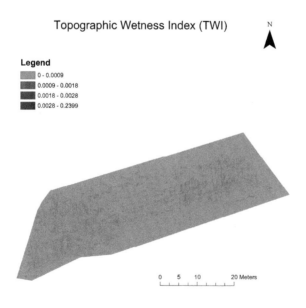

FIGURE 10.8   Map of the TWI.

Solar Map

The DTM also served as a base raster layer from which to calculate a solar map that expresses insolation at a certain location. The tool used for this computation was Area Solar Radiation (ESRI ArcGIS Spatial Analyst Extension: http://desktop.arcgis.com/en/arcmap/10.3/tools/spatial-analyst-toolbox/an-overview-of-the-solar-radiation-tools.htm) (Figure 10.9). Using data between April and October, the output was a group of raster maps where each image pixel contained a value representing the insolation (Wh/m^2).

Image Classification

The ortho-mosaic generated from the RGB images was used to separate the vineyard canopy from the ground using the Trimble eCognition developer software. In eCognition (www.ecognition.com), the RGB ortho-image was first segmented and pixels were grouped according to criteria of scale and colour (Figure 10.10).

Multi-resolution segmentation allows one to obtain homogeneous vectors from an image. The segmented layer was then classified using the nearest neighbour algorithm in order to separate the vineyard's canopy from the ground, and classification was based on colour and texture.

The nearest neighbour classification allows one to get samples from the feature classes and to define a "positive" class (in this case the vineyard's canopy) and a "negative" class (the ground).

The class "canopy" was then exported as in ArcGIS for visualisation.

SUMMARY AND CONCLUSIONS

The importance of spatial variability in vineyards is frequently often underestimated due to the lack of affordable tools to acquire and quantify the information as well as the expertise to process and interpret the information.

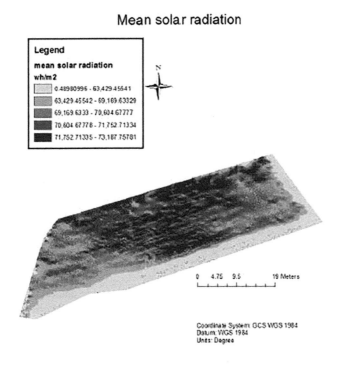

FIGURE 10.9 Solar map.

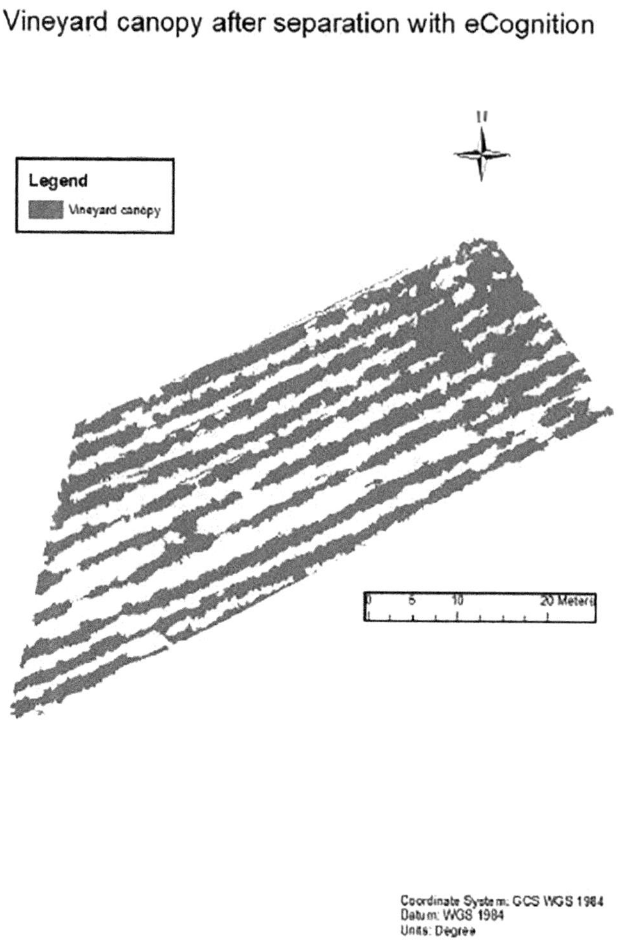

Vineyard canopy after separation with eCognition

FIGURE 10.10   Vine canopy isolation from the ortho-mosaic performed using the RGB image.

The information acquired from this project demonstrates how off-the-shelf hardware and software – in the form of UAVs and softcopy photogrammetry and image processing software – can successfully be used to create a 3D model of a vineyard and to add in useful layers of information to support decision-making by a vineyard manager.

In this example, low-cost equipment was used to demonstrate how affordable tools can successfully be used to gather data for small area vineyards with a limited budget and how it is possible to enhance knowledge about a complex environment such as a vineyard relatively easily.

Being able to acquire a DTM and derivatives such as the solar radiation layer has some important agronomic implications. For example, when replanting a vineyard, an agricultural implement such as a grade leveller, guided by a laser, could be used to modify the slope of the vineyard and in turn to enhance the solar radiation captured by the plants. This could help to optimise underperforming areas by improving the photosynthetic activity of the plants and possibly increase the sugar content of the grapes.

Whilst UAV and sensor hardware are now both more affordable and relatively easy to use, aside from some of the freely available apps and opensource software, softcopy photogrammetry and image processing software are both still quite expensive and also require specialist expertise to use, analyse, and interpret the results. This is gradually changing, however, as the tools and techniques now becoming available are increasingly popular, and companies such as Parrot are offering RTF hardware and sensor applications e.g. the Parrot Bluegrass multi-rotor designed to be flown by a vineyard manager with the imagery being processed into information using software in the Cloud.

Ideally RTK GPS-enabled UAVs e.g. using a Navspark (www.navspark.com.tw) or Piksi (www.swiftnav.com/piksi.html) unit would also help to improve the locational accuracy of the imagery acquired for processing and the resulting layers of information.

Whilst vineyard management has always relied upon the collection of data about the vine canopy, the terrain, and the environment, the tools and techniques developed in the context of PV have facilitated greater knowledge and understanding of the vineyard environment and its management for wine production.

Low-cost UAV-based remote sensing clearly has a lot of potential as the basis for collecting valuable data as and when it is needed, processing this into information, and increasingly as a cost-effective way to provide a means to 'hot-spot' areas that need management e.g. the application of irrigation, pesticides, and fertilisers to help reduce costs avoiding blanket applications. The use of such technologies is now just about becoming affordable for use in smaller vineyards.

As noted by Lamb et al. (2008), there are also many other factors limiting the wider use of the technologies including a reluctance to adopt technology, justifying the costs versus benefits, and the expertise perceived to be needed to make use of the many different tools and techniques now becoming available; some of which still need to be simplified to encourage their operational use and accompanied by opportunities for education and training.

But, as can now be seen from the quantity of information becoming available in the form of journal papers, webpages, and factsheets and books on PA and PV, and RTF products, it should not be long before more people become aware of the potential role and value of PV in vineyard monitoring and management.

REFERENCES

Berni, J.A.J., Zarco-Tejada, P. J., Suárez, L., González-Dugo, V., and Fereres, E., 2009. Remote sensing of vegetation from UAV platforms using lightweight multispectral and thermal imaging sensors. *Proceedings ISPRS Workshop. Hannover 2009. High-Resolution Earth Imaging for Geospatial Information Hannover*, Germany, June 2–5, 2009.

Fahrentrapp J., Hafele M., Schumacher P., Gómez C., and Green D., 2015. Identifying physiological differences in highly fragmented vineyards using NIR/RGB UAV photography. *Proceedings of the 19th International Meeting of Viticulture GiESCO Pech Rouge* – Montpellier May 31 – June 5, 2015, pp. 660–663.

Green, D.R., 2012a. Chapter 13: Geospatial tools and techniques for vineyard management in the 21st century. In, *Geography of Wine: Studies in Viticulture and Wine*. Dougherty, P.H. (Ed.). Springer, Berlin. p. 255.

Green, D.R., 2012b. Grape expectations—digital data in the vineyard. *GIS Professional*. (46):16–19.

Green, D.R., and Szymanowski, M., 2012a. Monitoring, mapping and modelling the vine and vineyard: Collecting, characterising and analysing spatio-temporal data in a small vineyard. *Proceedings of Ninth International Vitivinicultural Terroir Congress 2012 (ITC2012)*. 25–29 June 2012, France. p. 12.

Green, D.R., and Szymanowski, M., 2012b. Grape expectations: 'Terroir' explained—collecting, characterising and analysing Spatio-temporal data in a small vineyard. *GIS Professional*. (48):16–19.

Lamb, D.W., Frazier, P., and Adams, P., 2008. Improving pathways to adoption: Putting the right ps in precision agriculture. *Computers and Electronics in Agriculture*. 61:4–9.

Johnson, L.F., Roczen, D.E., Youkhana, S.K., Nemani, R.R., and Bosch, D.F. 2003a. Mapping vineyard leaf area with multispectral satellite imagery. *Computers and Electronics in Agriculture*. 38:33–44.

Johnson, L.F., Herwitz, S.R., Dunagan, S.E., Lobitz, B.M., Sullivan, D.V., and Slye. R.E., 2003b. Collection of ultra-high spatial and spectral resolution image data over California vineyards with a small UAV. *Proceedings of the 30th International Symposium on Remote Sensing of Environment*, Honolulu, HI.

Johnson, L.F., and Herwitz, S.R., 2004. UAVs for collection of farm imagery. *Online Document*. p. 3, http://www.uav-applications.org/gallery/doc/FieldTalk.pdf

Mathews, A.J., and Jensen, J.L.R., 2013. Visualizing and quantifying vineyard canopy LAI using an Unmanned Aerial Vehicle (UAV) collected high density structure from motion point cloud. *Remote Sensing*. 5(5), 2164–2183.

11 Forest Ecosystem Monitoring Using Unmanned Aerial Systems

Cristina Gómez
UCEMM – University of Aberdeen and
INIA-Forest Research Centre (CIFOR)

Tristan R.H. Goodbody
University of British Columbia

Nicholas C. Coops
University of British Columbia

Flor Álvarez-Taboada
Universidad de León

Enoc Sanz-Ablanedo
Universidad de León

CONTENTS

INTRODUCTION

Monitoring forest ecosystems is a complex task that requires data at multiple scales. UAV systems provide very fine scale, local information, acquired from both passive and active sensors, which can be used to complement the capacity of satellite and aerial remote sensing (RS), and to fill key information gaps (Klosterman and Richardson, 2017). UAV systems constitute a safe, flexible, and relatively low-cost technology for acquisition of frequent and very high spatial resolution data. UAV technologies can bridge the field and RS scales with above canopy perspective, providing data for calibration and validation of RS monitoring systems (Hall et al., 2016). UAVs may support forest inventory over small areas when used alone (Puliti et al., 2015) or complement inventory activities at broader scales (e.g., Puliti et al., 2018a). Current limitations to the use of UAVs are imposed by battery duration, payload weight, and local regulations (Manfreda et al., 2018), as well as massive data processing capability.

The scale of data acquisition with UAV ranges from the plot level (Goodbody et al., 2017b) to a few thousand hectares (Fernández-Guisuraga et al., 2018). Both fixed-wing and rotary-wing UAV platforms are well suited for forestry applications (Torresan et al., 2018). Fixed-wing UAVs are well positioned to cover large areas with a predefined flight at high altitude (e.g., 120 m), for example, to monitor broader stands canopy characteristics, whereas rotary-wing UAVs are preferred for manoeuvrability to fly lower (e.g., 20–50 m) and to acquire more detailed data, for example, for plot- or stand-level structural characterisation. To date, UAV technology has been used in forestry for health assessment (Dash et al., 2018), estimation of biomass (Kachamba et al., 2016), species identification and classification (Cao et al., 2018; Tuominen et al., 2018), characterisation of biodiversity (Zhang et al., 2016), assessment of invasive species (Mafanya et al., 2018; Baron et al., 2018), flowering assessment (Carl et al., 2018), forest recovery (Zahawi et al., 2015; Hird et al., 2017), regeneration (Feduck et al., 2018; Goodbody et al., 2017a), calibration of burn severity indices (Fraser et al., 2017), estimation of canopy fuel (Shin et al., 2018), detection of physiological stress (Smigaj et al., 2017), estimation of rut depth (Nevalainen P. et al., 2017) or characterisation of canopy gaps (Getzin et al., 2012, 2014). This diverse list reflects the versatility and utility of the UAV technology for forestry.

Typical sensors on-board UAVs for forestry applications are optical cameras including visible and infrared wavelengths, and increasingly hyperspectral and airborne laser scanning (ALS) sensors (Adão et al., 2017). Passive sensors have less power requirements than active sensors, enabling longer flights. Conventional photographic cameras are most popular for their low cost and for ease of data processing and interpretation. Combining two or more spectral bands into *indices* that correlate with phenological traits (Klosterman and Richardson, 2017) is a typical approach used for classification of species (Franklin and Ahmed, 2017) or for health assessment comparing data from repeated flights (Pádua et al., 2017; Dash et al., 2018). Image-matching algorithms (e.g., Structure from Motion (SfM [SFM]) multiview stereo (MVS) (Smith et al., 2016)), which require imagery acquired with high overlapping rate, facilitate producing dense point clouds to reconstruct forest three-dimensional structure and digital surface models (DSMs). UAV photos for forestry are typically acquired with a 70%–90% forward and lateral overlap, which is higher in abrupt terrains. These photogrammetric point clouds may be used for assessment of canopy height (Chen et al., 2017), estimation of biomass (Dandois and Ellis, 2013), and residual stem volume (Goodbody et al., 2017b), as well as for identification of individual trees (Nevalainen O. et al., 2017) and classification of tree species (Franklin et al., 2017). In comparison with ALS, UAV-based digital aerial photography (DAP) is inexpensive and the point cloud can easily match and surpass ALS densities. The larger point density does not necessarily yield greater vertical accuracy, as points do not penetrate vegetation and there is no ground-level reference (Guerra-Hernández et al., 2017). On the contrary, ALS penetrates the canopy (Wallace 2012a, b; Torresan et al., 2018), and in addition to direct measures of height for 3D structural reconstruction, it may provide understory and ground data to characterise the shrub layer and to retrieve terrain models. DAP technology has been shown valuable for the estimation of structural parameters with accuracies similar to those from ALS when an accurate DEM is available (Navarro et al., 2018). Precise comparisons between ALS and UAV photogrammetrically derived point clouds have shown similar capacity to detect individual-tree crowns (Thiel and Schmullius, 2017; Guerra-Hernández et al., 2018), with accuracy being dependent on the quality of the ground control (Sanz-Ablanedo et al., 2018) and the structural complexity of the forest (Jayathunga et al., 2018). The high cost and additional power required by UAV-ALS sensors – which in turn restricts the type of platform capable of carrying them – to date has limited the use of UAV-ALS for forestry applications. However, this is an area of rapid development and miniaturisation.

Processing UAV images remains challenging due to the large amount of data, the spatial location uncertainties, and the lack of calibration standards (Baron et al., 2018). Like the piloted counterparts, UAVs rely on an Inertial Measurement Unit (IMU) and Global Navigation Satellite Systems (GNSS) receivers for accurate location information. Geometric correction with ground control points (GCPs) is necessary for accurate identification of features and particularly for integration in workflows with multiple data sources. Although some UAVs are equipped with real-time kinematic (RTK) Global Positioning System (GPS) functionality, DSMs are more accurate and consistent when derived with support of GCPs (Forlani et al., 2018). When flying various sensors over the same area but at different times – typically needed with small UAVs – accurate registration of the imagery is required to facilitate analysis. Registration can pose difficulties because slight changes in illumination conditions may have significant impacts on the acquisition perspective and reflectance of target objects. Although in some circumstances dismissed (e.g., Pádua et al., 2017), radiometric correction of imagery is strongly recommended, particularly for multi-temporal comparison of datasets. In situ calibration of sensor bands with calibration reflectance panels is a frequent practice but leads to challenges related to banding noise and non-homogeneous radiometry. Empirical or sensor-based calibration approaches are applied for post-acquisition processing (Tu et al., 2018). Given the novelty of the UAV technology, there are still no established standards for UAV data acquisition and processing in forestry applications, but research is ongoing to identify the best parameterisation (Fraser and Congalton, 2018; Dandois et al., 2015). For example, flying altitude is a trade-off between spatial scale and coverage, and the most adequate is application and site dependent. While 100 m flying altitude was found optimal by Fraser and Congalton (2018) for a fixed-wing survey due to better image alignment, identification of tie points, and retrieval of planimetric models in complex forests, Perroy et al. (2017) found 30 m best for identification of sub-canopy individual invasive species in an open tropical rainforest environment. Forward overlap of images and ground sampling distance are the most relevant parameters for successful 3D structural reconstruction of forests (Frey et al., 2018). Time of day for data acquisition should allow controlling variation in illumination conditions and it is particularly relevant for applications based on spectral characterisations. Most applications aim for nadir image acquisitions, and the increasing automation of UAV systems only requires a choice of flight height and image overlap. Suboptimal UAV imagery can also be a source of information for certain applications (Puliti et al., 2017) although its use requires identification of limitations. Object-based image analysis frequently suits forestry applications, since the detailed UAV data enables identifying individual canopies (Franklin et al., 2017; Cao et al., 2018) and sub-canopy structures (Carr and Slyder, 2018) or other homogeneous spatial segments with special characteristics.

In the remaining chapters, we focus on four forestry applications in which UAV technologies contribute to get useful insights for sustainable management. An update of precision inventories of forest stands, assessing growth rates and volume increments near Williams Lake (British Columbia,

Canada); characterisation of coniferous regeneration in Quesnel (British Columbia, Canada); monitoring health of *Pinus radiata* in Fresnedo (León, Spain); and identification of invasive species in Viana do Castelo (Portugal).

PRECISION FOREST INVENTORY

UAV systems are pioneering a paradigm shift in how forest monitoring is conducted operationally, acquiring measurements for standard forest inventory attributes such as height or basal area but also novel attributes such as within-crown dimensions, individual-tree estimates, very high-resolution terrain surface data, and biodiversity metrics. These data are relevant to inform decision-making and meet product demand, to prevent supply shortages, and to capitalise on economic market conditions. Improved precision and data richness of operational inventories should translate into ecologically and economically sustainable forest management strategies. To date, multiple sensor types on-board UAVs have been explored as operational sources of data to inform forest management decisions (Table 11.1).

TABLE 11.1
Examples of Published Research Employing UAVs for Forest Management

Publication	Sensor Type	Forestry Application/Findings
Dunford et al. (2009)	Optical	Spatial quantification of riparian areas and determining vegetation composition
Lin et al. (2011)	ALS	Fine-scale mapping of tree height, pole detection, road extraction, and digital terrain model refinement
Wallace et al. (2012a)	ALS	Measurement of tree location, height, and crown width
Wallace et al. (2012b)	ALS	High-resolution forest change detection
Koh and Wich (2012)	Optical	Near real-time mapping of local land cover, monitoring of illegal forest activities, and surveying of large animal species
Dandois and Ellis (2013)	DAP	Observing phenological dynamics, spectral traits, monitoring changes, and measuring forested environment
Wallace et al. (2014a)	ALS	Tree detection and segmentation using ALS data
Wallace et al. (2014b)	ALS	Repeatable acquisition of forest inventory metrics
Zarco-Tejada et al. (2014)	DAP	Measuring canopy heights at operational scales
Tang and Shao (2015)	Review	Surveying, mapping canopy gaps, tracking wildfire, and supporting intensive forest management
Näsi et al. (2015)	Hyperspectral	Mapping bark beetle damage at the individual-tree level
Zahawi et al. (2015)	ALS	Monitoring tropical forest recovery
Puliti et al. (2015)	DAP	Estimating inventory attributes over a small forest area
Zhang et al. (2016)	DAP	Ultra-high-definition images of operational areas and a 3D point cloud in a single flight plan/long-term forest monitoring
Goodbody et al. (2017b)	DAP	Outline the location and estimate residual timber volume following selection harvesting operations
Michez et al. (2016)	Multi/hyperspectral	Classifying riparian species and health
Hassaan et al. (2016)	Optical	Counting urban trees
Nevalainen O. et al. (2017)	DAP/hyperspectral	Tree detection and classification

(Continued)

TABLE 11.1 (*Continued*)
Examples of Published Research Employing UAVs for Forest Management

Publication	Sensor Type	Forestry Application/Findings
Goodbody et al. (2017c)	Review/DAP	Updating ALS tree heights with UAS-DAP
Goodbody et al. (2017a)	DAP/optical	Assessing spatial, structural, and spectral characteristics of multi-age forest regeneration
Mohan et al. (2017)	DAP	Tree detection in open mixed-canopy forests
Guerra-Hernández et al. (2017)	DAP	Multi-temporal tree variable updates
Saarinen et al. (2018)	DAP/hyperspectral	Biodiversity assessment in boreal forests
Näsi et al. (2018)	Hyperspectral	Estimating bark beetle damage in urban forests
Goodbody et al. (2018)	DAP	Determining the influence of vegetation phenology on accuracy of DAP terrain models in low cover forests
Puliti et al. (2018a)	DAP	Using UAS imagery for hierarchical forest sampling
Puliti et al. (2018b)	Optical	Tree stump detection post-harvest

The sensor type employed in each study is indicated.

Goodbody et al. (2017b, c) examined the role UAV systems can play as a source of updating data for operational forest inventories, focusing on post-harvest inventories at the area-based and individual-tree scales. UAV-acquired data were used to update a component of a baseline ALS Enhanced Forest Inventory (EFI) by mapping harvest locations, estimating amounts harvested, and delineating the spatial distribution of timber volume remaining in the operational area. The effectiveness of UAV imagery and DAP point clouds for updating individual-tree level heights was also determined as a means to improve the frequency of forest inventories (Figure 11.1).

STUDY AREA

This research focused on a 31.6 ha compartment within the Alex Fraser Research Forest, south east of Williams Lake (British Columbia, Canada). The forest area is dominated by Douglas fir (*Pseudotsuga menziesii* var. *glauca*) and lodgepole pine (*Pinus contorta* var. *latifolia*), and it lies within a Cariboo Mule Deer Wintering Range (MDWR). As an MDWR area, harvesting restrictions include maintaining a minimum basal area, cutting cycles over 30 years, and leaving clumpy distributions of mature Douglas fir within the canopy openings. These managerial constraints require that operations be precise to balance the quality of habitat with timber values.

UAV IMAGERY AND ALS DATA

After group selection harvesting aligning with MDWR requirements, UAV imagery was acquired in March 2015 with a 16-megapixel RGB sensor mounted on an Aeryon SkyRanger®, a 2 kg vertical take-off and landing quadcopter. Flying at approximately 100 m above ground, images were acquired with ~85% forward and ~65% lateral overlap with average ground sample distance GSD = 0.03 m. Image acquisition took place over several hours in benign sun and wind conditions (Goodbody et al., 2017c). Pix4D SGM software was employed to derive DAP point clouds with horizontal and vertical accuracies of ±0.06 and ±0.09 m, respectively. The point cloud was normalised and co-registered to precise ALS data, allowing the creation of a canopy height model (CHM) that portrays detailed height of vegetation above ground (Figure 11.2). The use of ALS terrain information for normalisation of DAP point clouds is a standard approach, as ALS can provide

FIGURE 11.1 Schematic representation of the research goal: to evaluate the role UAV systems can play as a source of updating data for operational forest inventories, by assessing post-harvest structure and condition.

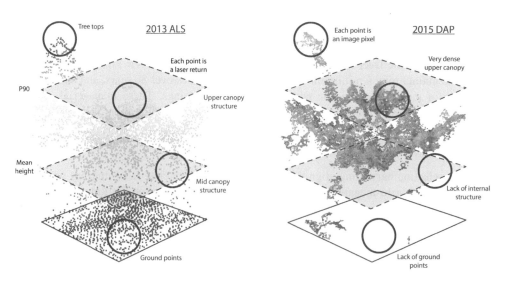

FIGURE 11.2 Visual comparison of the point clouds subsets: 2013 ALS (a) and 2015 DAP (b). Overlaid circles indicate areas of similarity and difference between point clouds. ALS ground points (dark blue) were used to generate the terrain model used for both ALS and DAP point cloud normalisation.

high-accuracy terrain information beneath forest canopies, whereas DAP is limited to upper canopy point cloud descriptions (see Figure 11.2). Normalisation of DAP data using co-existing ALS facilitates multi-temporal inventory analyses such as high-accuracy change detection.

Landscape-level ALS data was used for normalisation and co-registration of DAP point clouds. The ALS dataset was acquired with a Riegl VQ-580 laser scanner in August 2013 producing point clouds with 7–9 points $\times$ m^{-2}. These data were acquired $\pm30°$ from nadir and had an estimated vertical Root Mean Square Error (RMSE) of 0.048 m. ALS data, along with field measurements acquired in 20 systematically located sample plots, provided the means to generate a baseline EFI. Sample plots measured during ALS imagery acquisition (2013) were re-measured near the time of UAV flights (2015) to update plot-level inventories. All trees were measured providing reference data of height, diameter at breast height, and species.

AREA-BASED ESTIMATES

To produce area-based inventories, a statistical relationship between reference data (e.g., field measured height and diameter at breast height) and point cloud metrics (i.e., descriptive statistics of the point cloud) is necessary to estimate variables across data acquisition areas. Ideally, field measurements and flights would be conducted as close as possible to ensure the best matching between measurements. Mean height and the 90th percentile of height were generated from both ALS and DAP point clouds and clipped to the extent of co-located field-measured sample plots (Goodbody et al., 2017b). Direct relationships between each point cloud metric and the corresponding field measurements were established, that is, linear models for stem volume and basal area were generated independently for ALS and DAP point clouds.

Goodbody et al. (2017b) found these linear regression models accurate and effective for estimation of attributes. The DAP model for volume had an R^2 of 0.93 (Figure 11.3) and included three metrics describing height (90th percentile of height), canopy cover (cover >2 m), and cross-sectional density (mean height between 10 and 20 m). The application of linear models wall-to-wall allowed for the creation of maps that describe the spatial distribution of stem volume across the operation area (Figure 11.3, inset). Visual comparisons between the DAP CHM shown in Figure 11.4 and the wall-to-wall volume map in Figure 11.3 confirm that the volume models were accurate at delineating the location of harvest and representing residual volume.

INDIVIDUAL-TREE ESTIMATES

ALS and DAP CHMs detail the spatial distribution of standing timber before and after harvesting. Comparing both CHM, areas that had not been harvested between 2013 and 2015 were delineated (Goodbody et al., 2017c) and retained as virtual plots. A watershed detection algorithm was applied to the ALS and DAP CHMs for delineation of tree tops and crown extents in these virtual plots. Trees detected by the ALS were confirmed with trees identified in the DAP, and only trees matched in both datasets (171) were used for subsequent analysis of height and growth.

The maximum height (CHM height) and 95th percentile of height (P95) metrics corresponding to matching trees in both datasets were compared (Figure 11.5), indicating the growth of individual trees between 2013 and 2015. Height growth was validated using independent sample plots measured and averaged over 21 years. CHM heights and growth were slightly larger than P95 counterparts, regardless the source of data. Mean height growth estimated with CHM and P95 was 0.34 and 0.25 m, respectively, whereas field measured growths ranged between 0.25 and 0.30 m. Shorter trees had larger growth rates than taller trees, as expected in mature Douglas fir stands of the Williams Lake region.

Overall, these findings outline that DAP point clouds are capable of precisely and accurately updating a baseline ALS EFI at an individual-tree scale. Acquiring accurate estimates of height at individual-tree levels and monitoring growth through time provides a means of increasing

$$Volume_{DAP} = 1.971 \times P90 + 0.651 \times ln(Cover > 2\,m) + 0.114 \times MeanHt\,10{:}20$$

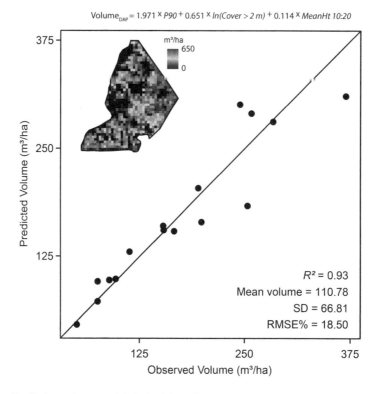

FIGURE 11.3 Predictive volume model derived from DAP. Block inlay shows spatial prediction of volume. (Adapted from Goodbody et al. 2017c.)

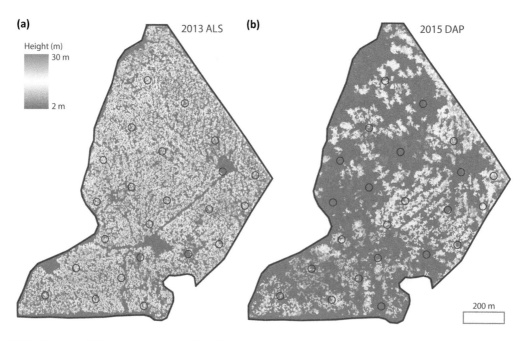

FIGURE 11.4 CHMs for pre-harvest ALS (a) and post-harvest DAP (b). Circles indicate locations of sample plots. (Adapted from Goodbody et al. 2017a.)

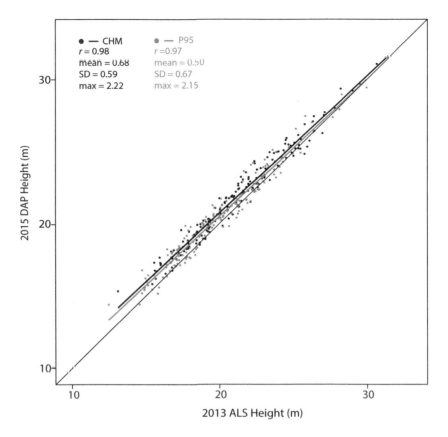

FIGURE 11.5 CHM (dark grey) and P95 (light grey) heights for 2013 ALS and 2015 DAP. (Adapted from Goodbody et al. 2017b.)

measurement samples while improving efficiency and cost-effectiveness. Precise data such as these can help to better inform estimates of growth and consequently better understand harvest rotation patterns. This methodology could be especially useful for naturally fast-growing species and productive plantations, where monitoring growth of individual trees is important for maximising harvest value.

Conclusion

This focus study demonstrates the capacity of UAV technology for inventory and operational management leading to reliable estimates, reducing economic risk in an increasingly volatile and competitive economic climate. In particular, it highlights the capacity of UAV-acquired DAP point clouds to update individual-tree heights. Given that UAVs have compatibility with a variety of sensor types, the ability to acquire multi-sensor data routinely greatly increases available managerial data to improve rationale for forest management decision-making. Establishing a precision inventory monitoring framework, especially in highly valuable stands, would help to consistently generate and update multi-source RS inventory data, refine estimates of interest, and ultimately monitor operational activities such as harvesting.

FOREST REGENERATION

Forest regeneration with little to no human intervention has low silvicultural costs but results in little control over the distribution, density, and content of the re-establishing forest (Duryea, 1987).

FIGURE 11.6 Schematic representation of the research goal: to determine the effectiveness of UAV imagery for assessment of regenerating forests.

Density and species composition are fundamental to effective planning of silviculture and play important roles in the future quality and quantity of timber and non-timber products, as well as in growth rates (Minore and Laacke, 1992). Artificial reforestation by planting nursery-grown trees or through dispersion of seeds requires silvicultural treatments like soil scarification to ensure forests regenerate emulating the pre-harvest state (Natural Resources Canada, 2016). To ensure the viability and success of forest regeneration, information on stocking density, health, and stand composition should be routinely acquired. These data are often manually collected through field sampling and reconnaissance, constituting costly efforts in vast areas like British Columbia (Canada) where surveys can exceed over 1 million hectares or ha annually (MFLNRO, 2016). More efficient and cost-effective ways to provide inventory data evaluating regeneration success are needed. To date, few works have explored the capacity of UAV to provide information of forest regeneration stages. Goodbody et al. (2018) determined the effectiveness of UAV imagery for detailing the distribution, height, and spectral information of regenerating forests from a Canadian perspective (Figure 11.6).

STUDY AREA

This research focused in an area located 150 km from Quesnel (British Columbia, Canada) where forestry is an economically and socially important industry. Intensive management takes place in this Sub-Boreal Spruce zone where lodgepole pine and hybrid spruce (*Picea engelmannii × glauca*)

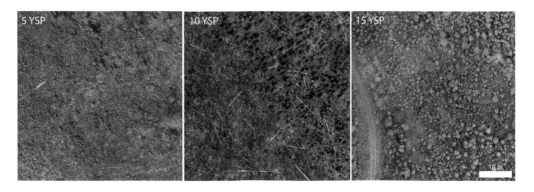

FIGURE 11.7   Sample UAV imagery from the 5 YSP, 10 YSP, and 15 YSP sites near Quesnel, BC, Canada.

are prevalent. Three stands similar in size, species composition, aspect, and slope, of approximately 5, 10, and 15 years since planting (YSP) after clear-cut were selected, covering a total area of 66.4 ha. Field measurements were acquired in 38 circular plots of $50\,m^2$, where coniferous and deciduous tree numbers were tallied per quadrant. The level of deciduous competition (high, moderate, low) and bare ground cover were visually estimated. One representative tree height within each plot was also recorded. Plot centre locations were averaged using a handheld Garmin GPSMAP 64s and marked with high-visibility red spray paint (Figure 11.7).

UAV IMAGERY AND PROCESSING

Following field measurements, wall-to-wall UAV RGB imagery was acquired with a 16-megapixel SUNEX MT9F002 sensor mounted on an Aeryon SkyRanger®, a vertical take-off and landing quadcopter of 2 kg. Flights at 100 m altitude were pre-determined in a grid pattern and took place in 26–30 June 2016 between 9.00 and 16.00 hours. Images acquired had 85% along-track and 70% across-track overlap and average GSD = 0.024 m. With Pix4D software, images were aligned using in-flight internal IMU and GNSS/GPS measurements, followed by matching of conjugate tie-point pixels in overlapping areas. Ortho-imagery and DAP point clouds were generated. The point clouds had average point densities of ~168 point $\times$ m^{-2} (Figure 11.8) with horizontal and vertical accuracies of ±0.06 and ±0.09 m, respectively. Point clouds were normalised using points previously classified as ground via LAStools software. The normalisation of DAP points was feasible given the low canopy density of these forests. Other studies have shown that as the level of ground occlusion from canopy cover increase, the reliability of DAP terrain information declines (Goodbody et al., 2018). CHMs of 0.05 m resolution were generated to provide high-resolution structural data.

CLASSIFICATION

To analyse the spatial distribution of forest regeneration in the three stands as well as their structure, and to infer temporal trends, an object-based image analysis (OBIA) approach was applied. Imagery were segmented based on spectral data, and polygons were attributed with spectral and structural data from the ortho-images and point cloud, respectively, and classified with random forest algorithms. Individual classifications at each stand avoided the impact of different illumination conditions. A suite of visible vegetation indices generated from the ortho-images were tested for differentiating regeneration cover, including the normalised green red vegetation index (NGRDI = $(G-R)/(G+R)$, where G and R are digital numbers for green and red channels, respectively). Training samples (an average of 1,578 from each stand) for the supervised classification were selected among segments located within the $50\,m^2$ sample plots. The classification scheme included conifer, deciduous, and bare ground classes (Figure 11.9).

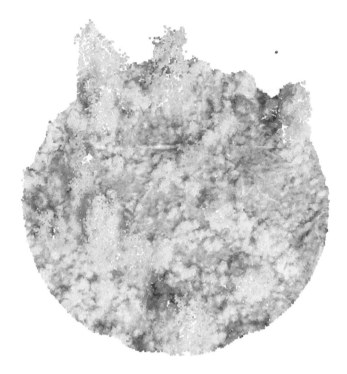

FIGURE 11.8 DAP visualisation of a 50 m² sample plot point cloud in the 15 YSP stand. The plot centre marked with red paint can be identified.

Classification accuracies for all stands were high (5 YSP: 95.6%, 10 YSP: 95.2%, 15 YSP: 93.2%) providing confidence in the spatial and structural analyses that followed (Figure 11.9). Considering the three stands as a chrono-sequence for analysis of structural trends, the proportion of coniferous coverage increased through time, with corresponding decrease in bare ground cover (Figure 11.9a). Deciduous cover also increased but at lower rate than conifer. The mean height of areas classified as conifer increased through time, with the 5 YSP site having the shortest trees on average (<0.5 m) and the 15 YSP stand having the tallest on average (2.2 m). CHM values in polygons classified as deciduous were highest in the 5 YSP site, indicating the dominance of deciduous species during the initial years of regeneration. Deciduous height was lower than conifer in the 10 YSP and 15 YSP sites (Figure 11.9b).

Conclusion

The potential of UAV-acquired imagery for quickly, accurately, and reliably providing detailed spatial, spectral, and structural information on forest regeneration has been shown (Goodbody et al., 2018). UAV data increase available information to determine the spatial distribution of regeneration wall-to-wall, while providing detailed structural information through time. The approach shown in this case study could be adopted to determine the effectiveness of silvicultural strategies and treatments such as deciduous brushing or replanting initiatives, and to ensure that legislative obligations are being upheld. This approach provides spectral and structural information that can be tracked through time, enabling multi-temporal inventories to improve our understanding of the early growth of trees and to improve growth projections, potentially reducing the number of necessary field samples. The provision of these data also facilitates improved tracking of regeneration success for diligence purposes. Given that forest regeneration obligations are auditable, the acquisition of imagery and structural data provides evidence to justify management decisions, reduces managerial risk, and improves quality and quantity of data for evidence-based decision-making.

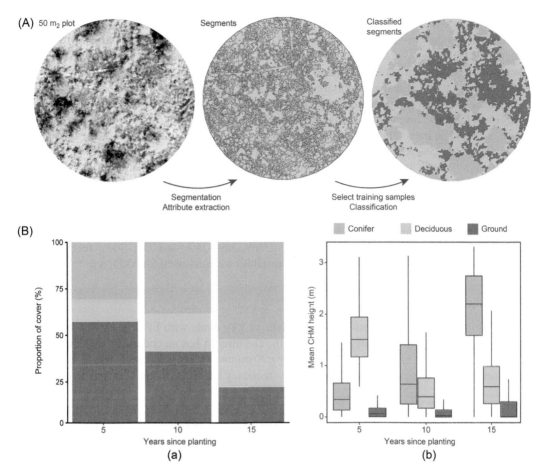

FIGURE 11.9 (A) Simplified object-based image analysis methodology. Spectral and structural data were initially segmented and attributes extracted. Sample segments were chosen for a supervised classification, and final classification was applied to all segments. (B) Proportion of spatial coverage (a). Mean height for conifer, broadleaved, and bare ground classes in the 5 YSP, 10 YSP, and 15 YSP stands (b).

FOREST HEALTH MONITORING

UAV systems can play an important role in multi-scale and multi-sensor systems to monitor forest health and the effects of pests on growth rate and tree mortality. Álvarez-Taboada et al. (2014) explored the capacity of a multi-sensor UAV to assess defoliation caused by *Lymantria dispar* L. (gypsy moth) in *P. radiata* D. Don (radiata pine) stands. UAV imagery was employed to map defoliation at both stand- and tree-level scales, and for comparison, Landsat OLI images were employed to map defoliation at the stand scale. The accuracy obtained with the UAV data was determined as well as the potential for operational use in decision-making, by establishing a relationship between the degree of defoliation and the growth rate and mortality of trees.

STUDY AREA AND FIELD DATA

The outbreak under study was located in a 150 ha area in Fresnedo (León, north-western Spain). Productive plantations of *P. radiata* are prevalent, and native broadleaved species stands are also present in this area. Although the gypsy moth is a particularly severe pest in other countries and over different tree species, this outbreak was unexpected and the first record of a large radiata pine area being attacked by this moth in Spain. Some trees were defoliated three times, causing a significant decrease in the growth rate and leading to a high mortality. The outbreak started after a thinning operation in a pure, even-aged 18-year-old stand of radiata pine with 800 trees/ha. During the 2 years of outbreak culmination (2012 and 2013), *ca.* 46 ha of radiata pine were severely defoliated, that is, with over 75% of leaf area removed (Castedo-Dorado et al., 2016).

In April 2014, before the season defoliation started, 21 rectangular field plots of 400 m² were established for forest inventory and to assess defoliation at tree and stand level. Each plot included between 26 and 35 trees. Three degrees of defoliation were classified at the stand scale in the field: non-defoliated (<25% leaf area loss); moderate defoliation (25%–75% leaf area loss); and severe defoliation (>75% leaf area loss). Six plots were assigned to the severe defoliation class, three to the intermediate defoliation, and twelve to the non-defoliated class. The position of each tree was measured with centimetre accuracy using a total station and a GPS receiver. Diameter at breast height, radial growth, total tree height, and degree of defoliation were measured for each tree at the time of the plot establishment and again in October 2014. Four classes of defoliation were established by visual inspection at tree scale: 1%–25%, 25%–50%, 50%–75%, and 75%–100% leaf area loss. Overall, the degree of defoliation and the forest variables were assessed for 650 trees.

UAV IMAGERY AND LANDSAT DATA

A small in-house-developed fixed-wing UAV plane of 2 kg made with EPO (expanded polyolefin) was flown over 1,000 ha in the area (Figure 11.10). The aircraft has an endurance of 40 minutes at 12 m × s⁻¹ cruise speed, and the actual flights were <20 minutes at altitudes over 500 m. The positioning system of the plane used a coded-base single-frequency L1, GPS U-Blox receiver. This receiver

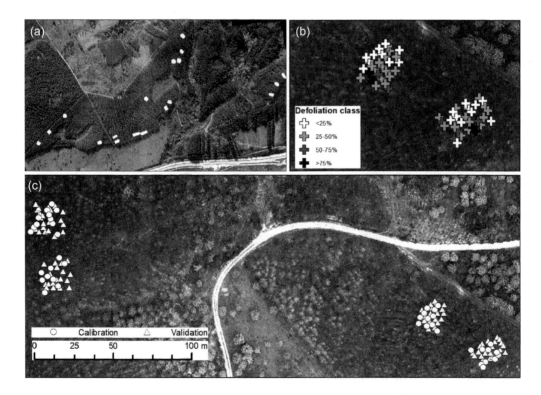

FIGURE 11.10 Design of study. (a) Location of the plots in the study area. (b) Sampled trees in two plots with different degrees of defoliation at the tree level. (c) Distribution of samples used for calibration and validation at the tree level in four plots.

was just used for aircraft navigation; therefore, its calculated position was not used during SfM (SFM) processing or georeferencing. The flight controller on the aircraft was an APM 2 autopilot, based on the open-source Ardupilot project. Ardupilot controller enables automatic mission flights based on waypoints of known latitude, longitude, and height. The firmware programmed into the flight navigates the UAV through all waypoints using information provided by the inertial system, precision barometer, electronic compass, and external GPS sensor. Two flights were planned with Mission Planner (open-source software) in east-west orientation (Figure 11.10), each one for carrying a single sensor (RGB or NIR). A Canon Powershot S100 12 MPix standard camera (4,000 × 3,000 pixels) modified with a low pass filter was employed to acquire infrared photography (ISO 160 F/2.8, T 1/640). The same camera with standard configuration acquired RGB photography (ISO 80 F/2.2 T 1/1000). The actual flights were carried out on 7 March 2014 at 12.30 (NIR) and on 8 April 2014 at 18.00 (RGB), over totally clear sky. Although the flights covered a larger area, the number of images used for this study was 89 RGB (GSD = 0.17 m) and 56 NIR images (GSD = 0.35 m) and had 80% forward- and 60% side-overlap rate. For georeferencing with Photoscan v.1, five GCPs measured on the ground with a Leica Viva GNSS receptor were employed, yielding residuals of 0.9 m (NIR imagery) and 0.3 m (RGB imagery). Mosaicking was based on the photo most perpendicular to the terrain. All images were acquired with the same exposition and were not radiometrically corrected.

Two Landsat OLI images (path/row: 203/030) (pre- and post-outbreak) acquired in two consecutive years (1 September 2013 and 13 April 2014) were downloaded from EarthExplorer (https://earthexplorer.usgs.gov/). Landsat images were radiometrically corrected and calibrated to at-sensor reflectance.

STAND-LEVEL DEFOLIATION

For identification of areas defoliated at the stand level, supervised classifications were applied both per-pixel and with OBIA approaches, and the results were more accurate using OBIA ($P < 0.05\%$). The RGB and NIR data gathered with the UAV system were segmented with eCognition Developer 8.9 using a range of scale parameters (150, 100, 50) and classified with the nearest neighbour algorithm. Independent calibration (ten plots per class) and validation (85 points) datasets were used. The most accurate results were obtained using RGB data (overall accuracy: 85.88%, KHAT: 0.77), while using RGB + NIR, the overall accuracy was 82.35% (KHAT: 0.71). More accurate results were obtained with the smallest scale parameter, pointing to a small size of defoliation patches. Errors were analysed via confusion matrix, finding lower omission errors for the extreme classes (<12%), whereas the moderate defoliation class incurred in high omission error (64%). The applicability of the approach is therefore discouraged when very fine classifications are needed (Table 11.2, Figure 11.11).

TABLE 11.2
Confusion Matrices of the OBIA Classification

UAV RGB	Validation					Landsat OLI	Validation				
Classification	1	2	3	Total	UA (%)	Classification	1	2	3	Total	UA (%)
No defoliation (1)	37	6	3	46	80.43	No defoliation (1)	35	0	0	35	100.00
Intermediate (2)	0	9	1	10	90.00	Moderate (2)	2	16	0	18	88.89
Total defoliation (3)	0	2	27	29	93.10	Total defoliation (3)	0	0	31	31	100.00
Total	37	17	31	85	-	Total	37	16	31	84	-
PA (%)	100.0	52.94	87.10		OA (%): 85.88	PA (%)	94.59	100.00	100.00		OA (%): 97.61

PA, producer's accuracy; UA, user's accuracy; OA, overall accuracy.

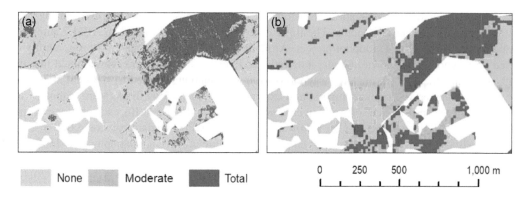

FIGURE 11.11   Classification results with UAV imagery (a) and Landsat OLI imagery (b).

For comparison with orbital imagery, the same three classes of defoliation were mapped using the multi-temporal Landsat 8 OLI at sensor reflectance data and an object-oriented supervised approach. Visible, NIR, and SWIR bands from the pre-outbreak Landsat OLI image and the ratio pre-/post-outbreak of band 5 (NIR) were input for classification, achieving an overall accuracy of 97.61% (KHAT: 0.96). In this case, the moderate defoliation class showed the lowest accuracy among the three (commission error: 11.11%, there was confusion with non-defoliated class), while the omission and commission errors were lower than 5% for the non-defoliated and the severe defoliation classes.

This work shows that it is possible to obtain accurate cartography of defoliation at stand level using UAV data; however, Landsat OLI data can be as convenient. The moderate defoliation class was the least common class but also the toughest to map. Moreover, using an object-oriented approach is preferred, and low-cost sensors and conventional RGB cameras are suitable to map these outbreaks.

TREE-LEVEL DEFOLIATION

To characterise defoliation at tree-level scale, the UAV imagery was segmented for identification of individual-tree crowns and these were classified into four, three, and two levels of defoliation for assessment of discriminative capacity. Trees measured on the ground were identified in the UAV imagery, designating 30% for training and 70% for validation of a nearest-neighbour supervised algorithm (Figure 11.10). The optimal combinations of segmentation parameters (including shape, scale, colour, and bands) yielded overall accuracies of 67.68% (KHAT: 0.53) for the four defoliation classes (1%–25%, 25%–50%, 50%–75%, and 75%–100% loss of leaf area), 71.72% (KHAT: 0.57) for three classes (1%–25%, 25%–75%, and 75%–100% loss of leaf area), and 92.93% (KHAT: 0.84) for two classes (1%–75% and 75%–100% loss of leaf area). Individual relationships were built between radial growth measured on the ground and level of defoliation, both estimated on the ground and estimated from UAV imagery.

Field data showed correlation between degree of defoliation at tree level and radial growth. In the 2014 growth season, the radial and height growth of individual trees was, on average, 45% and 40% lower, respectively, than that observed in non-defoliated trees (Lago-Parra et al., 2016). Lago-Parra et al. (2016) found that the radial growth of the non-defoliated trees or moderately defoliated (0%–25% and 25%–75% of leaf loss) was higher and significantly different than the radial growth of the severely defoliated trees (>75% of leaf loss). Similar results were obtained using the class assigned to each tree by the classification at tree level (two classes) and the radial growth measured in the field, indicating that mapped trees into these classes (0%–25% and 25%–75% of leaf loss) with UAV imagery can be used as a surrogate for mapping the impact in radial growth at tree scale.

From an operational perspective, considering three or four classes is not recommended, due to high confusion between the 1%–25% and 25%–50% leaf loss classes. Like the stand-level mapping, the intermediate defoliation classes (25%–50% and 50%–75%) were the most difficult to map. A smaller scale parameter (25 or 15) is recommended to map defoliation at the tree level. Using only two defoliation classes (loss of leaf area <75%, loss of leaf area >75%) is suitable from the management point of view and it reached an overall accuracy higher than 90%. In that case, including NIR information increases the classification accuracy. Regarding the estimation of defoliation at tree level, more accurate results were obtained when each stand was classified employing just its covering photos instead of the entire ortho-mosaic, pointing to the influence of illumination conditions in classification results (Figure 11.12).

CONCLUSION

Defoliation in *P. radiata* at the stand level can be successfully mapped using UAV imagery, being the high and low classes of defoliation severity more accurately identified than the moderate defoliation class, which was often confused and underestimated. At the stand level, using UAV and multi-date Landsat OLI imagery as input showed similar classification accuracies, although the single-date UAV imagery misclassified some areas without vegetation as severely defoliated, whereas OLI multi-date imagery successfully discriminated this class. UAV technology provides easiness for flight repetition and should be exploited for forest health monitoring. UAV imagery can be used as a surrogate for mapping the impact in radial growth at tree scale.

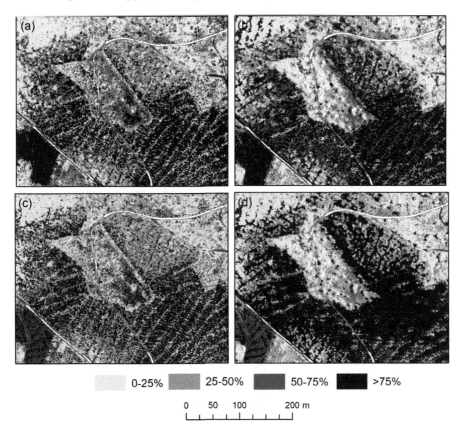

FIGURE 11.12 Visualisation of the classification results with different parameter combinations (segmentation scale and input bands). (a) Scale parameter 25, bands RGB. (b) Scale parameter 25, bands RGBNIR. (c) Scale parameter 10, bands RGB. (d) Scale parameter 10, bands RGBNIR.

INVASIVE SPECIES

Invasive species threaten ecosystem equilibrium and reduce diversity and provision of services (Morais et al., 2017). The expansion of invasive species should be controlled and their effects mitigated. Monitoring the geographical location, extension, and evolution is a first control step in which UAV systems may play a significant role. Álvarez-Taboada et al. (2017) explored the capacity of UAV to identify the extension of *Hakea sericea* in Northern Portugal, comparing its ability to identify and quantify the areas covered by *H. sericea* through an object-oriented approach with a similar approach employing WorldView-2 data.

STUDY AREA

H. sericea Schrad. & J.C.Wendl is an invasive prickly shrub from south-east Australia, characterised by its winter flowering. *H. sericea* (needlebush) is colonising extensive areas in the north of Portugal and threatening the ecological interest of Serra d'Arga, a region protected for its unique landscape by the Portuguese law. Needlebush was added to the European and Mediterranean Plant Protection Organization (EPPO) list of invasive alien plants developed in 2012. In Viana do Castelo, a district in north Portugal and where most land is forest, shrubs, or uncultivated fields, numerous areas have been colonised by needlebush, which flourishes earlier than most other local species.

UAV IMAGERY AND WORLDVIEW-2 DATA

UAV data consisted of RGB and NIR/R/G ortho-images covering *ca.* 160 ha (Figure 11.13). These images were obtained from data gathered on 1 August 2013 in two consecutive flights. An eBee® fixed-wing UAV was flown twice along the same flight lines carrying different cameras on each flight. RGB data was acquired with a Canon IXUS 220 HS camera and NIR/R/G data was captured by a Canon PowerShot ELPH 300HS (NIR-R'-G') camera. The images captured in each flight were ortho-rectified to the WGS84 UTM Zone 29 with 50 GCPs and the SenseFly software (accuracies of 0.95 m for XY and 0.89 m for Z), and mosaicked. Final imagery consisted of R-G-B and a NIR-R'-G' ortho-images with GSD = 0.07 m and 8 bits of radiometric resolution (Álvarez-Taboada et al., 2017).

MAPPING OF THE INVASIVE SPECIES

An OBIA was implemented in eCognition 8.9 (Benz et al., 2004). The three RGB image bands and the NIR band of the NIR-R'-G' image were input to the multiresolution segmentation algorithm, with scale = 50, shape = 0.2, and compactness = 0.5. A supervised classification with the nearest

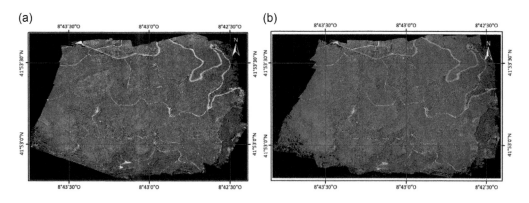

FIGURE 11.13 Ortho-images of the study area (datum: WGS84). (a) RGB ortho-image visualisation. (b) NIR-R'-G' ortho-image visualisation.

neighbour algorithm was then applied to the segmented imagery. Training and validation samples for seven land covers in the study area (*H. sericea*, woodlands, infrastructures, bare soil, shrubs, agricultural lands, and water) were visually identified in the images and verified in the field in August 2013. For training, 226 areas were located in the UAV data (54 polygons for the *H. sericea* class). A random stratified sampling design was followed to choose the validation points: 212 for the UAV ortho-imagery (50 validation points for the *H. sericea* class).

Two NDVIs were derived: (i) NDVI-1: using the red and NIR bands from the NIR-R′-G′ ortho-image; and (ii) NDVI-2: using the red band from the RGB ortho-image and the NIR band from the NIR-R′-G′ ortho-image. Regarding the textural features, we computed the local variance and the 3×3, 5×5, and 7×7 pixel window sizes (T3R, T3G, T3B, T5R, T5G, T5B, T7R, T7G, and T7B) for the RGB image, following Fernández-Luque et al. (2014). The final feature space from the UAV ortho-images for classification comprised 15 bands: 4 multispectral bands (R, G, B, and NIR), 2 vegetation indices, and 9 textural features. To test the suitability of spectral and textural data for classification of *H. sericea*, eight combinations of variables from the feature space were tested (Table 11.3) and their relative performance was evaluated as per Congalton and Green (2009).

The highest accuracies for the classification of *H. sericea* were obtained with the *Basic 1* set (R, G, B, and NIR), with overall accuracy 75.47% (KHAT: 0.68), (user's and producer's accuracy for *H. sericea* of 72.90% and 76.92%, respectively). Nevertheless, none of the classifications could be considered more accurate than the others, since the test to compare the accuracies for the *H. sericea* class (Congalton and Green, 2009) indicated that there were no significant differences among classifications ($P < 0.10$). The dataset *Texture 1* achieved the lowest omission error for the *H. sericea* class (19.23%), followed by *Basic 1*, *Index 1*, and *Index 2* (23.08%), while the highest value was obtained when using *Texture 5* (28.85%). This result showed that, when textural features with an unsuitable window size (in this case, ≥5 pixels) were included, the omission error for the target class increased. The spatial representation of the classification using *Texture 1* was not suitable, since large areas of *H. sericea* were misclassified as woodlands (Figure 11.14). The estimated area with needlebush presence was 63.48 ha (48.58% of the study area).

For comparison and to assess the accuracy of the UAV technology, orbital data from the very high spatial and radiometric resolution sensor WorldView-2 (16 bits) was employed. Imagery acquired on 15 February 2012 covered 2,550 ha. The multispectral image (GSD = 2 m) comprised eight bands (coastal blue (CB), blue (B), green (G), yellow (Y), red (R), red-edge (RE), NIR1, and NIR2) and the panchromatic image had GSD = 0.5 m. Imagery were acquired in ortho-ready standard 2A format and were geometrically corrected to WGS84 UTM Zone 29.

TABLE 11.3

Set of Spectral Bands, Indices, and Textures Which Define Each Feature Space (FS)

FS	UAV Ortho-Images
Basic 1	R + G + B + NIR
Index 1	Basic 1 + NDVI-1
Index 2	Basic 1 + NDVI-2
Texture 1	Basic 1 + (T3R + T3G + T3B) + (T5R + T5G + T5B) + (T7R + T7G + T7B)
Texture 3a	Index 1 + T3R + T3G + T3B
Texture 3b	Basic 1 + T3R + T3G + T3B
Texture 5	Index 1 + T5R + T5G + T5B
Texture 7	Index 1 + T7R + T7G + T7B

For Unmanned Aerial Vehicle (UAV) imagery, bands R, G, B belong to the RGB ortho-image and the NIR band to the NIR-R′-G′ ortho-image.

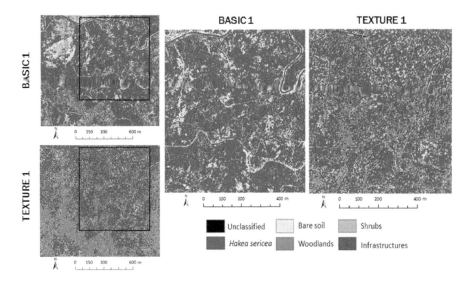

FIGURE 11.14 Visual comparison of classification results of UAV imagery with 'Basic 1' and 'Texture 1' variables sets.

A similar classification approach was applied to the WV2 imagery. The image was calibrated to at-sensor reflectance and pan-sharpened using the panchromatic band, to make the spatial resolution of the WV2 and the UAV ortho-images comparable. The spectral dimensionality of the WV2 data was also increased by calculating vegetation indices and textural features. In this case, the most suitable feature space to map *H. sericea* was the dataset including the CB, B, G, Y, R, RE, NIR1, and NIR2 bands with overall accuracy of 80.98% (KHAT: 0.77), and a user's and producer's accuracy for *H. sericea* of 94.81% and 93.59%, respectively (KHAT: 0.81). In this case, the higher spectral resolution of the UAV data did not compensate the more detailed spectral signature derived from the WV2 data. In addition, the WV2 image gathered at the time of needlebush flowering (February) made its detection and discrimination from still dormant vegetation easier. The results obtained with UAV and orbital technologies were consistent and rather similar in accuracy. The difference in the extension estimated by both sources of imagery was less than 15%.

CONCLUSION

This research over an area heavily infested by *H. sericea* tested the UAV and orbital technologies for operationally mapping and monitoring the presence of invasive species. The kind of information provided by both sources of data is useful not only for mapping the state of invasion but also for mitigation plans. Furthermore, the insights provided about the temporal and spatial patterns of invasion are also a valuable tool for risk assessment. Since the detection and control phases of invasive species have different spatial and temporal requirements, UAV and WV2 data can complement each other at different spatio-temporal scales of observation. WV2 is adequate for detection and identification at regional scale, and UAV is better suited for controlling the effectiveness of mitigation actions at smaller scale, due to its finer spatial resolution.

FINAL REMARKS AND CONCLUSION

UAVs are increasing in utility and popularity as source of data in forestry applications due to the flexibility of the technology, relatively low cost, and capacity to provide a link between field measures and airborne or satellite RS. Equipped with one or more sensors, UAVs have been shown to provide useful data for applications where other RS technologies are less well suited, such as health status, tree- or stand-level inventories, or land cover mapping due to the very high spatial and temporal resolutions. Ortho-mosaics are generally employed for detailed spectral classification, and 3D point clouds derived from overlapped imagery are employed for structural characterisation. Although acquisition parameters and processing workflows remain application dependent, ongoing research should work towards standards and development of geometrically and radiometrically accurate products.

REFERENCES

Adão, T., Hruška, J., Pádua, L., Bessa, J., Peres, E., Morais, R., Sousa, J.J. 2017. Hyperspectral imaging: A review on UAV-based sensors, data processing and applications for agriculture and forestry. *Remote Sensing*, 9, 1110.

Álvarez-Taboada, F., Sanz-Ablanedo, E., Rodríguez Pérez, J.R., Castedo-Dorado, F., Lombardero, M.J. 2014. Multi-sensor and multi-scale system for monitoring forest health in *Pinus radiata* stands defoliated by *Lymantria dispar* in NW Spain. *Proceedings of the ForestSAT Open Conference System*, http://ocs.agr.unifi.it/index.php/forestsat2014/ForestSAT2014/paper/view/245

Álvarez-Taboada, F., Paredes, C., Julián-Pelaz, J. 2017. Mapping of the invasive species *Hakea sericea* using unmanned aerial vehicle (UAV) and WorldView-2 imagery and an object-oriented approach. *Remote Sensing*, 9, 913.

Baron, J., Hill, D.J., Elmiligi, H. 2018. Combining image processing and machine learning to identify invasive plants in high-resolution images. *International Journal of Remote Sensing*, 39(15–16), 5099–5118, doi:10.1080/01431161.2017.1420940

Benz, U.C., Hofmann, P., Willhauck, G., Lingenfelder, I., Heynen, M. 2004. Multi-resolution, object-oriented fuzzy analysis of remote sensing data for GIS-ready information. *ISPRS Journal of Photogrammetry and Remote Sensing*, 58, 239–258.

Cao, J., Leng, W., Liu, K., Liu, L., He, Z., Zhu, Y. 2018. Object-based mangrove species classification using unmanned aerial vehicle hyperspectral images and digital surface models. *Remote Sensing*, 10, 89, doi:10.3390/rs10010089

Carl, C., Landgraf, D., van der Maaten-Theunissen, M., Biber, P., Pretzsch, H. 2018. *Robinia pseudoacacia* L. flowers analyzed by using an unmanned aerial vehicle (UAV). *Remote Sensing*, 9, 1091, doi:10.3390/rs9111091

Carr, J.C., Slyder, J.B. 2018. Individual tree segmentation from a leaf-off photogrammetric point cloud. *International Journal of Remote Sensing*, 39(15–16), 5195–5210, doi:10.1080/01431161.2018.1434330

Castedo-Dorado, F., Lago-Parra, G., Lombardero, M.J., Liebhold, A.M., Álvarez-Taboada, M.F. 2016. European gypsy moth (*Lymantria dispar dispar* L.) completes development and defoliates exotic radiata pine plantations in Spain. *New Zealand Journal of Forestry Science*, 46(18).

Chen, Sh., McDermid, G.J., Castilla, G., Linke, J. 2017. Measuring vegetation height in linear disturbances in the boreal forest with UAV photogrammetry. *Remote Sensing*, 9, 1257, doi:10.3390/rs9121257

Congalton, R.G., Green, K. 2009. *Assessing the Accuracy of Remotely Sensed Data*, 2nd edition. Boca Raton, FL: CRC Press.

Dandois, J.P., Ellis, E.C. 2013. High spatial resolution three-dimensional mapping of vegetation spectral dynamics using computer vision. *Remote Sensing of Environment*, 136, 259–276.

Dandois, J.P., Olano, M., Ellis, E.C. 2015. Optimal altitude, overlap, and weather conditions for computer vision UAV estimates of forest structure. *Remote Sensing*, 7, 13895–13920, doi:10.3390/rs71013895

Dash, J.P., Pearse, G.D., Watt, M.S. 2018. UAV multispectral imagery can complement satellite data for monitoring forest health. *Remote Sensing*, 10, 1216, doi:10.3390/rs10081216

Dunford, R., Michel, K., Gagnage, M., Piégay, H., Trémelo, M.L. 2009. Potential and constraints of unmanned aerial vehicle technology for the characterization of Mediterranean riparian forest. *International Journal of Remote Sensing*, 30, 4915–4935.

Duryea, M.L. 1987 *Forest Regeneration Methods: Natural Regeneration, Direct Seeding and Planting*. Florida Cooperative Extension Service, Institute of Food and Agricultural Sciences, University of Florida.

Feduck, C., McDermid, G., Castilla, G. 2018. Detection of coniferous seedlings in UAV imagery. *Forests*, 9, 432.

Fernández-Guisuraga, J.M., Sanz-Ablanedo, E., Suárez-Seoane, S., Calvo, L. 2018. Using unmanned aerial vehicles in postfire vegetation survey campaigns through large heterogeneous areas: Opportunities and challenges. *Sensors*, 18, 586.

Fernández-Luque, I., Aguilar, F.J., Aguilar, M.A., Álvarez, M.F. 2014. Extraction of impervious surface areas from GeoEye-1 and WorldView-2 VHR satellite imagery using an object-based approach. *Journal of Selected Topics in Applied Earth Observations and Remote Sensing*, 7, 4681–4691, doi:10.1109/ JSTARS.2014.2327159.

Forlani, G., Dall-Asta, E., Diotri, F., Morra di Cella, U., Roncella, R., Santise, M. 2018. Quality assessment of DSMs produced from UAV flights georeferenced with on-board RTK positioning. *Remote Sensing*, 10, 311, doi:10.3390/rs10020311

Franklin, S.E., Ahmed, O.S. 2017. Deciduous tree species classification using object-based analysis and machine learning with unmanned aerial vehicle multispectral data. *International Journal of Remote Sensing*, 39(15–16), 5236–5245.

Franklin, S.E., Ahmed, O.S., Williams, G. 2017. Northern conifer forest species classification using multi-spectral data acquired from an unmanned aerial vehicle. *Photogrammetry Engineering and Remote Sensing*, 83(7), 501–507.

Fraser, R.H., van der Sluijs, J., Hall, R.J. 2017. Calibrating satellite-based indices of burn severity from UAV-derived metrics of a burned boreal forest in NWT, Canada. *Remote Sensing*, 9, 279, doi:10.3390/ rs9030279

Fraser, B.T., Congalton, R.G. 2018. Issues in unmanned aerial systems (UAS) data collection of complex forest environments. *Remote Sensing*, 10, 908, doi:10.3390/rs10060908

Frey, F., Kovach, K., Stemmler, S., Koch, K. 2018. UAV photogrammetry of forests as a vulnerable pro-cess. A sensitivity analysis for a structure from motion RGB-image pipeline. *Remote Sensing*, 10, 912, doi:10.3390/rs10060912

Getzin, S., Wiegand, K., Scho, I. 2012. Assessing biodiversity in forests using very high resolu-tion images and unmanned aerial vehicles. *Methods in Ecology and Evolution*, 3, 397–404. doi: 10.1111/j.2041-210X.2011.00158.x

Getzin, S., Nuske, R.S., Wiegand, K. 2014. Using Unmanned Aerial Vehicles (UAV) to Quantify Spatial Gap Patterns in Forests. *Remote Sensing*, 6, 6988–7004; doi:10.3390/rs6086988

Goodbody, T.R.H., Coops, N.C., Hermosilla, T., Tompalski, P., Crawford, P. 2017a. Assessing the status of forest regeneration using digital aerial photogrammetry and unmanned aerial systems. *International Journal of Remote Sensing*, 39(15–16), 5246–5264.

Goodbody, T.R.H., Coops, N.C., Marshall, P., Tompalski, P., Crawford, P. 2017b. Unmanned aerial systems for precision forest inventory purposes: A review and case study. *The Forestry Chronicle*, 93, 71–81.

Goodbody, T.R.H., Coops, N.C., Tompalski, P., Crawford, P., Day, K.J. 2017c. Updating residual stem volume estimates using ALS- and UAV-acquired stereo-photogrammetric point clouds. *International Journal of Remote Sensing*, 38(8–10), 2938–2953.

Goodbody, T.R.H., Coops, N.C., Hermosilla, T., Tompalski, P., Pelletier, G. 2018. Vegetation phenology driv-ing error variation in digital aerial photogrammetrically derived terrain models. *Remote Sensing*, 10, 1554.

Guerra-Hernández, J., González-Ferreiro, E., Monleón, V.J., Faias, S.P., Tomé, M., Díaz-Varela, R.A. 2017. Use of multi-temporal UAV-derived Imagery for estimating individual tree growth in *Pinus pinea* stands. *Forests*, 8, 300.

Guerra-Hernández, J., Cosenza, D.N., Rodríguez, L.C.E., Silva, M., Tomé, M., Díaz-Varela, R.A., González-Ferreiro, E. 2018. Comparison of ALS- and UAV (SfM [SFM])-derived high-density point clouds for individual tree detection in Eucalyptus plantations. *International Journal of Remote Sensing*, 39(15–16), 5211–5235, doi:10.1080/01431161.2018.1486519

Hall, R.J., Castilla, G., White, J.C., Cooke, B.J., Skakun, R.S. 2016. Remote sensing of forest pest damage: A review and lessons learned from a Canadian perspective. *Canadian Entomologist*, 148, 1–61.

Hassaan, O., Nasir, A.K., Roth, H., Khan, M.F. 2016. Precision forestry: Trees counting in urban areas using visible imagery based on an unmanned aerial vehicle. *IFAC-PapersOnLine*, 49, 16–21.

Hird, J.H., Montaghi, A., McDermid, G.J., Kariyeva, J., Moorman, B.J., Nielsen, S.E., McIntosh, A.C.S. 2017. Use of unmanned aerial vehicles for monitoring recovery of forest vegetation on petroleum well sites. *Remote Sensing*, 9, 413, doi:10.3390/rs9050413

Jayathunga, S., Owari, T., Tsuyuki, S. 2018. Evaluating the performance of photogrammetric products using fixed-wing UAV imagery over a mixed conifer–broadleaf forest: Comparison with airborne laser scan-ning. *Remote Sensing*, 10, 187, doi:10.3390/rs10020187

Kachamba, D.J., Ørka, H.O., Gobakken, T., Eid, T., Mwase, W. 2016. Biomass estimation using 3D data from unmanned aerial vehicle imagery in a tropical woodland. *Remote Sensing*, 8, 968, doi:10.3390/ rs8110968

Klosterman, S., Richardson, A.D. 2017. Observing spring and fall phenology in a deciduous forest with aerial drone imagery. *Sensors*, 17, 2852.

Koh, L.P., Wich, S.A. 2012. Dawn of drone ecology: Low-cost autonomous aerial vehicles for conservation. *Tropical Conservation Science*, 5, 121–132.

Lago-Parra, G., Castedo-Dorado, F., Álvarez-Taboada, M.F., Lombardero, M.J. 2016.Estudio del brote epi-démico de lagarta peluda (*Lymantria dispar* L.) en masas de *Pinus radiata* en Cubillos del Sil (El Bierzo, León). *Cuadernos de la Sociedad Española de Ciencias Forestales*, 43, 315–328.

Lin, Y., Hyyppä, J., Jaakkola, A. 2011. Mini-UAV-borne LIDAR for fine-scale mapping. *IEEE Geoscience on Remote Sensing Letters*, 8, 426–430.

Mafanya, M., Tsele, Ph., Botai, J.O., Manyama, Ph., Chirima, G.J., Monate, T. 2018. Radiometric calibra-tion framework for ultra-high-resolution UAV-derived orthomosaics for large-scale mapping of invasive alien plants in semi-arid woodlands: *Harrisia pomanensis* as a case study. *International Journal of Remote Sensing*, 39(15–16), 5119–5140.

Manfreda, S., McCabe, M.F., Miller, P.E., Lucas, R., Pajuelo Madrigal, V., Mallinis, G., Ben Dor, E., Helman, D., Estes, L., Ciraolo, G., Müllerová, J., Tauro, F., de Lima, M.I., de Lima, J.L.M.P., Maltese, A., Frances, F., Cylor, K., Kohv, M., Perks, M., Ruiz-Pérez, G., Su, Z., Vico, G., Toth, B. 2018. On the use of unmanned aerial systems for environmental monitoring. *Remote Sensing*, 10, 641.

MFLNRO, 2016. *Silviculture Survey Procedures Manual.* Victoria, BC: Ministry of Forests, Lands and Natural Resources Operations.

Michez, A., Piégay, H., Lisein, J., Claessens, H., Lejeune, P. 2016. Classification of riparian forest species and health condition using multi-temporal and hyperspatial imagery from unmanned aerial system. *Environmental Monitoring and Assessment*, 188, 1–19.

Minore, D., Laacke, R.J. 1992. Natural regeneration. In S.D. Hobbs (Ed.), *Reforestation Practices in Southwestern Oregon and Northern California* (pp. 258–283). Corvallis, OR: Forest Research Laboratory, Oregon State University.

Mohan, M., Silva, C.A., Klauberg, C., Jat, P., Catts, G., Cardil, A., Hudak, A.T., Dia, M. 2017. Individual tree detection from unmanned aerial vehicle (UAV) derived canopy height model in an open canopy mixed conifer forest. *Forests*, 8, 1–17.

Morais, M., Marchante, E., Marchante, H. 2017. Big troubles are already here: Risk assessment protocol shows high risk of many alien plants present in Portugal. *Journal for Nature Conservation*, 35, 1–12.

Näsi, R., Honkavaara, E., Lyytikäinen-Saarenmaa, P., Blomqvist, M., Litkey, P., Hakala, T., Viljanen, N., Kantola, T., Tanhuanpää, T., Holopainen, M., Wang, C., Wynne, R.H., Thenkabail, P.S. 2015. Using UAV-based photogrammetry and hyperspectral imaging for mapping bark beetle damage at tree-level. *Remote Sensing*, 7, 15467–15493.

Näsi, R., Honkavaara, E., Blomqvist, M., Lyytikäinen-Saarenmaa, P., Hakala, T., Viljanen, N., Kantola, T., Holopainen, M. 2018. Remote sensing of bark beetle damage in urban forests at individual tree level using a novel hyperspectral camera from UAV and aircraft. *Urban Forestry & Urban Greening*, 30, 72–83.

Natural Resources Canada, 2016. *The State of Canada's Forests: Annual Report 2016.*

Navarro, J.A., Fernández-Landa, A., Tomé, J.L., Guillén-Climent, M.L., Ojeda, J.C. 2018. Testing the quality of forest variable estimation using dense image matching: A comparison with airborne laser scanning in a Mediterranean pine forest. *International Journal of Remote Sensing*, 39(14), 4744–4760.

Nevalainen, O., Honkavaara, E., Tuominen, S., Viljanen, N., Hakala, T., Yu, X., Hyyppä, J., Saari, H., Pölönen, I., Imai, N.N., Tommaselli, A.M.G. 2017. Individual tree detection and classification with UAV-based photogrammetric point clouds and hyperspectral imaging. *Remote Sensing*, 9, 185, doi:10.3390/rs9030185

Nevalainen, P., Salmivaara, A., Ala- Ilomäki, J., Launiainen, S., Hiedanpää, J., Finér, L., Pahikkala, T., Heikkonen, J. 2017. Estimating the rut depth by UAV photogrammetry. *Remote Sensing*, 9, 1279, doi:10.3390/rs9121279

Pádua, L., Jonáš, H., Bessa, J., Adão, T., Martins, L.M., Gonçalves, J.A., Peres, E., Sousa, A.M.R., Castro, J.P., Sousa, J.J. 2017. Multi-temporal analysis of forestry and coastal environments using UASs. *Remote Sensing*, 10, 24, doi:10.3390/rs10010024

Perroy, R.L., Sullivan, T., Stephenson, N. 2017. Assessing the impacts of canopy openness and flight param-eters on detecting a sub-canopy tropical invasive plant using a small unmanned aerial system. *ISPRS Journal of Photogrammetry and Remote Sensing*, 125, 174–183.

Puliti, S., Ørka, O., Gobakken, T., Næsset, E. 2015. Inventory of small forest areas using an unmanned aerial system. *Remote Sensing*, 7, 9632–9654.

Puliti, S., Saarela, S., Gobakken, T., Ståhl, G., Næsset, E., 2018a. Combining UAV and Sentinel-2 auxiliary data for forest growing stock volume estimation through hierarchical model-based inference. *Remote Sensing of Environment*, 204, 485–497.

Puliti, S., Talbot, B., Astrup, R. 2018b. Tree-stump detection, segmentation, classification, and measurement using unmanned aerial vehicle (UAV) imagery. *Forests*, 9, 102.

Puliti, S., Ene, L.T., Gobakken, T., Næsset, E. 2017. Use of partial-coverage UAV data in sampling for large scale forest inventories. *Remote Sensing of Environment*, 194, 115–126.

Saarinen, N., Vastaranta, M., Näsi, R., Rosnell, T., Hakala, T., Honkavaara, E., Wulder, M., Luoma, V., Tommaselli, A., Imai, N., Ribeiro, E., Guimarães, R., Holopainen, M., Hyyppä, J. 2018. Assessing biodiversity in boreal forests with UAV-based photogrammetric point clouds and hyperspectral imaging. *Remote Sensing*, 10, 338.

Sanz-Ablanedo, E., Chandler, J.H., Rodríguez-Pérez, J.R., Ordóñez, C. 2018. Accuracy of unmanned aerial vehicle (UAV) and SfM (SFM) photogrammetry survey as a function of the number and location of ground control points used. *Remote Sensing*, 10, 1606, doi:10.3390/rs10101606

Shin, P., Sankey, T., Moore, M.M., Thode, A.E. 2018. Evaluating unmanned aerial vehicle images for estimating forest canopy fuels in a ponderosa pine stand. *Remote Sensing*, 10(8), doi:10.3390/rs10081266

Smigaj, M., Gaulton, R., Suárez, J.C., Barr, S.L. 2017. Use of miniature thermal cameras for detection of physiological stress in conifers. *Remote Sensing*, 9, 957, doi:10.3390/rs9090957

Smith, M.W., Carrivick, J., Quincey, D. 2016. Structure from motion photogrammetry in physical geography. *Progress in Physical Geography*, 40(2), 247–275. ISSN 0309-1333

Tang, L., Shao, G. 2015. Drone remote sensing for forestry research and practices. *Journal of Forestry Research*, 26, 791–797.

Thiel, Ch., Schmullius, Ch. 2017. Comparison of UAV photograph-based and airborne LIDAR based point clouds over forest from a forestry application perspective. *International Journal of Remote Sensing*, 38(8–10), 2411–2426, doi:10.1080/01431161.2016.1225181

Torresan, Ch., Berton, A., Carotenuto, F., Di Gennaro, S.F., Gioli, B., Matese, A., Miglietta, F., Vagnoli, C., Zaldei, A., Wallace, L. 2018. Forestry applications of UAVs in Europe: A review. *International Journal of Remote Sensing*, 38(8–10), 2427–2447.

Tu, Y.-H., Phinn, S., Johansen, K., Robson, A. 2018. Assessing radiometric correction approaches for multispectral UAS imagery for horticultural applications. *Remote Sensing*, 10, 1684, doi:10.3390/rs10111684

Tuominen, S., Näsi, R., Honkavaara, E., Balazs, A., Hakala, T., Viljanen, N., Pölönen, I., Saari, H., Ojanen, H. 2018. Assessment of classifiers and remote sensing features of hyperspectral imagery and stereophotogrammetric point clouds for recognition of tree species in a forest area of high species diversity. *Remote Sensing*, 10, 714.

Wallace, L., Lucieer, A., Watson, C., Turner, D. 2012a. Development of a UAV-LIDAR system with application to forest inventory. *Remote Sensing*, 4, 1519–1543.

Wallace, L., Lucieer, A., Watson, C.S. 2012b. Assessing the feasibility of UAV-based LIDAR for high resolution forest change detection. *ISPRS International Archives of Photogrammetry, Remote Sensing and Spatial Information Science*, 38, 499–504.

Wallace, L., Lucieer, A., Watson, C.S. 2014a. Evaluating tree detection and segmentation routines on very high resolution UAV LIDAR data. *IEEE Transactions on Geoscience and Remote Sensing*, 52, 7619–7628.

Wallace, L., Musk, R., Lucieer, A. 2014b. An assessment of the repeatability of automatic forest inventory metrics derived from UAV-borne laser scanning data. *IEEE Transactions on Geoscience and Remote Sensing*, 52, 7160–7169.

Zahawi, R.A., Dandois, J.P., Holl, K.D., Nadwodny, D., Reid, J.L., Ellis, E.C. 2015. Using lightweight unmanned aerial vehicles to monitor tropical forest recovery. *Biological Conservation*, 186, 287–295.

Zarco-Tejada, P.J., Diaz-Varela, R., Angileri, V., Loudjani, P. 2014. Tree height quantification using very high resolution imagery acquired from an unmanned aerial vehicle (UAV) and automatic 3D photoreconstruction methods. *European Journal of Agronomy*, 55, 89–99.

Zhang, J., Hu, J., Lian, J., Fan, Z., Ouyang, X., Ye, W. 2016. Seeing the forest from drones: Testing the potential of lightweight drones as a tool for long-term forest monitoring. *Biological Conservation*, 198, 60–69.

12 Monitoring Oil and Gas Pipelines with Small UAV Systems

Cristina Gómez
UCEMM – University of Aberdeen and
INIA-Forest Research Centre (CIFOR)

David R. Green
UCEMM – University of Aberdeen

CONTENTS

INTRODUCTION

Pipelines are the safest means for transport of oil and gas. Millions of cubic metres of hydrocarbons are transported daily by pipelines. Globally there are approximately three million km of pipelines (CIA, 2013) valued almost 9,000 million dollars in 2014 (Markets and Markets, 2014). Pipeline networks are of all sizes and lengths and may have above- or below-ground configurations and diametric size up to more than a metre. Despite the safety provided by pipelines, equipment failure may occur with a risk to spill large amounts of oil and gas, damaging the environment through contamination and pollution and affecting ecological health and human security. Typical reasons for failure of equipment are over-age (structures become more prone to corrosion), natural ground movement, accidental hot tap, and third-party interference (CONCAWE, 2015). Minor incidents and failures are more frequent than catastrophic accidents and can also cause important environmental damage and economic losses. According to the Energy Resources Conservation Board, the number of pipeline breaks per 1,000 km year exposure (pipeline length×duration) in Alberta (Canada) was 1.5 in 2011 and 2012 (ERCB, 2013). This rate is estimated to be 110–140 per 1,000 km per year in Russia. In Europe, these figures decreased from 1.2 incidents per 1,000 km×year in the 1970s to 0.23 incidents per 1,000 km×year in 2013 (CONCAWE, 2015) for oil pipelines, and from 0.87 to 0.33 in the period 1970–2013 for gas pipelines (EGIG, 2015). Theft incidents have increased in the last years in both oil and gas pipeline networks (CONCAWE, 2015; EGIG, 2015), becoming one of the most important causes of spillage.

Regardless of their size, placement, or location, the safety and security of pipelines is of paramount importance to stakeholders and to the public. Safety guidelines and regulations for installation and management of pipelines exist worldwide (IPLOCA, 2003; GL, 2010), commonly including an 'inspection and maintenance of pipelines integrity plan'. To prevent failures and detect problems on time, a monitoring system providing regular information of the structural state and functional conditions of the pipelines is necessary. Furthermore, monitoring oil and gas pipeline networks involves acquiring knowledge of the impact pipelines produce on the environment over time (i.e. how vegetation and wildlife is affected).

METHODS FOR MONITORING OIL AND GAS PIPELINES

The most widely used methods for monitoring oil and gas transmission pipelines are foot patrols along the pipeline route and aerial surveillance using light aircraft or helicopters. Patrols are carried out at regular intervals throughout the year and regardless of weather conditions. These methods ensure a high level of security but are expensive and with a main disadvantage of detecting failures late, when the output (oil or gas) has been reduced or the environment has already been affected and damaged. Airborne solutions also bring their own difficulties in terms of safety and operation: manned aircraft using pilot and/or operator for detection and identification are forced to fly very close to the terrain; frequently can only detect visible effects (i.e. not gas leak detection); and require expensive aircraft limiting the frequency and duration of flight. Alternative approaches that do not rely on human interventions utilise real-time monitoring systems based on a network of small sensors (e.g. pressure, acoustic, and temperature) designed to measure real-time flow, enabling detection of leakages or to identify changes in the wall thickness through temperature or noise measurements (El-Darymli et al., 2009). These sensors are vulnerable to damage at any point along the network and can produce ambiguous data, providing incomplete or inaccurate information.

Satellite data (e.g. radar, optical) are used operationally to detect oil spills in marine environments (Brekke and Solberg, 2005), where the hydrocarbons' spectral signature is very distinctive. However, the potential of satellite remote sensing to reach the demands of pipeline monitoring required by pipeline operators has been investigated in numerous research projects (e.g. PRESENSE, PIPEMOD, and GMOSS) without satisfactory conclusions. Progress in high-resolution remote sensing and image processing technology has provided the basis for designing pipeline monitoring systems using remote sensors and context-oriented image processing software (Hausamann et al., 2005; van der Werff et al., 2008).

DETECTION OF LEAKS FROM HYDROCARBON PIPELINES

Hydrocarbon leaks can be identified by trained operators with visual observation (interpreting colour, texture, and pattern) on real-time inspection or analysing a recorded image (still or video). Beyond the visible, other electromagnetic wavelengths are practical to detect hydrocarbon leaking. The thermal infrared (TIR) wavelengths are particularly useful due to the temperature differences between the fluids (i.e. hydrocarbons) and the soil. Draining liquids affect the thermal conductivity of soils, and whilst an oil leak creates a warmer area, gas escapes show colder than the ground substrate. The rationale for detecting hydrocarbon leakages from pipelines using TIR data surveys is based upon thermal differences, either on a single image, where leakage points look remarkably different to the surroundings, or by comparison of images of the same area captured on different days. Detection of small temperature differences is possible with dedicated image analysis techniques, but care should be placed on other factors affecting the soil temperature (e.g. water content). The availability of frequent and regular images increases the sensitivity of the data through the use of averaging techniques, enabling small differences in heat capacity of the soil to be detected, on a day-by-day basis. Leakages can also be identified through a reduction in the vigour of vegetation eventually leading to death (Mishra et al., 2012). Repetitive imagery is needed to identify this change as well as the use of some kind of automated threshold to provide the alert.

Fine spectroscopy of 0.05–0.1 nm spectral resolution has also proven to identify the specific spectral signature of hydrocarbons, as well as the early effects of oil and gas pollution on vegetation. Measurements of solar-induced fluorescence (F) could be used as an early indicator of the health and status of vegetation, although the quantitative estimation of F from the air is complicated by the absorption of the atmosphere *en route* to the sensor (Meroni et al., 2009). Currently, laser fluorosensors are the most useful and reliable instruments to detect oil on various backgrounds, including water, soil, weeds, ice, and snow. In fact, they are the only reliable sensors to detect oil in the presence of snow and ice, and they do not detect false positives. A different and unequivocal means for the detection of specific gases (e.g. methane) from a certain distance is gas detection, which works in day and night conditions. Despite gas diffusion and dispersion of gas contamination into the atmosphere, particularly in windy conditions, the highest concentration of gas is a reliable indicator of the leakage location (Allen et al., 2015).

EMERGING OPPORTUNITIES WITH UAV REMOTE SENSING

Progress in remote sensing technology – including sensors and platforms – and data processing software provides new opportunities for the development of spatially precise and comparatively inexpensive monitoring systems to inspect pipelines and identify hydrocarbon leaks. Unmanned aerial vehicles (UAVs) offer an important option with advantages such as improved mission safety, flight repeatability, the potential for reduction in operational costs, and fewer weather-related flying limitations. However, these advantages are dependent upon the type and size of airborne platform, sensor type, mission objectives, and the regulatory requirements imposed. Small UAVs, both *fixed-wing* and helicopters with *rotary wings*, are increasingly considered as reliable platforms for capturing data for environmental applications, and a low-cost alternative to larger-scale platforms (e.g. ARC, 2003). With technological developments such as sensor miniaturisation, stabilisation, and navigation systems, small UAVs provide a powerful basis to gather data and produce information that can be tied to in situ ground data and to large-scale satellite imagery, providing a link between multiple spatial scales.

Small UAVs can provide a very flexible means to acquire unique data and information. Currently there is a wide range of commercial UAVs available for the acquisition of low-cost aerial remotely sensed data. The rapid development of microelectronics and microprocessors, battery technology, GPS and navigation systems, together with reduced costs over the last five years have all helped in triggering an unprecedented demand for, and growth in, the use of UAV platforms for many civilian applications (Watts et al., 2012; Colomina and Molina, 2014).

UAVs are now evolving as highly effective tools for tackling the requirements of oil and gas pipeline monitoring, a specific environmental application of the UAV technology. This chapter describes the current use of UAV platforms and sensors, as well as the foreseen potential of small UAV systems for monitoring oil and gas pipelines. The remainder of this chapter will include sections covering the following: (i) an overview of the characteristics of UAVs; (ii) the use of UAVs for oil and gas pipeline monitoring to date with particular attention paid to the strengths and successes, as well as the weaknesses; (iii) considerations and developments in the technology of small-scale aerial platforms and sensors specifically tailored to oil and gas pipeline monitoring applications, including battery, sensor, navigation, software, and platform; and (iv) future prospects for UAV development and application.

UNMANNED AERIAL SYSTEMS (UAS)

A UAV is flown without a pilot on-board and is either remotely and fully controlled from another place (e.g. ground, another aircraft, space) or programmed and fully autonomous (ICAO, 2011). An unmanned aerial vehicle or system comprises the flying *platform*, the elements necessary to enable and *control* navigation, including taxiing, take-off and launch, flight and recovery/landing, and the elements to accomplish the mission objectives: *sensors* and equipment for data acquisition and transfer. A brief description of each of the main elements (platforms, sensors, and auxiliary equipment) now follows:

UAVs, also called remotely piloted aircraft systems (RPAS), can be classified under different schemes, using criteria such as flying height and range, size, and weight (frequently referred to as Mean take-off-weight – MTOW). A strict categorisation of UAVs is not however possible because certain characteristics in the various classes overlap (Skrzypietz, 2012). Table 12.1 provides an overview of the types of flying platform as considered by UVS International; a non-profit association dedicated to promote unmanned aerial systems.

UAV PLATFORMS

The very small platforms, micro and mini aerial vehicles (Table 12.1) can fly for less than 1 hour at an altitude below 250 m. Micro platforms are considerably smaller than mini platforms (i.e. <5 kg versus 20–150 kg), but both have a similar flying range. An example of micro UAV is the Phantom 1 or 2, with MTOW below 1.3 kg and a very light payload capacity; Aibot-X6 is another micro UAV with 3.4 kg MTOV and maximum payload of 2 kg. Mini is the most abundant type of platform produced for civilian applications, doubling the number of micro- and medium-range UAV platforms (UVS, 2014). An example of mini UAV is the Camcopter, with an MTOV of 68 kg and maximum

---

TABLE 12.1
Characteristics of Non-Military Remotely Piloted Aircraft Systems (RPAS)

Name	Acronym	Mass (kg)	Range (km)	Altitude (m)	Endurance (h)
Micro	MAV	<5	<10	250	1
Mini	Mini	<20–150[a]	<10	150[a]	<2
Close range	CR	25–150	10–30	3,000	2–4
Short range	SR	50–250	30–70	3,000	3–6
Medium range	MR	150–500	70–200	5,000	6–10
MR endurance	MRE	500–1,500	>500	8,000	10–18
Low alt. deep penetration	LADP	250–2,500	>250	50–9,000	0.5–1
Low alt. long endurance	LALE	15–25	>500	3,000	>24
Medium alt. long endurance	MALE	1,000–1,500	>500	5/8,000	24–48
High alt. long endurance	HALE	2,500–5,000	>2,000	20,000	24–48
Stratospheric	Strato	>2,500	>2,000	>20,000	>48
Exo-stratospheric	EXO	TBD	TBD	>30,500	TBD

Source: Adapted from UVS international (2008).

MAV, micro air vehicles; VTOL, vertical take-off and landing; LASE, low altitude, short endurance; LALE, low altitude, long endurance; MALE, medium altitude, long endurance; HALE, high altitude, long endurance.

[a] According to national legal restrictions.

payload capacity of 25 kg. On the other side of the scale, medium-altitude, long-endurance (MALE) platforms (e.g. Talarion, Predator) and high-altitude, long-endurance (HALE) platforms (e.g. Global Hawk, Euro Hawk) have a flying endurance of several days at an altitude of 20,000 m. The latter aerial platforms are comparable in size to manned aircraft.

Nanodrones are miniature UAVs able to carry small still and video cameras. These UAVs can fly in all directions and perform manoeuvres and mid-air stunts. For example, the palm-size Micro Drone 2 weighs 0.034 kg and has a flying range of 120 m and endurance of 6–8 minutes. Other small UAVs are now flown as tethered aerial vehicles to circumvent the risks associated with flying. The Pocket Flyer by CyPhy Works is a 0.080 kg tethered platform that can fly continuously for 2 hours or more, sending back high-quality HD video the entire time. With improved tether technology, all data, control, and endurance can be built into the tether, providing long endurance. Furthermore, ZANO operates on a virtual tether connected to a smart device, allowing simple gestures to control it.

Small powered UAV platforms based on airframes can be grouped into two categories: rotary-wing and fixed-wing UAVs. Fixed-wing UAVs have a relatively simple structure, making them stable platforms that are relatively easy to control during autonomous flights. Their efficient aerodynamics enables longer flight duration and higher speeds, making them ideal for applications such as aerial survey (requiring the capture of georeferenced imagery over large areas). On the down side, fixed-wing UAVs need to fly forward continuously and need space to both turn and land. These platforms are also dependent on a launcher (person or mechanical) or a runway to facilitate take-off and landing, which can have implications on the type of payloads they carry.

Typical lightweight fixed-wing UAVs currently in the commercial arena have a *flying wing* design with wings spanning between 0.8 and 1.2 m, and a very small fin at both ends of the wing. In-house vehicles tend to have slightly longer wings to enable carrying the required heavier sensors (Petrie, 2013a, b). A second type of design is the *conventional fuselage*. The dimensions are around 1.2–1.4 m length for the fuselage and 1.6–2.8 m wing length (Figure 12.1). In the UK, there are around 20 companies operating commercial airborne imaging services using fixed-wing UAVs (Petrie, 2013a, b).

Rotary-wing aircraft (N-copter or N-rotor) have complex mechanics, which translates into lower speeds and shorter flight ranges. Among their main strengths, rotary-wing UAVs can fly vertically, take-off and land in a very small space, and can hover over a fixed position and at a given height. This makes rotary-wing UAVs well suited for applications that require manoeuvring in tight spaces and the ability to focus on a single target for extended periods (e.g. structural inspections). On the downside, rotary-wing UAVs can be less stable than fixed-wing counterparts and also more difficult to control during flight.

Single-rotor and coaxial rotor platforms (with two counter-rotating rotors on the same axis) are similar to conventional helicopters, with a single lifting rotor and two or more blades. These platforms maintain directional control by varying blade pitch via a servo-actuated mechanical linkage. Single-rotor and coaxial rotor UAVs are typically radio-controlled and powered by electric motors, although some of the heaviest examples use petrol engines. They require cyclic or collective pitch control.

Multi-copters have an even number of rotors and utilise differential thrust management of the independent motor units to provide lift and directional control. As a general rule, the more the rotors, the higher the payload they can take and are functional in strong wind conditions, as the redundant lift capacity provides for increased safety, and more control in the event of a rotor malfunction or failure. Some platforms now available are capable of autonomous flight, which significantly improves the capability to undertake repeat aerial video and photography.

SENSORS ON-BOARD UAV FOR MONITORING OIL AND GAS PIPELINES

The information reachable by the survey mission is determined by the type and quality of the sensors carried on-board the flying platform. Despite an increasing range of sensors available for small-scale UAVs, thanks to miniaturisation and advancements in battery technology, for some of the most adequate oil or gas leak detection techniques, (e.g. fluorescence) there is still no sensor adapted to

FIGURE 12.1 Examples of lightweight commercial UAVs. Fixed-wing flying-wing design: (a) Trimble Gatewing X100; (b) swinglet CAM; (c) smartone. Conventional fuselage design: (d) MAVinci Sirius; (e) Composites Pteryx; (f) CropCam. Single rotor designs: (g) AT-10 (Advanced UAV Technology); (h) Syma S107C. Multi-rotor designs: (i) Parrot AR Drone quadcopter; (j) Cinestar 8 octocopter.

UAV conditions (i.e. laser fluorosensor). Selecting a combination of platform and sensor to provide the necessary data in adequate conditions for monitoring and mapping oil and gas pipelines remains a challenge. The main sensor types with commercial adaptations to UAV mechanics that can be used for monitoring oil and gas pipelines are listed in Table 12.2.

The essential difference between *active* and *passive* sensors originates from the source of energy illuminating the target objects (passive sensors rely on the sun; active sensors emit some kind of radiation themselves). Therefore, missions with an active sensor require higher lifting and carrying capacity UAV platforms. Active sensors emit some kind of radiation measuring the fraction reflected by the target objects and the difference in time between emission and reception. Active sensors require power supplied by a source, adding weight to the aerial system which makes active equipment less versatile for use on UAVs than passive equipment. The capacity to perform a particular monitoring task and to work under certain environmental conditions (e.g. topography, weather) is sensor dependent (Table 12.2).

Some examples of commercial sensors from most of the techniques listed in Table 12.2 which are adapted to small UAVs can be found in Gómez and Green (2015, report) and Colomina and Molina (2014).

AUXILIARY EQUIPMENT

A series of systems and elements support the aircraft and sensors to make an UAV mission successful. The most relevant systems are those dedicated to position and navigation, to autonomous flight, and for communications. The need of elements for the launch, recovery and retrieval, and the mechanics and payloads are UAV and mission dependent.

TABLE 12.2

Selection of Sensors Suitable for Monitoring Oil and Gas Pipelines; Strengths and Weaknesses for the Purpose and Typical Performing Tasks

Type		Strengths	Weaknesses	Typical task
Passive	Visible (wavelength: 0.38–0.76 µm)	• Visual interpretation	• Only suitable in daylight conditions • Limited by atmospheric effects such as clouds, haze, or smoke	• Infrastructure inspection • Spill detection
	Multispectral (multiple bands)	• Visual interpretation • Vegetation indices	• Only suitable in daylight conditions • Limited by atmospheric effects such as clouds, haze, or smoke	• Characterisation and monitoring of environmental condition
	SWIR (wavelength: 0.9–1.7 µm)	• Very sensitive in low-light conditions • Low power consumption (thermoelectric cooler) • Identification of materials and substances	• Not visible for human eye but sensed with indium gallium arsenide (InGaAs) sensors • Scarce production of detector material (InGaAs)	• Night time characterisation and monitoring of environmental condition
	Thermal IR (8–14 µm)	• Enables detection of leaks • Night vision • Vision through smoke, haze, cloud	• Reference for comparison is needed	• Leak detection • Leak monitoring
	Near infrared (NIR) (wavelength: 0.76–14 µm)	• Sensitive to vegetation condition	• Reference for comparison is needed	• Characterisation and monitoring of environmental condition
	Hyperspectral (hundreds of bands)	• Identification of materials and substances • Flexible/customisable number and resolution of spectral bands	• Library needed	• Characterisation and monitoring of environmental condition
	Video	• Life monitoring if video downlink enabled • Enables generation of 3D imagery	• Redundant information • Typically lower spatial resolution than stills	• Monitoring leakage /spill
	Stereo cameras	• Enables generation of 3D imagery • Can be used as the basis for navigation systems	• Augments weight	• Infrastructure inspection
	Gas IR camera			• Leak detection • Leak monitoring

(Continued)

TABLE 12.2 (*Continued*)

Selection of Sensors Suitable for Monitoring Oil and Gas Pipelines; Strengths and Weaknesses for the Purpose and Typical Performing Tasks

Type		Strengths	Weaknesses	Typical task
Active	LIDAR	• Enables 3D measures • High precision	• Power consumption • Dependable on Inertial Navigation System • Lack of commercial sensors • Difficulties for miniaturisation (size and weight)	• Background characterisation (3D) • Infrastructure inspection
	Radar	• Detection of oil spills in water • All weather conditions • Day and night conditions	• Power consumption • Differential imagery needed • Lack of commercial sensors • Difficulties for miniaturisation (size and weight)	• Leak detection • Leak monitoring
	Laser gas detector	• Measurement of gas emissions (methane concentration) • Early detection of pipeline misfunction • Underground pipeline leak detection • Day and night conditions • No false detections	• Power consumption • Limited range of action (~100 m; <500 m) • Imprecision in windy conditions	• Leak detection • Leak monitoring
	Laser fluorosensor	• Day and night conditions • Reliable detector of oil in snow and ice	• Power consumption • Lack of commercial sensors • Requires clear atmosphere (no fog) • Specialised processing	• Leak detection • Leak monitoring

Source: Gómez and Green (2015).

The position and navigation systems play a crucial role in UAV missions, controlling the location of the UAV at all times by remote operator, or by the autonomous pre-programmed flight plan. Currently available Global Navigation Satellite System (GNSS) equipment for small UAVs is lightweight, compact, and capable of receiving signal from multiple satellite systems (e.g. GPS, Galileo, BeiDou) providing high-accuracy location information (Colomina and Molina, 2014), especially when operated as differential GPS, and facilitates all UAV navigation. For remote control from the ground, air, or sea, where non-autonomous operations are necessary, radar or radio tracking solutions are required.

Autonomous capacity is useful to take-off and land, as well as to repeat the same area survey at regular intervals, flying along a path defined with n-waypoints pre-programmed by the UAV operator. Another desirable capacity of UAVs, to assure safety operations, is the ability to navigate amongst obstacles in the flight path and to sense and avoid other vehicles. Ideally control software (e.g. Mission Planner, APM Planner, Droid Planner) enables switching between pre-programmed autonomous flight and manual control.

Communication between small UAVs and the control system (CS) is usually through radio frequency, commonly in the 900 MHz, 1.2 GHz, 2.4 GHz, and 5.8 GHz bands. Uplink transmission (i.e. CS to aircraft) consists of a flight plan, real-time flight-control commands, control commands to the different payloads, and updated positional information. The downlink information (i.e. aircraft to CS) consists of the payload data (e.g. imagery), positional data, and aircraft housekeeping data (e.g. battery or fuel state). Two different frequencies are necessary to keep the transmission of command information and sensor-acquired data independently, avoiding interference.

Fixed-wing vehicles require additional equipment to assist with launch and recovery. Launch equipment is typically an acceleration ramp with a trolley. Recovery equipment can be a parachute installed within the aircraft, combined with a means to absorb the impact energy (e.g. airbag or an easily replaceable piece of material). To secure the sensor and control the pointing direction and orientation during the flight, solutions go from a simple bracket and rubber mounts in between the UAV and the sensor to elaborated 3D gimbals.

UAV REGULATIONS

Rules for operation of small-scale UAV are a limiting factor for success of the UAV technology. The capacity and responsibility to regulate UAVs rely on different bodies internationally (e.g. European Aviation Safety Agency (EASA) in Europe, Federal Aviation Administration (FAA) in the USA, Civil Aviation Administration (CAAC) in China, Directorate General of Civil Aviation (DGCA) in India). The International Civil Aviation Organisation (ICAO) aims to coordinate global interoperability and harmonisation of UAVs rules (Hayes et al., 2014). The rapid development of small UAV civil applications worldwide, coupled with a fair number of incidents, calls for tightening up the rules and regulations for small UAV flights. Specific rules are being established for light vehicles, and the responsibility to regulate flights of vehicles below 150 kg concerns national or state authorities. Regulations typically require education and training of operators and pilots. In Europe, beyond national level, there is a consortium funded by the European Commission (the Unmanned Aerial Systems in European Airspace (ULTRA)) working to develop a master plan for the insertion of RPAS in the European air transport system. Despite differences in essential aspects concerning platform categories (EASA, 2015), national rules typically impose some kind of restriction to fly the UAVs in certain localities (e.g. airports) and over certain heights (~125–150 m) and distances (i.e. line of sight). Some sort of certification and insurance is also required for commercially based flights.

The expansion of UAV applications will impact the airspace, but other industry areas will also be affected and need regulations and adaptation too. Radio frequencies for communication of UAV with the ground CS require sufficient band width. The International Telecommunication Union (ITU) has not yet allocated such bandwidth to UAVs, and they may have to use different radio frequencies in every country (Everaerts, 2008), something to be considered by international operators and manufacturers. Since regulations are these days changing fast, updated information should be consulted in the regulators' webpage.

USE OF UAVs FOR OIL AND GAS PIPELINE MONITORING

UAVs are expected to play an important role in the inspection, monitoring, and maintenance of oil and gas pipelines, because unmanned vehicles are likely to do anything the energy companies don't want to send people to do (e.g. dangerous operations). Oil and gas companies are looking into incorporating UAVs as part of their Intelligent Pipeline Management initiatives. A number of feasibility studies have taken place, but only a few examples are already in an operational phase.

Considerations for Specifications of a UAV System for Monitoring Oil and Gas Pipelines

To configure a UAV system for a specific monitoring mission, a range of factors have to be considered (Table 12.3). First of all, the type of task (e.g. inspection of infrastructure, detection of leakage, follow-up of spillage) and the kind of information to be acquired (e.g. images of the pipelines, environmental indicators like temperature or vegetation status). Furthermore, the weather and day or night conditions may limit the sensor capability (Table 12.2); the physical characteristics of the environment (e.g. accessibility, terrain roughness, distance to target pipeline) also impose limitations to certain systems or configurations.

The characteristics of the mission flight – time, distance, area, and height – require thorough planning and are crucial for selection of a platform. Although capable of higher flights (400–600 m, depending on system of communication with the CS), small UAVs fly at lower altitude (100–150 m) due to law and regulation restrictions and during a time (~20–25 minutes) limited by the power provided by the battery life. When the pipeline monitoring mission requires surveillance and mapping of an extensive area, fixed-wing UAV platforms capable of long flights in a reasonable time, such as Trimble Gatewing X100, are more useful. In practice, it is advisable to maintain constant altitude during the data capture phase (either stills or video), in order to get consistent imagery and to minimise the need for complex post-processing of the data. Most UAV platforms are now very stable, and the height or altitude can easily be monitored through the use of an altimeter, where a reading can be displayed on a first-person view (FPV) monitor and controlled with a standard controller and

---

TABLE 12.3
Considerations for Selection of UAV System for Oil and Gas Pipeline Monitoring

Observations needed	The pipeline system can be monitored by direct detection of hydrocarbon leaks or indirectly by monitoring indicators or surrogates of leaks (e.g. change in soil or vegetation condition).
Terrain conditions	Flat terrain simplifies UAV navigation.
	Constant height and speed are the best option for easiness of data processing. Alternatively, to monitor pipelines in heterogeneous conditions, an adaptable navigation system is necessary.
Flight distance	A strategic design of the flight aims for efficiency and cost savings. Flying distance and flying time depends on the characteristics of the network of pipelines to monitor (e.g. length, connections, risk points).
	One-way flight along the pipeline route with recovery stations at both ends (or in intermediate stations) is superior to return flights.
Platform	The type of platform to choose depends on what is required of the exercise: • Flying the entire pipeline from one end to the other on a regular basis would require an autonomous fixed-wing UAV carrying one or more sensors or a video camera. • Hot-spot inspection or monitoring would be better suited to an N-copter deployed at intervals along the pipeline.
Sensor	The sensor or sensors should be optimised for the monitoring task. These should be functional or adaptable regardless of weather conditions.
Payload weight	The platform must carry the sensor and auxiliary equipment (e.g. GPS and INU for navigation).
Data processing	Processing of data acquired to generate useful information usually involves: • Geometric correction. An exact spatial correspondence of features captured in images with reality and other data sets is crucial. • Radiometric calibration. Reliance of repetitive surveys is based on perfect radiometric calibration of measurements.
Legislation	National regulations control the options for use of one type of UAV or another. Currently the use of UAVs is still relatively restricted. As a result of a more pressing demand for applications, it is expected to be developed further.

even with a mobile phone app. For hot-spot inspection missions, e.g. to monitor the state of a pipe joint or a new structure, where high-resolution imagery is required, the flying height may be only around 5–10 m over the ground. In those cases, requiring manoeuvrability, rotary wing platforms with hover capacity are preferred.

Environmental conditions (e.g. strong wind, extreme temperatures) may restrict the flight of some UAV platforms. Further, most platforms are not waterproof, and consequently, their sensors can only be flown in dry conditions. This is especially true for most electric powered UAVs; if a UAV gets caught in the rain, it is generally best to land as soon as possible. Small N-rotor platforms are recommended to fly only up to a light breeze, and although they can be flown in a moderate breeze, the aerial imagery captured may be compromised by the instability of the yaw, pitch, and roll of the platform. Light rain and drizzle are tolerable although the rain may spoil the shots.

The data acquired by UAVs, most times images, can be visually interpreted with aerial photo-interpretation techniques. Digital image processing software, much of it now low cost, can also be used to geo-correct and mosaic the photographic prints or images together as the basis for onscreen interpretation and the mapping of thematic information for subsequent input to a GIS. Image processing (e.g. geometric and radiometric corrections) can be performed with standard or specialised DIP software. Furthermore, there is now a small range of UAV-dedicated software packages available, specifically aimed at UAV image acquisition and correction, which have similar functionality and better price. A few examples of UAV-dedicated software are listed in Table 12.4.

ADVANTAGES AND LIMITATIONS OF UAVS FOR MONITORING PIPELINES

UAV technology has some advantages over other methods (e.g. manned airborne platforms, ground surveys) for oil and gas pipeline monitoring tasks (Table 12.5). The economic cost, operational safety, and freedom of use are some important factors. Ground surveys are much more expensive, and aerial surveys are also less secure and flexible. Other important benefits of small UAVs are

TABLE 12.4
Some Software Solutions for Processing of UAV-Captured Data

Software (Company Website)	Description
Pix4D http://pix4d.com/solutions	Automatically combines raw images captured by lightweight UAVs to produce accurate measurements and visualisation of the environment, enabling timely, on-demand local 3D mapping. Automatically turns many images into accurate, 2D maps (orthomosaics) and 3D models (digital elevation models)
MosaicMill http://www.mosaicmill.com/products.html	Provides photogrammetric tools for both UAV operators and conventional manned aircraft operators. EnsoMOSAIC creates orthomosaics, 3D models, XYZ point clouds from images, GPS positions and camera calibration parameters. Calibration of camera. Output products are ready to open in any GIS software.
AirPhotoSE http://www.uni-koeln.de/~al001/airphotose.htmlv	Free and open source software. Geometric rectification of oblique aerial images and generation of orthophotos. Provides high flexibility of input /output formats and it is compatible with GIS software.
AgiSoft Photoscan Pro http://www.agisoft.com	Allows generation of high-resolution georeferenced orthophotos (5 cm accuracy with GCP) and detailed DEMs. Fully automated workflow enables processing thousands of aerial images on a desktop computer.

TABLE 12.5

Main Advantages and Disadvantages of Using UAVs for Monitoring Oil and Gas Pipelines

Advantages	Limitations
Safety in operations. Operational risk is reduced	Legal constraints: • lack of regulation • restriction of use in certain areas • restricted size for free flight
High temporal and spatial resolution data	Specialist expertise required
Programmatic flexibility: • use when convenient (weather) • on-the-fly change of schedule	Small scale of operation: • only small platforms permitted for civil use • limitation to carry specialised sensors due to weight
Access to difficult areas and perspectives	
Economic cost: • inexpensive insurance • reduced human expenses	Lack of standards
Environmentally friendly: less noise, emissions, pollution, and disturbance	Lack of collision avoidance technology
Imagery of very high spatial resolution	

related with weather conditions. Surveys based on conventional aerial platforms are restricted by wind, clouds, and other climatological agents, while UAVs are very flexible, benefiting from below cloud flying altitude and short-term change of plan. The low-height flight provides for high spatial resolution images. Small UAVs can fly in temperatures of –33°C and in 50 km/h winds offering a safe alternative to manned flights in storms and arctic climates and can also make observations in difficult environments where traditional aircraft cannot access. The cost per hour of UAVs can be low enough to make the return on investment very favourable. Since there are no crew members on-board, the vehicle safety concerns are alleviated and insurance costs are reduced, further improving the benefits of this approach. There is a wide range of geospatial and geophysical sensing and imaging technology (Table 12.2) that can now potentially be mounted onto UAVs, its use depending on the platform.

Currently the greatest limitation for use of UAVs lies in the absence of legislation and regulations to operate in airspace non-segregated from manned aircrafts airspace (Skrzypietz, 2012). This restriction is supported by security reasons, based on a UAV's lack of on-board capability to sense and avoid other aircraft.

Operational Cases

To date, most UAVs systems in use or under development for oil and gas monitoring are very costly and sophisticated systems based on large UAV platforms flying at altitudes that permit repeat coverage of large areas; the mission task is to survey areas in conflict to assure security. However, the rapid development of small-scale platforms (mini, micro UAV) and sensors as part of the UAV technology provides big potential for monitoring pipelines at the more local scale. Table 12.6 illustrates some real-case scenarios of current pipeline monitoring systems. For a different monitoring goal in each case, there is a corresponding appropriate strategy and related combination of platform and sensor.

In June 2014, the British Petroleum (BP) and AeroVironment agreement to inspect the Prudhoe Bay oil field area in search of deteriorated infrastructure and to identify flood areas vulnerable to flood (Table 12.6) represented the first large-scale, government-approved commercial use of unmanned aircraft in the US AeroVironment's UAS operators performed photogrammetric (visible

TABLE 12.6

Examples of Operational UAVs Systems for Monitoring Oil and Gas Pipelines

BP and AeroVironment	**Goal:** inspection of the oil field area for detection of deteriorated infrastructure and areas vulnerable to flood	**Platform:** Puma AE **Sensor**: LIDAR or EO/IR
	Technique: production of 3D maps of the Prudhoe Bay oil field roads, pipelines, and well pads	
ConocoPhillips and Boeing	**Goal:** surveying marine mammals and ice areas in the Arctic, necessary to meet environmental and safety rules before drilling on the sea floor	**Platform:** Scan Eagle X200 **Sensor:** EO/IR imagers and video
	Technique: offshore surveys taking off from a vessel. Controlled by a pilot on the Westward Wind, the ScanEagle sends real-time video and telemetry to the ground CS on the vessel	
Aeronautics	**Goals:** 1. Patrolling offshore fields for security 2. Undertaking leak detection in buried oil and water pipelines	**Platform:** Aerostar **Sensor:** IR camera
	Technique: differential thermal imaging, ultra-wideband, or differential RF, subsurface probing.	
BP and University of Alaska Fairbanks	**Goal**: pipeline inspection for potential leak detection in Alaska and change detection	**Platform:** Aeryon Scout **Sensor:** high-resolution visual and IR cameras
	Technique: observation from near ground level, with top and side pipeline flights. Detection of hot-spots with thermal images allowing closer inspection; change detection through repetitive flights.	

Source: After Gómez and Green (2016).

and IR) and light detection and ranging (LIDAR) data capture and analysis to monitor Prudhoe Bay infrastructure, including the gravel roads, pipelines, and a gravel pit. The hand-launched aircraft used was the Puma AE, which has a 3 m wingspan and 6 kg weight. It is capable of up to 3.5 hours flight time per battery and has the ability to fly low (~120–150 m above ground level), and slowly (<40 knots), and provided BP with highly accurate location maps to help manage the complex. Soon after, on 24 September 2014, ConocoPhillips announced it had completed the country's first commercial UAV flight off of Northwest Alaska in the Chukchi Sea (Table 12.6) with the aim to survey marine mammals and ice areas in the Arctic, something necessary to meet environmental and safety rules before drilling on the sea floor. The survey was performed with visible and IR cameras and video mounted on the ScanEagle X200 UAV which was launched from Fairweather's Westward Wind research vessel during a week of flights. The ScanEagle X200 is a 3.1 m wingspan vehicle weighing ~18 kg. The Puma AE and ScanEagle X200 are the first two UAVs approved by the US FAA for commercial applications. They both were given a Restricted Category type certificate, enabling operation in certain sectors of the Arctic area, at a maximum altitude of 600 m, 24 hours a day for research and commercial purposes. This certification represents a precursor of open UAV commercial operations expected to be approved by US Congress.

Aeronautics employs the Aerostar platform carrying an IR camera to patrol offshore fields for security (Table 12.6) in Angola. The Aerostar platform has a 12-hour flight endurance and slow loitering speed, allowing it to remain in the air the whole night long and inspect each rig thoroughly with a fully programmed flight path, making it ideal for surveillance. Aerostar is also employed to detect leaks in buried oil and water pipelines by repeatedly surveying the same area and applying differential thermal imaging. The Aerostar is a 230 kg (MTOW) platform, operates at the range of 250 km, and carries 50 kg of payloads.

In a different mission, BP employs the Aeryon Scout carrying high-resolution visual and IR cameras to inspect pipelines for potential leaks in Alaska (Table 12.6). The Aeryon Scout is a small quadcopter (~1.4 kg without payload) able to hover and fly close to the pipes in short missions of up to 25 minutes. It is particularly useful for observation from near ground level (flying altitude is below 150 m), with top and side pipeline flights. The thermal sensors on-board enabled the detection of hot-spots with thermal images. The flexibility of operations to carry on repetitive flies provides for a good change detection tool.

OIL AND GAS PIPELINES MONITORING SCENARIOS

Current inspection and monitoring needs of the oil and gas pipeline networks, coupled with existing UAV technology, suggest the possibility of a few scenarios (Table 12.7). The basics of each mission and hypothetical UAV configurations are described. Examples of commercial vehicles and sensors are provided to demonstrate the actual feasibility of the mission proposed, but other solutions are of course possible.

Scenario 1: Proximity Survey with Visual Identification of Pipe Damage

Inspection of proximate pipelines requiring detailed observation of difficult positions may require short distance flights of a micro N-copter with high manoeuvrability (e.g. Phantom). The flying altitude required is low, as the mission focus on observation of pipelines on or near the ground. This scenario represents a typical operation for assessment of risk points like junctions or for evaluation of new features in the pipeline network. A hypothetical UAV system powered by batteries for the

---

TABLE 12.7
Proposed Scenarios for Monitoring Oil and Gas Pipelines

Proximity survey/visual identification of pipe damage	UAV scenario 1: small and lightweight low-altitude UAV	
	Flying altitude	Very low (<50 m)
	Payload	<7 kg
	Endurance	<1 h
	Platform	N-copter with hovering capacity and high manoeuvrability
	Sensor	High-resolution video camera with on the fly transmission
Short-distance survey/visual identification of leak	UAV scenario 2: small and lightweight low-altitude UAV	
	Altitude	Low (<100 m)
	Payload	<25 kg
	Endurance	5–6 h
	Platform	Fixed-wing
	Sensor	Optical or NIR camera, LIDAR
Long-distance survey/automatic sensing of oil properties	UAV scenario 3: swarm of small and lightweight low-altitude UAV	
	Altitude	Low (<150 m)
	Payload	<25 kg
	Endurance	<1 h
	Platform	Fixed-wing
	Sensor	Optical or NIR camera, LIDAR

Source: After Gómez and Green (2015).

---

flight and to supply the sensor calls for the operator to be equipped with extra batteries for replacement (Figure 12.2).

Scenario 2: Short Distance Survey with Visual Identification of Leak

For monitoring inspection of a short to medium length pipeline of up to various km (depending on local legislation), an altitude of around 100 m – typically below clouds – might be appropriate. This scenario represents a repetitive and periodic monitoring mission, and will benefit from a fixed flying plan, determined by a number of way points. An adequate platform is a fixed-wing UAV (e.g. swinglet CAM), equipped with visible and IR still/video camera and possibly a LIDAR. Data should be recorded for comparison with previous and future surveys and for change detection with automatic algorithms. As part of this scenario, an additional and complementary task is the characterisation of the environmental condition for monitoring the effect of the energy infrastructure. The repetitive coverage resulting from planned flight makes for a perfect case for assessment of change (Figure 12.3).

Scenario 3: Long-Distance Survey with Automatic Sensing of Oil Properties

Beyond the current capacity of single small UAV platforms, a more difficult stage is the monitoring of long pipelines in remote areas, needing periodical observation for detection of damage or malfunction. With existing UAV technology, this monitoring scenario can be performed by larger UAV systems operated in the controlled airspace and must, therefore, be implemented in a full Air Traffic Control environment. A large UAV can be powered by a wide variety of engines and motors, as well as fuel and battery. Since the UAV is operated above 1,000 m i.e. in general not below clouds, a radar (SAR) sensor is desirable, which could be complemented by an optical/IR sensor system. This equipment in turn requires an appropriate payload capacity. The image processing and feature extraction efforts are more complicated than previous scenarios, since radar data require more

Scenario 1:
Proximity survey for visual identification of damage

FIGURE 12.2 Monitoring scenario 1: proximity survey for visual identification of damage.

Scenario 2:
Short distance survey with visual identification of leak

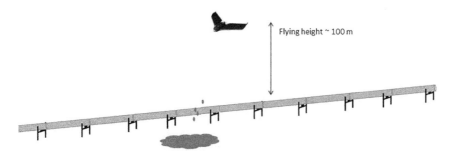

FIGURE 12.3 Monitoring scenario 2: short-distance survey with visual identification of leak.

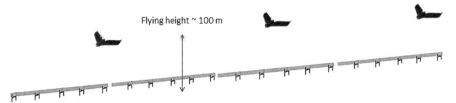

FIGURE 12.4   Monitoring scenario 3: long-distance survey with automatic sensing of soil properties.

sophisticated processing steps and is better exploited if combined with data from other sensors. We envision for the not too far future a swarm of small UAVs capable of performing this monitoring task, with individual UAVs in charge of monitoring a pipe legs, in the form of a relay system. Some technological improvements are required to this end: charging base stations at the end of each pipe leg, autonomous decision capacity for recharge or return to base in case of too bad weather conditions (Figure 12.4).

DEVELOPMENTS IN UAV AND RELATED TECHNOLOGIES

Technological improvements now underway will presumably affect the development and suitability of small UAVs and the related technology for oil and gas pipeline monitoring (Table 12.8).

The capacity to carry different sensors enables operational versatility of aerial platforms; the mounting mechanism that allows rapid change-over or exchange of different sensors on the UAV platform is known as *hot swappable sensors* and provides multi-sensor capability. For example, Aeromapper UAVs (http://www.aeromao.com/aeromapper_uav) using a hot-swappable mount can carry a wide range of sensors either on their own or in multiple configurations. Other UAV manufacturers such as QuestUAV have developed hot-swappable sensor pods designed to contain one or more sensors for ease of exchange. Similarly, Tekever Autonomous Systems (http://autonomous. tekever.com/ar4/specs.html) have developed a hot-swappable sensor payload capability for their AR4 Light Ray system for electro-optic, infrared, night vision, and low-visibility sensors.

UAV platforms are continually being developed, with main improvements including the nature, size, durability, and materials used for the construction of the platform, the capacity of the motors and batteries, and the load capacity. Some manufacturers are now issuing upgrades and new models within a very short period of time, including significant developments of the technology. DJI and 3D Robotics e.g. are revealing new and more capable platforms with the capacity to carry higher payloads. Specific developments in relation to platforms are those that combine new technologies in a safe package which provides mechanical protection for target collisions, protecting the electronics, rotors, and on-board sensors e.g. WorkFly's Flying Sensor S3 (www.workfly.net) and Riegl's RiCopter with an integrated Riegl VUX-1 laser scanning system (www.riegl.com). More recently, with the popularity of GoPro cameras on many UAV platforms, GoPro are reported to be developing their own UAV platform.

Using several UAVs simultaneously to perform a survey task may improve the efficiency of the system. A group of UAVs or swarm, inter-communicated via on-board wireless LAN may operate over the same area or in the form of a relay, reducing the time taken to perform some surveys. The approach also increases tolerance to fault, because if a sensor on a UAV fails it can return to base and another UAV could continue with the work without much interruption. Any swarm configuration (i.e. number of UAVs and distances between them) can be set up to fulfil the application requirements, aiming to optimise the coverage in adequate time and with fine data provision. For oil and gas pipeline monitoring, a swarm of UAVs could be configured, each carrying a different type

TABLE 12.8
Developments in UAV and Related Technologies

Technology	Description
	Advantage/Improvement for Oil and Gas Monitoring
Hot swappable sensors	Mechanisms to allow for the rapid change-over or exchange of different sensors on the UAV platforms • Operational versatility of platform
Platform development	Continuous improvement of design, size, durability, and materials • Improved security and reliability
Swarm (multi-UAV configuration)	Simultaneous use of several UAVs for a survey task • Reduction of survey time • Different sensors used simultaneously to gather complementary data
Cloud-based data storage	Remote data storage • Increases the value of data acquisition • Reduces need of on-board storage capacity
Battery technology	Reduction of weight speed of recharge, discharge rate, and cost • Increases UAV capacity: longer flights and heavier payload
3D printing	Creation of three-dimensional objects by layering materials under computer control • Fast and versatile production of prototypes • Cost reduction
GIS	Data storage, retrieval, analysis, and visualisation • Facilitates integration of data from numerous sources • Makes the most of data
Security systems	Improvement of emergency landing, radio connections, sense, and avoid technology • Facilitate access of UAVs to shared airspace

of sensor (e.g. visible camera, LIDAR) for acquisition of complementary data. Small UAVs with reduced storage capacity, currently an SD or MicroSD card, will benefit from Wi-Fi connections to external networks and cloud-based data storage.

Perhaps the most awaited improvements in the UAV technology are related with battery power and security systems; both will indeed contribute to the expansion and liberalisation of small UAV technology for civilian applications. More efficient batteries will enable improved capacity of small UAVs (e.g. longer missions, heavier payload including multiple sensors). Electrochemical batteries (e.g. Lithium Polymer, Lithium Sulphide) are being particularly developed, as well as metal–air batteries (e.g. Zinc–Air, Aluminium–Air, Lithium–Air). Furthermore, graphene is a promising material for battery development, with very low weight and high conductivity.

The security of UAV flights sharing airspace with manned vehicles is crucial for civilian use of these platforms. Security will become increasingly important as more of these platforms, both civilian and commercial take to the skies. Multiple efforts are now being directed to improve emergency landing procedures, radio connections, sensor function, and the availability and use of *sense and avoid* technology that will facilitate access of UAVs to shared airspace.

CONCLUSIONS

For oil and gas pipeline monitoring, UAV technology constitutes an alternative method to more traditional airborne or patrol survey, with main advantages based on cost, security, and operational flexibility. Multiple UAV systems are commercially available offering myriad possible configurations. The choice of UAV system should be led by the specific task and a precise identification of the information needs. Extensive areas to survey systematically are better covered with fixed-wing platforms and automatic flight design, whilst N-rotor platforms provide more flexibility in shorter missions, thanks to the hover capacity and high manoeuvrability. An important decision concerns the type of sensor to be carried by the aerial platform, since the sensor determines the type of data acquired and the obtainable information, as well as the need for specific mechanical designs (e.g. gimbal). Miniaturisation of sensors (e.g. LIDAR, hyperspectral) is an ongoing technological goal, being payload weigh and battery power key limitations for full capability of UAV systems in mapping and monitoring projects.

UAVs have demonstrated, through research and operational cases, the capacity to make easier some oil and gas pipelines monitoring tasks, in remote and difficult areas, as well as in repetitive operations. A number of low-cost and high-cost off-the-shelf technological solutions are available, all of which have potential for undertaking the monitoring missions, by surveying and mapping the pipelines and the environment.

REFERENCES

Allen, G., Pitt, J., Hollingsworth, P., Mead, I., Kabbabe, K., Roberts, G., Percival, C. 2015. Measuring land-fill methane emissions using unmanned aerial systems: Field trial and operational guidance, Report – SC140015. Environmental Agency, Horizon House, Deanery Road, Bristol, BS1 5AH, UK

Ames Research Center. 2003. http://www.spaceref.com/news/viewpr.html?pid=12357 (accessed 30/09/14)

Brekke, C., Solberg, A.H.S. 2005. Oil spill detection by satellite remote sensing. *Remote Sensing of Environment*, 95, 1–13

CIA. 2013. https://www.cia.gov/library/publications/the-world-factbook/ (accessed 08/10/14)

Colomina, I., Molina, P. 2014. Unmanned aerial systems for photogrammetry and remote sensing: A review. *ISPRS Journal of Photogrammetry and Remote Sensing*, 92, 79–97

CONCAWE. 2015. Performance of European cross-country oil pipelines. Statistical summary of reported spillages in 2013 and since 1971. Report No 4/15. Brussels. https://www.concawe.eu/wp-content/uploads/2015/01/Spillage-descriptions-2005-2016.pdf (accessed 28/07/15)

EASA. 2015. Advance Notice of Proposed Amendment 2015-10. Introduction of a regulatory framework for the operation of drones. http://easa.europa.eu/system/files/dfu/A-NPA%202015-10.pdf (accessed 21/12/15)

EGIG. 2015. Gas pipeline incidents. 9th Report of the European Gas Pipeline Incident Data Group (period 1970–2013). http://www.egig.eu/uploads/bestanden/ba6dfd62-4044-4a4d-933c-07bf56b82383 (accessed 28/07/15)

El-Darymli, K., Khan, F., Ahmed, MH. 2009. Reliability modeling of wireless sensor network for oil and gas pipelines monitoring. *Sensors & Transducers*, 106(7): 6–26

ERCB. 2013. *ST57–2013: Field Surveillance and Operations Branch – Field Operations Provincial Summary 2012.* Calgary, AB, p. 18

Everaerts, J. 2008. The use of unmanned aerial vehicles (UAVs) for remote sensing and mapping. *International Archives of the Photogrammetry, Remote Sensing and Spatial Information Sciences*, XXXVII, part B1

GL. 2010. Review and comparison of petroleum safety regulatory regimes for the Commission for Energy Regulation. Report Number: AA/73-01-01/03. https://www.cer.ie/docs/000458/cer11015.pdf (accessed 29/07/15)

Gómez, C., Green, D. 2015. Small-scale airborne platforms for oil and gas pipeline monitoring and mapping. Unpublished Report – University of Aberdeen. Redwing Ltd., p. 56.

Hausamann, D., Zirnig, W., Schreider, G., Strobl, P. 2005. Monitoring of gas pipelines—a civil UAV application. *Aircraft Engineering and Aerospace Technology: An International Journal*, 77(5), 352–360

Hayes, B., Jones, C., Töpfer, E. 2014. Eurodrones Inc. Transnational Institute and Statewatch. https://www.tni.org/es/search?search=eurodrones&sort_by=search_api_relevance&=Buscar (accessed 21/12/2015)

ICAO. 2011. ICAO circular 328, Unmanned Aircraft Systems (UAS). Technical Report. International Civil Aviation Authority, Montreal, Canada

IPLOCA. 2003. Safety guidelines for international onshore pipeline construction. The World Federation of Pipeline Industry Associations. http://iploca.com/platform/content/element/6885/Safety-Manual.pdf (accessed 29/07/15)

MarketsandMarkets. 2014. Pipeline transportation market by solution (security solutions, automation and control, integrity and tracking solutions, network communication solutions, and other), by modes (oil and gas, coal, chemical, water, and other)—global forecast to 2019. http://www.marketsandmarkets.com/Market-Reports/pipeline-transportation-market-110375125.html (accessed 10/10/14)

Meroni, M., Rossini, M., Guanter, L., Alonso, L., Rascher, U., Colombo, R., Moreno, J. 2009. Remote sensing of solar-induced chlorophyll fluorescence: Review of methods and applications. *Remote Sensing of Environment*, 113, 2037–2051

Mishra, D.R., et al. 2012. Post-spill state of the marsh: remote estimation of the ecological impact of the Gulf of Mexico oil spill on Louisiana Salt Marshes. *Remote Sensing of Environment*, 118, 176–185

Petrie, G. 2013a. Commercial operation of lightweight UAVs for aerial imaging and mapping. *GEOInformatics*, 1, 28–38

Petrie, G. 2013b. Current developments in airborne laser scan: Progress is being made! *GEOInformatics*, 12, 16–22

Skrzypietz, T. 2012. Unmanned aircraft systems for civilian operations. Big policy paper no 1, p. 28. http://www.bigs-potsdam.org/images/Policy%20Paper/PolicyPaper-No.1_Civil-Use-of-UAS_Bildschirmversion%20interaktiv.pdf (accessed 03/02/16).

UVS. 2014. *RPAS Yearbook. Remotely piloted aircraft systems: The global perspective.* 12th edition. Paris, Blyenburgh &Co.

van der Werff, H., van der Meijde, M., Jansma, F., van der Meer, F., Groothuis, G.J. 2008. A spatial-spectral approach for visualization of vegetation stress resulting from pipeline leakage. *Sensors*, 8, 3733–3743.

Watts, A.C., Ambrosia, V.G., Hinkley, E.A. 2012. Unmanned aircraft systems in remote sensing and scientific research: Classification and considerations of use. *Remote Sensing*, 4, 1671–1692; doi:10.3390/rs4061671

13 Drone-Based Imaging in Archaeology
Current Applications and Future Prospects

Stefano Campana
University of Siena (Italy)

CONTENTS

THE BACKGROUND

A BRIEF OUTLINE OF THE HISTORY OF UAV APPLICATIONS IN ARCHAEOLOGY

Leaving aside the very first acquisition of aerial photographs from an Unmanned Aerial Vehicle (UAV), the first applications in the field of archaeology took place around the turn of the millennium. By UAV applications we mean in this context structured projects ranging in their scope from flight planning to the flights themselves, along with data acquisition, post-processing, and ultimately the interpretation and discussion of the results (Nex and Remondino, 2013).

One of the most significant pioneering initiatives was that promoted by ETH Zurich under the leadership of Professor Armin Gruen. He first developed, virtually independently, a reliable pipeline based on UAV data acquisition for achieving accurate and detailed 3D models and digital cartography (Eisenbeiss et al., 2005; Eisenbeiss and Zhang, 2006; Lambers et al., 2007; Pueschel et al., 2008; Remondino et al., 2009; Eisenbeiss, 2009). Another important case study, especially for its attention to the sensors and to a wide range of potential archaeological applications, was initiated by the University of Gent in Belgium. In addition to the development of an interesting aerial platform (Verhoeven et al., 2009) within the traditional category of the Helikite, the developers at Gent carried out experiments in the field of archaeological diagnostics, using optical sensors that operated in the non-visible portions of the electromagnetic spectrum (Verhoeven and Loenders, 2006).

Other pioneering initiatives followed in the wake of the innovative work undertaken by ETH Zurich, characterised in particular by the first use of commercially developed aerial platforms (Campana et al., 2008; Remondino et al., 2011; Lo Brutto et al., 2013; Sordini et al. 2016) or of systems assembled within university laboratories and research institutes (Oczipka et al., 2009; Seitz and Altenbach, 2011, Rinaudo et al., 2012) or small commercial companies (Piani, 2013). At that stage, everything was still highly experimental, and the development work required significant technical skills. Nowadays the main applications of UAVs in archaeology are aimed at the photogrammetric survey of landscapes for the creation of digital surface and digital terrain models (DSMs and DTMs) and at the 3D recording of excavations and historic monuments. The case studies cited in this context have amply demonstrated the effectiveness and advantages of these novel systems and procedures compared with traditional instruments; at the same time they have involved the procedural development of cost-effective and reliable pipelines, free of obstructive bottlenecks (Chiabrando et al., 2011; Nocerino et al., 2013; Sordini et al., 2016).

There have of course been many problems to overcome, starting with the aerial platforms themselves. At the 4th International Congress on Remote Sensing in Archaeology, held at Beijing in October 2012, Armin Gruen, in his keynote address on UAV systems defined this technology as potentially extraordinary in its impact but still unripe in its reliability, its capacity to operate in critical condition, and in its safety of operation (considerations repeated by Professor Gruen in 2014: https://www.youtube.com/watch?v=LurNobNVlEM). Similar problems were encountered with the software for image processing and the generation of high-quality 3D models. In recent years, however, there has been a substantial improvement in all of the key segments of the procedural pipeline: the platform, the sensors, and the software for post-processing. In particular, the platforms have benefitted from important developments involving systematic improvements to all of the main evaluation parameters listed in Figure 13.1. Costs have also fallen drastically but high-end UAV systems still require investments in the order of tens, up to hundreds and thousands of euro. That said, there remains ample room for further improvement. For instance, the software systems on the platforms themselves are still fairly basic, good enough for following predetermined routes but incapable as yet of responding independently to unforeseen events or circumstances during flight. The relatively short flight endurance of most platforms (rarely more than about 30 minutes before a necessary change of battery) often reduces their efficiency in use, forcing the user to undertake several flights in the same working day.

The final segment in the procedural workflow involves the post-processing software, in which development has in recent years advanced in leaps and bounds. Until the first decade of the present millennium, image-based modelling (IBM) software procedures were for the most part manageable only by specialists in the pure and applied sciences, using in-house software subject to severe limitations in the implementation of automatic processing (Remondino and El-Hakim, 2006). Nowadays, however, there are open-source, low-cost, and semi-automated commercial software packages that can be operated by competent users without necessarily demanding the involvement of specialists (Nex and Remondino, 2013; Sordini et al., 2016).

THE MAIN TYPES OF UAV PLATFORMS CURRENTLY EMPLOYED IN ARCHAEOLOGY

The cost of a UAV for archaeological purposes can vary enormously, from less than €500 up to €30,000 or even as much as €250,000. The variation of course reflects substantial differences in the on-board instrumentation (sensors and Global Positioning System – GPS), the payload, the flight range, the platform type, and the degree of automation and therefore versatility in use.

Low-cost platforms have improved significantly in recent years. It is possible today to find within the commercial market systems that can perform autonomous flight, with a payload sufficient to carry, or more often to embed, a good quality camera, with good reliability, and an acceptable capacity to deal with wind and other environmental factors.

For archaeological purposes, it is possible to draw a distinction between two different types of aerial platform. The first, the so-called 'fixed-wing' variety, consists of small- and medium-sized aircraft ranging from about 60 cm to 2 m in approximate wingspan; these can provide coverage at landscape scale, making it possible in optimal conditions to scan within a day's work an area of up to about 600 hectares, flying at a height of about 150 m with a ground resolution of about 2 cm/pixel. The second main category consists of 'multi-rotor' systems equipped with three, four, six, eight, or more propellers, sometimes in pairs rotating in opposite directions. These can be further subdivided according to their size, payload, and level of sophistication into 'low' and 'medium' brackets for the kind of UAV generally used in archaeological work, along with 'higher-end' systems involving the use of more sophisticated and innovative sensors, those enabling the collection of Light Detection and Ranging (LIDAR) data or geophysical data (radar and magnetic measurements) being of particular interest in the present-day context. Despite the differences between these sub-categories, it is possible to identify some common characteristics that apply with reasonable consistency to all multi-rotor systems: accuracy in positioning of the platform, an acceptable level of automation, a high geometric resolution of data acquisition, an ability to record vertical structures such as the facades of buildings and other structures as well as more or less horizontal surfaces such as archaeological excavations in progress (Figure 13.1). The fixed-wing and multi-rotor types of UAV are thus complementary in the functions that they can serve.

DRONE-BASED SENSORS CURRENTLY USED IN ARCHAEOLOGY

Remote sensing traditionally identifies three key segments: platform, sensor, and data processing. In archaeology, as in other fields of application, the platform/sensor/base-station se-tup is not an end in itself but simply the means of acquiring of the desired information. In this process, the sensor clearly plays a fundamental role. Having defined the objectives of a hypothetical research programme and established the need for UAV platform to undertake aerial survey, it is necessary to define the required characteristics of the sensor; in combination with other factors this will in turn affect the choice of aerial platform. In recent years, the rapid development of the UAV market

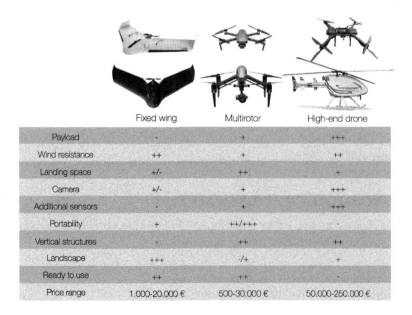

	Fixed wing	Multirotor	High-end drone
Payload	-	+	+++
Wind resistance	++	+	++
Landing space	+/-	++	+
Camera	+/-	+	+++
Additional sensors	-	+	+++
Portability	+	++/+++	-
Vertical structures	-	++	++
Landscape	+++	-/+	+
Ready to use	++	++	-
Price range	1.000-20.000 €	500-30.000 €	50.000-250.000 €

FIGURE 13.1 Summary of the main features of the three main types of UAV platforms used in archaeology.

has stimulated manufacturers to design sensors specifically adapted for this type of platform. The published literature on remote sensing identifies two main types of sensor: 'active' and 'passive'.

'Passive' sensors are in essence those used in digital cameras for recording the visible portion of the electromagnetic spectrum. This type of sensor is widely available, either separately or in combination with cameras that offer a wide and constantly evolving functionality. The choice of the technical characteristics of the sensor should be based on the research objectives. There is a progressive move towards the documentation of structures, sites, and landscapes in three dimensions through the use of multi-image photogrammetry (IBM). For this kind of work, the quality and resolution of the sensor must be medium to high. Generally, the higher the quality of the camera (and especially the lens), the greater is its weight. However, weight is a parameter that must be kept under control in aerial work since it has a direct impact on the type of platform that will be used, the autonomy of the flight, the skill of the intended pilot, and ultimately, of course, the overall cost. A very important factor in the choice of the camera is the possibility to save and output files in *.raw or other uncompressed formats (for subsequent conversion, for example, to *.tiff and *.bmp formats). Indeed, compressed files make it much more difficult, and sometimes even impossible, to create high-quality 3D models. In addition to the production of 3D models and maps, digital cameras can be used in exploratory aerial reconnaissance and the monitoring of sites and landscapes.

Another interesting category of 'passive' sensors has been developed for recording information in the near, medium, and thermal infrared and near-ultraviolet ranges (Verhoeven, 2009b; Verhoeven and Schmitt, 2010). Within the last year or two (2015–2016), several cameras specifically designed for use on UAVs have made their appearance. For a few examples, see Tetracam ADC Lite (http://www.tetracam.com/Products-Micro_MCA.htm), MAIA (http://www.salengineering.it/public/it/p/camera_multispettrale_maia.asp), MicaSense RedEdge (http://www.micasense.com/rededge), and FLIR Vue Pro (http://www.flir.com/suas/content/?id=70728). These sensors have great potential value, especially for exploratory aerial reconnaissance. An interesting recent development is the great interest of some major manufacturer (e.g. DJI and Parrot) to develop multi-sensor low-cost drones including good-quality embedded thermal cameras.

Among the 'active' group of sensors, those designed for recording LIDAR data have dominated the scene so far, for reasons that will be discussed later in connection with the exploratory recording of wooded or partially wooded areas. An obvious indication of the high demand for, and interest in, 3D data in many areas of landscape study lies in the wide range of miniaturised systems for UAV observation that are currently available (Figure 13.2). However, LIDAR sensors have cost and weight implications that demand the use of high-end aerial platforms, thereby increasing the overall cost of any intended survey work.

The latest drone-based category of sensors currently under development and of great value for archaeology is airborne geophysics. Particularly magnetic (http://www.gemsys.ca/uav-magnetometers/) and radar sensors (http://www.radarteam.se) are currently under development also for archaeological applications.

PRINCIPAL ARCHAEOLOGICAL APPLICATIONS

In the past 15 years or so, the main UAV applications, at varying levels of intensity and with encouraging degrees of success, have played fundamental roles in:

a. 3D documentation of archaeological excavations
b. 3D recording of historic building and standing monuments
c. 3D survey of sites and landscapes
d. Aerial exploration and monitoring
e. Detection and documentation of under-canopy archaeology in forested areas

FIGURE 13.2 Some examples of drone-based sensors applied in archaeology. Clockwise from top-left: commercial low-cost microdrones with embedded thermal cameras; Riegl multi-rotor (Ricopter) equipped with LIDAR and high-resolution camera; Swiss drone (Dragon 35) equipped with Riegl LIDAR, high-resolution camera and thermal camera; DJI drone equipped with Cobra plug-in GPR antenna.

The rapid development of platforms, sensors, and cloud-based data processing (Amazon AWS, Google Cloud Platform, etc.) has been matched by the pioneering of new applications by archaeologists, engineers, geophysicists, and others. Drones are no longer the esoteric curiosity that they seemed in the past: for today's archaeologists, they are an everyday presence, a piece of equipment ready to hand at any time, a basic tool for any researcher who ventures into the field to explore, examine or record, a fly-day utility (e.g. DJI Mavic series, GDU O2, Parrot Anafi, Levetop). At the time of writing, exciting experiments are in hand with new and potentially revolutionary applications, both in fundamental research and in well-established fields of study such as magnetometry, terrestrial radar, field-walking survey, and aerial reconnaissance.

There can be little doubt that in the next few years drone platforms will revolutionise basic methods and practice within archaeological survey at both the point and local scale (Campana, 2018), radically expanding the opportunity to deploy these innovatory tools and procedures within an ever-widening range of environmental and cultural contexts.

Documenting Archaeological Excavations through
the Use of Drones and 3D Recording

Examples of ground-based experience in the 3D documentation of excavations through the use of laser scanners and photogrammetry have increased in numbers and sophistication in recent years: see for instance (Doneus et al., 2003). By contrast, the development of new standards and procedures for the use of UAVs in 3D documentation is still in its relative infancy, the published discussions of this topic being as yet few in number. In addition to those discussed below, however, it is worth mentioning the contributions of Sauerbier and Eisenbeiss (2010), Seitz and Altenbach (2011), and Rinaudo et al. (2012). Interesting ideas have also been put forward as a result of some earlier researches, notably by Tokmakidis and Skarlatos (2000) and Shaw and Corns (2008).

Archaeological excavation is a fundamental but essentially unrepeatable process in the exploration and understanding of the past. These two objectives, exploration and understanding, are the main reasons why archaeologists consider it imperative to seek constant improvement in their excavation techniques and in refining the quality of their documentation. The attention to excavation techniques and the achievement of increasingly accurate documentation inevitably involve the expenditure of time and attention on these aspects of the process. In essence, these are the circumstances which in 2004 led the writer in his official role at the Laboratory of Landscape Archaeology and Remote Sensing (LAP&T) at the University of Siena to begin experiments with new ways of documenting excavation evidence. In collaboration with Leica Geosystems, a series of tests were carried out using laser scanning in the years between 2004 and 2006. The context was a long-running archaeological excavation at the site of a Roman villa that later developed into a large early Christian church at Pava in the Asso Valley in southern Tuscany. The initial results were positive, but as the excavation grew in size, it was soon realised that this system was unsustainable in the longer term for a variety of reasons, including the relatively low acquisition rate, the high number of stations necessary to limit the incidence of occlusions, and the time-consuming post-processing routines.

With the aim of overcoming these limitations, a collaboration was initiated in 2007 with Fabio Remondino of FBK Trento and with Zenith, a commercial company, to test the potential of UAV photogrammetry in recording the progressively increasing excavation area (Campana et al., 2008; Remondino et al., 2011). The ultimate objective was to develop a complete workflow, from flight planning to 3D modelling, layers and volumes mapping, so as to achieve the semi-automatic creation of 3D and 4D documentation of the archaeological excavation. The essential characteristics of the process were envisages as being speed, reliability, accuracy, standardisation, and user-friendliness. In addition to the potential applicability to any archaeological context, it was taken as granted from the outset that any effective workflow established in the context of the Pava excavations could then play a significant role in the wider context of planning-led and rescue archaeology, within which there is often a need to document large surface areas in a relatively brief period of time.

In addition to a reduction of working time in the field, the fundamental 'added value' of such a pipeline, in qualitative terms, would lie in the development of easily applicable standard procedures to a wide variety of archaeological contexts, with a consequent gain in the expected quality and homogeneity of the resulting documentation. The UAV platform made available by the Zenith company was a Microdrones md4–200 system equipped with a Pentax compact camera (Figure 13.3). Unfortunately, the problems encountered in the early years of testing tended to outweigh the benefits of the results. Difficulties were encountered in almost every part of the process. The aerial platform had major problems of reliability and operability in response to even moderate wind, while the sensor was limited by the relatively low lifting power of the UAV to a few lightweight compact cameras, none of which was capable of recording uncompressed (RAW) files. These limitations greatly complicated the processes of data processing and 3D modelling.

Nevertheless, these initial experiments were important in making it possible within the next few years to implement the procedure in its entirety through the advent of the next generation of aerial

FIGURE 13.3 Clockwise from top left. Microdrones md4–200 UAV; DTM of the Pava excavation following the processing of the drone photography, created by Fabio Remondino using in-house ETH Zurich software; aerial recording in progress at the Pava excavations; and detail of the camera used in the recording work.

platforms, cameras, and processing software. The context for this next phase of experimentation, carried out in collaboration with the ATS srl, a spin-off company of the University of Siena, was provided by archaeological excavation at Santa Marta, a large Roman villa situated in the Ombrone Valley that spans the provinces of Grosseto and Siena, again in the southern part of Tuscany. The new aerial platform was an Aibot X6 hexacopter, a system with six rotors protected by a carbon fibre cowling, capable of carrying a payload of up to 2 kg (Figure 13.4). The drone can be flown either manually or autonomously, including take-off and landing, with the aid of an integrated GPS navigation system. A significant advantage of the Aibot drone is its enhanced tolerance of variations in atmospheric condition, in particular wind, making the system as a whole far more versatile for use in various kinds of archaeological fieldwork. The carbon fibre cowling protects the six rotors from potential damage, making it possible for the aircraft to operate in spatially restricted environments and at low levels with a very high degree of reliability.

Within this set-up, it is possible to choose any type of camera (within the 2 km payload limit of the UAV). The nature of the camera mount minimises any vibration that might be transmitted to the sensor, allowing the capture of digital images that have accurate focus and a good depth of field. The setting up of the system is very fast; in no more than a few minutes, it is ready to fly. The payload and the stability of the UAV allow the use of either reflex or mirrorless cameras, producing very high-quality images.

This system had been used in 2013, 2014, and 2015 to record the progress of the Santa Marta excavations. The documentation of the open areas, amounting to about 700 m² of the overall 2,000 m² extent of the excavation, was achieved in less than 2 hours of work on each occasion, including related topographic survey on the ground. The level of detail in the images, the repeatability of

FIGURE 13.4 Clockwise from top -left. Aibot X6 Hexacopter; processing of images with AgiSoft PhotoScan Pro software; overview of the vector drawing overlaid on the orthophoto image in the project's GIS; detail of the vector drawing; aerial photo from the drone recording the state of excavation at the end of the 2013 season of fieldwork.

successive flights conforming to the same basic parameters, and the high speed in collecting the photogrammetric coverage made this complex of instrumentation particularly suitable for repeated documentation of the excavation area as the work progressed.

In order to optimise, the process of 3D recoding is of course necessary to integrate the aerial documentation derived from the drone with that created by total station survey at ground level; otherwise it would be necessary to undertake a fresh flight whenever a new feature or stratigraphical unit was identified, making the process unsustainable in practical terms. This is not a problem in reality since any present-day archaeologist can quickly learn how to collect (and geo-locate) ground-based photographs for photogrammetric purposes, using a camera and lens of relatively mid-quality specification.

The images collected from the drone were processed through the low-cost and semi-automatic photogrammetry package AgiSoft Photoscan Pro (http://www.agisoft.ru). The software, as would be expected, requires total coverage of the survey area with at least 60% overlap between the frames (ensuring that every part of the area is covered in a minimum of three overlapping images), along with geo-location of relevant reference points at ground level through total station survey.

In this kind of recording work, it is important to execute the successive stage in the workflow with a consistent level of accuracy if one is to achieve reliable documentation in which the spatial relationships between layers and features are properly expressed, especially when dealing with successive episodes of on-site recording. Given this precondition, the PhotoScan software provides a powerful tool for integrating different point clouds to create diachronic models, a very useful feature within archaeological research and dissemination. For example, the part of the bath area that had been excavated at Santa Marta in 2013 was then covered over for conservation purposes, making it possible in the following year to survey only the area that was newly excavated during the 2014 campaign. Removing the coverage from the 2013 excavation was considered too time consuming and expensive, so the archaeologists did not have the possibility of seeing or recording the whole of the excavated area at the same time. However, it proved possible to create a model of the whole area uncovered during the successive season of fieldwork by integrating the different datasets acquired in 2013 and 2014. Each point cloud was first 'cleaned' by removing points from the areas that overlapped with one another. A single mesh for the whole area was then created through the PhotoScan software. The resulting orthophoto is shown in Figure 13.5.

The next step of the workflow involved texture mapping on the polygonal meshes and the creation of orthophotos of each area (Figure 13.6).

The documentation achieved at each stage was used for the creation of vector drawing of each stratigraphical and structural context. Orthophotos were imported into the (2D) Geographic Information System (GIS) used for storage and management of information collected during the successive stages of excavation. The high precision, accuracy, and resolution of the aerial images allowed drawings of each layer and feature to be made in great detail and in less time than would have been possible by traditional on-site measurement and drawing. It is important to emphasise here, however, that the work at Pava, and at Santa Marta more specifically, has always envisaged a close integration between work on-site and work in the laboratory. This should in no sense be a linear and sterile relationship but one that involves a continuous feedback in both directions between fieldwork and laboratory. At present, the project GIS at Santa Marta uses MeshLab (http://meshlab. sourceforge.net/) to manage 3D models of stratigraphic contexts. This makes it possible to manage a large amount of data about individual layers and features, displaying or hiding them as and when necessary, implementing real-time measurements and (for instance) varying the direction of lighting within the digital 3D environment. The archaeologists involved in the project can thus explore the archaeological data in 3D as well as consulting a traditional GIS (for a detailed description of

FIGURE 13.5 Orthophoto mosaic of two adjacent survey areas made by combining the results of UAV photography collected during separate excavation campaigns at Santa Marta. Note the comparison with the image at bottom-left in Figure 6, in which the apse structure at upper-left in that image had not yet been excavated.

FIGURE 13.6 Photomosaic produced from UAV images taken during two different excavation campaigns at Santa Marta. Note the comparison with the image at bottom-left in Figure 6, in which the apse (towards the top of that image) had not yet been excavated.

these operations see Sordini et al., 2016). This kind of workflow, based on both UAV and ground-based photogrammetry, can also be used to facilitate traditional 2D mapping and GIS data management as well as 3D vector drawing (see for instance Dell'Unto, 2014).

3D Survey of Monuments and Historic Buildings

The UAV-based survey of archaeological monuments and historic buildings has been among the most prevalent applications from the early days of these application in archaeology and heritage management. The research team at ETH Zurich, for instance, concentrated much of its early experimentation on cultural heritage in general, and individual sites and monuments in particular, notably at the Maya site of Copan in Honduras, where Temple 22 was recorded within its surrounding context (Remondino et al., 2009). Another relevant project concerned recording of the medieval castle at Landenberg in Switzerland (Pueschel et al., 2008). There have been many subsequent examples of this kind of application, for instance, by LAP&T in Tuscany (Campana et al., 2008, 2009), by 3DOM and the Fondazione Bruno Kessler (Remondino et al., 2011; Fiorillo et al., 2013), by the Polytechnic of Turin (Chiabrando et al., 2011), and by various private companies.

From the experience accumulated so far, it is possible to identify some general feedback. A first aspect concerns the choice of aerial platform for initiating the recording work. While fixed-wing UAVs are very efficient at surveying large areas in order to produce orthophotos, maps, and 3D landscape models, multi-rotor systems are more suitable for the recording of individual sites and monuments because of their inherent characteristics. Multi-copters can record both horizontal and vertical or sloping surfaces with equal facility, as required in the modelling of facades and architectural details. Another common feature of working with these applications is the need to integrate the results of UAV survey with ground-based photogrammetry and laser scanning. Among the main benefits remarked on in the scientific literature so far are the relative speed, economies of cost, levels of accuracy, and ability to record elements that are not readily visible from the ground or from available vantage points (roof tops, architectural details, and stratigraphical relationships, etc.).

A very important feature, explored more fully below, is the inherent capacity of all forms of aerial photography to place individual monuments and historic buildings within a record of their immediate surroundings, thereby recognising and communicating the essential relationship between architectural structures and the urban or rural landscape within which they lie (Figure 13.7). No doubt

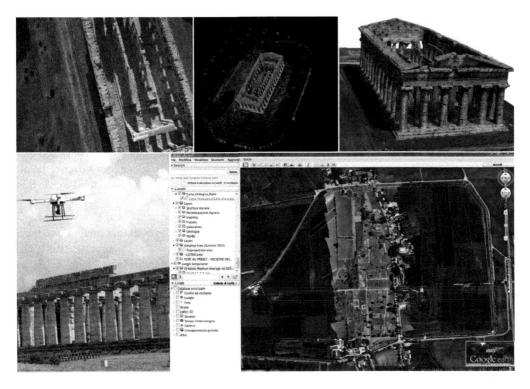

FIGURE 13.7   UAV survey of temples and the ancient city of Paestum, Italy. (Courtesy of Fabio Remondino and 3DOM FBK Trento.).

many readers will have experienced the uncomfortable sense of alienation that can be aroused by otherwise excellent 3D models presented in a virtual space totally devoid of information about their original (or even present) surroundings. This should surely act as a spur to go beyond the simple measurement and 3D representation of the collected data (with all its well-known advantages in monitoring, measurement, and the automatic extraction of plans, profiles, and cross-sections) by switching to a more analytical and critical approach in the use and interpretation 3D documentation, not just for individual structures but also for the immediate surroundings which form an essential part of their existence (Figure 13.7).

THE SURVEY OF ARCHAEOLOGICAL SITES, RURAL AND URBAN LANDSCAPES

It has already been mentioned that the topographical survey and digital representation of sites and landscapes figure strongly among the outputs of UAV applications within heritage studies. The final goal in these instances is to achieve detailed maps and 3D models of local landscapes and the archaeological sites that lie within them. This scale of landscape has proved to be an elusive target until quite recent times. In the past, first total station survey, then GPS-based survey, and finally laser scanning have provided successive approaches to the recording of such landscapes. However, the mapping of landscapes using these devices proved to be highly time-consuming and expensive; moreover, the outcomes often lacked sufficient detail and the tools used did not provide any capacity to add texture to the results achieved. On the other hand, traditional aerial photogrammetry, in addition to being very expensive, was not always capable (depending on the size of the sites involved) of providing sufficient detail for their detection and depiction. Against that background, and without citing specific examples, there is no doubt that drone-based 3D recording fills a methodological gap within the archaeological mapping process. In the literature, numerous case studies have been reported from around 2005 onwards. Originally, their primary aim was to test new technological developments and to verify the technical quality and accuracy of the survey results (Bendea et al., 2007; Lambers et al., 2007; Remondino et al., 2011; Fiorillo et al., 2013). In the author view, however, recent publications show a mature approach, most of all in the composition of the research teams, which no longer consist almost exclusively of scientists and engineers but also benefit from the presence/input of archaeologists. In addition, present-day implementations of UAV systems now tend to form parts of a complex strategy aimed primarily at answering archaeological questions (Brenningmeyer et al., 2016; Sonnemann et al., 2016). For instance, around the end of 2018, the author's specialist team of archaeologists carried out a drone-based survey of Nineveh (Mosul). The demanding local situation allowed only about 5 hours of work per day but in 3 days it proved possible to collect 750 hectares of 2 cm resolution data using a simple DJI Phantom 4 drone, the aim being the detailed recording and assessment of the damage caused by warfare and the Islamic State of Iraq and Syria (ISIS) occupation within the archaeological area.

In conclusion, the results achieved in recent years have clearly demonstrated the extraordinary speed and effectiveness of drone-based survey in the creation of high-resolution images and maps; other aspects include the high levels of accuracy that can be achieved and the benefits that accrue from having access to 3D images with all of the analytical potential that these provide. The advantages of this type of relief mapping, at this scale of detail, are by now manifest and convincing, in many cases rendering anachronistic the systems that were used until just a few years ago.

Exploratory Survey and Aerial Reconnaissance

While survey work aimed at the better understanding of already-known sites and features is a highly developed field of work, drawing on a wide range of methods and platforms (Verhoeven, 2009a), exploratory survey through the use of UAVs is an innovation that is directly related to the dynamic characteristics of this particular form of platform (Brenningmeyer et al., 2016). The main novelties compared with the use of light aircraft in traditional aerial reconnaissance, are enhanced cost-effectiveness, high geometric resolution, unprecedented opportunities to acquire data in a wide range of differing environmental and field conditions (lighting, seasonality, land use, morphology, etc.), and coverage at a 'new' scale of detail, filling a real gap in the previous methodological workflow. Indeed, in this kind of survey, the main cost lies in the initial purchase of the UAV, there being few subsequent management or maintenance costs apart from the occasional replacement of moving parts affected by wear (or occasional accidental damage).

Fixed-wing UAVs are particularly suitable for this kind of exploratory reconnaissance, being capable of covering relatively large areas, up to as much as 300 hectares per day. Technically, there is no reason why this scale of coverage could not be increased to two or three times as much, but present aviation regulations in most countries compel the pilot to keep direct eye contact with the UAV. However, even in this case, the choice of the most appropriate platform should be weighted according to research objectives; successful applications in this area have also been achieved through the use of multi-rotor platforms (Oczipka et al., 2009; Sconzo, 2014; Lang et al., 2016). The experience gained so far, though still fairly limited in its scope, confirms some of the readily imagined potential of the method, highlighting further aspects of particular or more detailed interest. For example, research at the huge Celtic settlement of Heidengraben in Germany, as well as in Oman at Al Khashabah and in the Orkhon Valley of Mongolia, has shown that in addition to traditional diagnostic capabilities such as the detection of cropmark and soilmark evidence, there is enormous potential for the collection of high-resolution photogrammetric data, and the extraction of high-quality DTMs can assist the discovery, mapping, and subsequent survey and monitoring of micro-morphological features of archaeological or related significance (Lang et al., 2016; Oczipka et al., 2009).

Micro-morphological features of this kind fall within well-established categories of aerial survey interpretation, but the environmental conditions in which such features become visible, in particular the incidence of low oblique lighting, are limited to certain seasons of the year, especially autumn and winter, and specific times of the day, soon after dawn and just before sunset (Piccareta and

Ceraudo, 2000; Musson et al., 2005). Conversely, operating on a digital 3D model acquired in less specific environmental conditions, it becomes possible to simulate whatever conditions, angles, or directions of lighting that will give emphasis to the micro-morphological features. This can also be done in such a way as to allow the angle of view to be varied in real time, just as it would be if the observer were able to move at will around the chosen target – a key element in enhancing the capacity of eye and mind to comprehend the physical characteristics and potential archaeological interpretation of the site or object under study. It should be acknowledged, of course, that the creation and examination of 3D models of this kind is also possible (admittedly at considerably greater cost), through 'traditional' aerial survey from light aircraft, or even from existing historical air photographs if these are available in sufficient numbers (for an example involving the mapping of the Roman-period mine workings in Spain see Verhoeven, 2012).

Traditional aerial photography (using optical sensors operating within the visible electromagnetic spectrum) has among its main limitations the need to work within fairly short time windows, combined with the inability to control all (or any) of the variables that determine success or failure: geology, pedology, micro-climate, agricultural patterns, farming practices, and historical landscape development. Archaeologists have long been conscious of their inability to control the conditioning factors that influence the visibility or invisibility of archaeological features buried beneath the soil (Jones and Evans, 1975). As a result, the discovery of archaeological traces through aerial photography is to a significant extent subject to the incidence of chance and serendipity (Campana, 2016).

An important contribution to expanding the current time windows and reducing the effect of serendipity on the outcome of aerial reconnaissance (thereby strengthening the diagnostic potential of the photography) could come from the measurement and analysis of non-visible portions of the electromagnetic spectrum. UAV systems offer the archaeological community the possibility of equipping platforms not only with optical systems operating in the visible range but also with sensors collecting data within the non-visible parts of the spectrum. Key experiments in this field have been conducted by the Department of Archaeology at the University of Gent, in Belgium. In this case, the data were collected through the medium of a Helikite. The research was undertaken mainly within the Potenza Valley Project in central Italy, a long-term collaboration that has been integrating a wide variety of research methodologies, including surface survey, geo-archaeological survey, geophysical prospection, and archaeological excavation (Vermeulen et al., 2006). In this case, the sensor was a standard single-lens reflex (SLR) camera that had been modified to record the near-infrared portion of the electromagnetic spectrum (Verhoeven, 2008). Overall, the results showed that the use of the differing bands, in combination, resulted in a considerable improvement in the visibility of the archaeological features, in some cases even allowing the recovery of traces that were otherwise completely invisible. This situation has been verified not only for cropmark evidence but also for soil marks (Verhoeven, 2012). Recently, an interesting experience has been developed in New Mexico by a team of the University of Arkansas and North Florida (Casana et al., 2014). The case study has been conducted at the Chaco-period Blue J community in north-western New Mexico. It presents a method for collection of high-resolution thermal imagery using an unmanned aerial system (UAS), as well as a means to efficiently process and orthorectify imagery using photogrammetric software. Results show clearly the size and organisation of most habitation sites to be readily mapped and also reveal previously undocumented architectural features. That said, the use of UAVs for this kind of experimental work with infrared or multispectral imagery is still in an early stage of development though a number of innovatory projects currently in progress will no doubt be brought to publication in the coming years.

ARCHAEOLOGY BENEATH THE WOODLAND CANOPY

From the second half of the 20th century, topographical studies within the field of archaeology have increased in number and sophistication, in the Mediterranean region as well as in other parts of Europe. Millions of hectares of open landscape have now been investigated, and our understanding of settlement patterns, rural populations, productive systems, and both large-scale and local trading patterns has improved substantially compared with the past. There have been several published discussions about the advantages and limitations of this kind of survey work (for instance Campana, 2009 or more recently Campana, 2016). However, wooded areas continue to present a very challenging environment in which to operate for purposes of archaeological survey.

The proportion of the landscape concealed beneath tree cover of one density or another is very considerable – about 45% of the total landmass, for instance, in Italy and the Mediterranean world more generally (FAO, 2006). This issue is exacerbated by the fact that these areas are not equally distributed in all kinds of habitats but are mainly concentrated in upland and mountainous areas (Blondel et al., 2010). In this context, Fernand Braudel (1949), in the first pages of his monumental work on the Mediterranean world, complains about the uneven geographical distribution of historical research. More than 30 years later, Graeme Barker, in the introduction to his monograph on the Biferno Valley (1995), echoes Braudel in reiterating that in the intervening years, landscape survey had still tended to concentrate on lowland areas at the expense of the higher ground. Today, a further 20 years after the Biferno Valley publication, this remains very much the case in the Mediterranean area, despite a few notable exceptions. The reasons, from a strictly archaeological perspective, must be sought in the physical overlap between 'highland' and 'woodland' and in the relative lack of archaeological survey methods for tackling the intrinsic characteristics of wooded areas. Indeed, in these more or less impenetrable environments, unsuited to arable cultivation and hence undisturbed by soil movement in recent centuries, field-walking survey is completely ineffective, as are most other approaches used by archaeologists in more open areas. Only aerial photography can claim to have had some limited success in these upland areas and even then only for the highest areas where the woodland cover is less dense or in some cases absent altogether. Metaphorically, forested areas constitute a massive 'black hole' in the firmament of landscape archaeology, creating a void of similar proportions in our understanding of the landscapes of the past. From around the beginning of the present millennium, however, the increasing use of the LIDAR sensor in topographical mapping and archaeological survey has introduced new opportunities for the exploration of previously unrecorded archaeological features preserved beneath the woodland canopy.

In the last 10 years or so, a fair number of LIDAR surveys have been implemented for archaeological purposes, and it is now widely accepted that this technique represents the most efficient system for the exploration of wooded areas (and in some respects for open pastureland too). This has led the writer, within his own research work as a Marie Curie Fellow at the University of Cambridge in the UK, to compile a fairly detailed survey of projects that have made active use of LIDAR survey for specifically archaeological purposes, not only around the Mediterranean but also across other parts of Europe more generally (Figure 13.8).

The result shows some interesting trends and some equally obvious limitations in the application of LIDAR technology within archaeology. Particularly noticeable is the lack of case studies in the Mediterranean area, which accounts for only 25% of research projects, the remainder lying in the alpine zone, in central or northern Europe or in Britain. The overall picture is even worse if the analysis is expressed in terms of extent of the areas investigated; on this basis, the Mediterranean contribution drops to just over 2% (5,292,301 hectares in Continental Europe and the UK against 118,660 in the Mediterranean area). The main reason for the lack of case studies around the Mediterranean – where most projects have in practice failed to achieve their stated objectives – lies not in the matter of archaeological schools of thought or other forms of cultural bias, nor even in the availability or otherwise of research funding. The reason stems from the character and density of the vegetation and the presence of dense scrub in many areas. Traditional LIDAR systems based on aerial platforms have so far been substantially ineffective in such situations, wherever they lie in Europe, with the exception of recent research conducted with high-resolution acquisition in Croatia (Doneus et al., 2015).

It is fair to point out in this context that LIDAR data has been captured for specifically archaeological purposes in only a few instances; even when this is the case, the process rarely lies under the

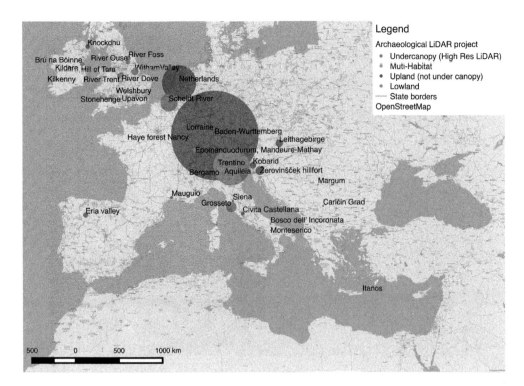

FIGURE 13.8 The distribution around the Mediterranean Sea, in Continental Europe and in the UK of LIDAR prospection within archaeological projects. The size of the circles is proportional to the size of the survey in km².

full control of the archaeologist, who in most instances is only able to specify the desired ground resolution (which is influenced by factors such as the kind of sensor used, the speed of the aerial platform, and the flight altitude) while technicians retain control of all of the other parameters, such as the scanning angle, the number of return pulses, the type of sensor, the flying speed, and even the season when the scanning work is undertaken (Opitz and Cowley, 2012; Opitz, 2016).

Despite these caveats, it is undeniable that UAV platforms represent an innovation of the very greatest interest, both for undertaking low-altitude aerial photogrammetry and for the rapid development of lightweight LIDAR systems designed specifically for such platforms. Low-altitude photography cannot of course penetrate the dense Mediterranean vegetation to provide a true record of the under-canopy landscape. There are, however, several intermediate situations in which it could perhaps play an interesting role. Some types of staple crops may be 'open' enough to offer useful results, for instance, olive groves and vineyards where in addition to colour differences (not always easily perceived because of the typically discontinuous pattern of the ground cover) photogrammetric processing may be able to play a significant role through the creation of high-resolution DTMs suitable for the detection of micro-morphological features.

In the case of the thicker Mediterranean tree cover, the combination of UAV platforms and lightweight LIDAR sensors could mark a significant turning point. The main innovation in this case is the archaeologist's capacity to determine a spatial resolution of anything up to 200 points/m² (as well as to influence all of the other acquisition parameters). Recent considerations have suggested 8 points/m² as the desirable minimum resolution for archaeological applications, with 16 points/m² as the ideal resolution for bare soil or very sparse vegetation canopies (Opitz, 2016). The highest resolution achieved so far in an archaeological context has been the 64 points/m² reached through the use of a slow-moving helicopter at the Hill of Tara and in the offshore island of Skellig Michael in the Republic of Ireland (Corns and Shaw, 2012).

FIGURE 13.9   Clockwise from top-left. The general character of the surveyed area between Roselle and Grosseto in southern Tuscany; UAV Ri-copter Riegl during LIDAR survey; detail of Riegl lightweight LIDAR device; the Ri-copter preparing for take-off; and the UAV being carried to the take-off area.

At the moment, there are no informative publications on the topic of UAV-based LIDAR acquisition for the detection of archaeological features beneath a dense woodland canopy. Recently, the author within his Emptyscapes research work at the University of Cambridge has initiated a test flight in collaboration with Microgeo (Italy, Florence) and RIEGL Laser Measurement Systems (Austria, Horn) over a small area of dense woodland in the Maremma region of southern Tuscany. The area is characterised by typical Mediterranean vegetation with a dense coverage of mostly evergreen trees and an under-canopy of younger trees, scrubs, and herbaceous plants (Figure 13.9). Although the results are still under detailed analysis, it appears likely that in this case it has proved possible to penetrate the vegetation and reveal a number of potential archaeological features beneath the obscuring canopy. The potential of this kind of drone-based LIDAR survey seems great, but the path to its regular and successful application may still be a long one.

SUMMARY AND CONCLUSIONS

Although UAV systems of various kinds have been available for less than a decade, and are still in an experimental stage, archaeologists and applied scientists have already identified numerous archaeological applications and many more are under testing in which aerial platforms of this kind could play an important and innovative role. A key opportunity offered by drones is about the scale of the detail in which UAVs operate which has always been somewhat problematic – relatively restricted in comparison with traditional systems based on conventional air photography or aerial photogrammetry but relatively large by contrast with terrestrial detection systems such as total station survey, global navigation satellite system (GNSS), and ground-based laser scanning. In a sense, however, drones offer the opportunity to fill a gap in the effective range and detail of low-altitude survey, with an effective coverage of between about 20 and 200 m fling altitude and the capacity to acquire data for landscape areas ranging from less than a hectare to as much as 600 hectares per day.

The geometric resolution that can be obtained is unprecedented, allowing the acquisition of images of excellent quality for both air photo interpretation and 3D modelling. A further aspect, of great interest, lies the capacity of the UAV to take to the air at short notice, almost anywhere and at almost any time of day and year (according to domestic rules), increasing what might be termed the 'temporal resolution' of the instrument. This facility, compared with the difficulties involved in the hire of traditional light aircraft from perhaps distant airfields, introduces completely new opportunities for high-resolution survey, exploration, and landscape monitoring, in some cases also providing access to areas or features that are inaccessible from the ground. These characteristics open up new scenarios not only for the monitoring of individual sites or monuments but also for archaeological conservation more generally in response to the many activities and development proposals that daily threaten the cultural heritage. The opportunity to develop highly accurate 3D models of monuments and archaeological sites, even including those of very significant size (as at Paestum) offers new perspectives for World Heritage recording and conservation. In the case of environmental or other dramatic events and disasters, UAV-based data collection will make it possible to create digital 3D representations of the actual environmental conditions in the affected locality, assisting rescue or reconstruction work and serving as agents for memory preservation. It should be emphasised that these are complex issues that require specific analysis. The risk factors are not limited to war scenarios and terrorist attacks, of course. An even more incisive impact can come from other factors, including intensive agriculture, new construction work, industrial activity, and infrastructure development as well as from natural erosion and systematic looting, for instance. Another significant aspect is the cost of the equipment. In recent years, the initial costs have been in sharp decline, and nowadays the purchase of a professional-grade system equipped with good-quality sensors and offering high reliability and good tolerance of variable environmental conditions has become an affordable financial investment for most archaeologists. Given appropriate training, UAV systems have now become easy to use, though the acquisition of the necessary permits is mandatory. The fact that the survey work can in many cases be carried out effectively and safely by just a single person is another factor in the balance between initial capital investment and subsequent operating costs. Nevertheless, the world of remote sensing in archaeology (as in environmental monitoring, aviation, and aerospace applications in general) is on the cusp of a revolution based on autonomous flight and drone technology. That said, several important problems have yet to be overcome.

Drones are in most cases significantly weather dependent and are especially affected by strong wind and rain/snowfall. In many instances, the lifting capacity needs to be improved, as does the tolerance of non-ideal weather conditions and the relatively poor on-board 'intelligence' of many of the available platforms – few drones as yet have any significant capacity to respond independently to variable wind conditions or the presence of stationary or moving obstacles that lie outside the direct sight-line of the pilot. In this sense, autonomy and reliability, both of which vary greatly with the type and capital cost of the UAV, represent key considerations for potential users, especially in the initial phases of research design. Reliability must also be seen as a factor which is directly related to the search for safety of operation: serious accidents are thankfully rare, but they *do* occur, and any form of unreliability could become a contributory factor in future events of this kind. In addition to responsible behaviour and the observance of professional ethics, meticulous attention must also be paid to the regulations in force at the time and place of operation – the regulations, unfortunately, can be quite uneven and inconsistent between one country or context and another.

Ultimately the major methodological novelty of UAVs, beyond mere technical innovation, lies in their capacity to provide archaeologists with the opportunity to exercise direct and independent control over all aspects of the survey process: the platform, the sensors, and the subsequent processing of the collected data. This capacity to control the process as a whole will give the researcher the freedom to develop applications and programmes of work that are directly related to the framing and answering of specifically archaeological questions, without having to deal (apart from the regulatory framework) with any non-archaeological intermediary. This kind of challenge has been a constant in the history of archaeology; the advent of drones and their increasing capacity to carry varying kinds of sensors represents a major breakthrough that could in important respects revolutionise the future potential of archaeological survey, interpretation, and problem-solving.

ACKNOWLEDGEMENTS

The research for this chapter would not have been possible without the collaboration of many research institution and private company. Financial support comes from many research grants, particularly, the Marie Curie action for the Emptyscapes project (FP7-PEOPLE-2013-IEF n. 628338) and the Culture 2007 ArchaeoLandscpes Europe project (Grant Agreement nr. 2010/1486/001-001).

The author is indebted to the many people who have collaborated with the Laboratory of Landscape Archaeology and Remote Sensing (LA&T) by sharing resources and know-how. Particular thanks go to Dr. Fabio Remondino and 3DOM (FBK), who guided our first steps in 2007 and shared in our long and continuing journey of research.

Sincere thanks also go to Daniele Sarazzi and Zenit Ltd. who generously provided the Microdrones md4–200 UAV which gave us the opportunity to develop innovative documentation systems for archaeological excavations between 2007 and 2012. Our gratitude goes also to the constant support of Leica Geosystems of Florence.

The spin-off company of the University of Siena, ATS Ltd., has played a crucial role in the development of the workflow for excavation recording. The company and in particular Dr. Francesco Pericci and Dr. Matteo Sordini shared in the piloting of the Aibot X6 Hexacopter, DJI Phantom 4, Delair UX11, and performed the post-processing of the collected data, participating actively in the discussion of results and the improvement of pipeline.

Many thanks go also to Geostudi Astier (Livorno, Italy), FENIX AIR (Pisa-Italy), and Radarteam (Sweden) for offering the opportunity to test the lightweight cobra plug-in GPR (Radarteam) for UAV.

Thanks goes, too, to Microgeo Ltd. and in particular to Simone Orlandini, as well as to RIEGL Laser Measurement Systems for offering the opportunity to test the Riegl lightweight LIDAR for UAV.

We have been indebted to the Italian Civil Aviation Authority (ENAC), and in particular to Eng. Alessandro Cardi and Eng. Riccardo Delise for their ever-present availability, support, and expertise.

Finally, I must congratulate my friend and colleague Chris Musson, from Wales, for turning my initial text into fluent English prose.

REFERENCES

Barker G. 1995. *A Mediterranean Valley: Landscape Archaeology and Annales History in the Biferno Valley.* Leicester: Leicester University Press.

Bendea H., F. Chiabrando, F.G. Tonolo and D. Marenchino. 2007. Mapping of archaeological areas using a low-cost UAV. The Augusta Bagiennorum test site. In *AntiCIPAting the Future of the Cultural Past. Proceedings of the CIPA 2007 XXI International Symposium*, 01–06 October 2007, Athens, Greece.

Blondel J., J. Aronson, J.Y. Bodiou and G. Boeuf. 2010. *The Mediterrenean Region. Biological Diversity in Space and Time.* Oxford: Oxford University Press.

Braudel F. 1949. *La Méditerranée et le monde méditerranéen à l'époque de Philippe II.* Paris: Armand Colin.

Brenningmeyer T., K. Kourelis and M. Katsaros. 2016. The Lidoriki project: low altitude, aerial photography, GIS, and traditional survey in rural Greece, In Campana S., R. Scopigno, G. Carpentiero and M. Cirillo (eds). *CAA 2015 Keep the Revolution Going. Proceedeings of the 43rd CAA*, 30 March–2 April 2015, Siena-Italy. Oxford: Archaeopress Publishing. pp. 979–988. Available in open-access ISBN 9781784913380, e-PDF: https://www.academia.edu/23671738/CAA2015_KEEP_THE_REVOLUTION_GOING_ Proceedings_of_the_43rd_Annual_Conference_on_Computer_Applications_and_Quantitative_ Methods_in_Archaeology

Campana S. 2009. Archaeological site detection and mapping: Some thoughts on differing scales of detail and archaeological 'non-visibility'. In Campana S. and S. Piro (eds.), *Seeing the Unseen*. Leiden: Taylor & Francis. pp. 5–26.

Campana S. 2016. Archaeology, remote sensing. In Gilbert A.S. (ed). *Encyclopedia of Geoarchaeology*. Dordrecht, Heidelberg, New York, London: Springer.

Campana S., 2018. *Mapping the Archaeological continuum. Filling 'empty' Mediterranean Landscapes*. New York: Springer.

Campana S., M. Sordini and F. Remondino. 2008. Integration of geomatics techniques for the digital documentation of heritage areas. In *Proceedings of the 1st European Association of Remote Sensing Laboratories (EARSeL) International Workshop on "Advances in Remote Sensing for Archaeology and Cultural Heritage Management"*, Rome, September 30–October 4, 2008.

Campana S., M. Sordini and A. Rizzi. 2009. 3D Modeling of a Romanesque church in Tuscany: archaeological aims and geomatic techniques. In *Proceedings of the 3rd International Workshop on the 3D Virtual Reconstruction and Visualization of Complex Architectures* (25–28 February 2009, Trento, Italy).

Casana J., Kantner J., Wiewel A. and Cothren J. 2014. Archaeological aerial thermography: a case study at the Chaco-era Blue J community, New Mexico. *Journal of Archaeological Science* 45: 207–219.

Chiabrando F., F. Nex, D. Piatti and F. Rinaudo. 2011. UAV and RPV systems for photogrammetric surveys in archaeological areas: two tests in the piedmont region (Italy). *Journal of Archaeological Sciences* 38: 697–710. doi:10.1016/j.jas.2010.10.022.

Corns A. and R. Shaw. 2012. LIDAR and World Heritage Sites in Ireland: why was such a rich data source gathered, how is it being utilised, and what lessons have been learned? In Opitz R.S., D.C. Cowley (eds). *Interpreting archaeological topography. Airborne Laser Scanning, 3D Data and Ground Observation*. Oxford: Oxbow books. pp. 146–160.

Dell'Unto N. 2014. The use of 3D models for intra-site investigation in archaeology. In Remondino F., and S. Campana (eds.), *3D Surveying and Modeling in Archaeology and Cultural Heritage Theory and Best Practices*. BAR International Series. Oxford: BAR Publishing. pp. 151–158.

Doneus M., W. Neubauer and N. Studnicka. 2003. Digital recording of stratigraphic excavations. In Enter the Past. The E-way into the Four Dimensions of Cultural Heritage, In Ausserer, K.F., W. Börner, M. Goryany, N. Karlhuber, N. Vöckl (eds), Proceedings of the 31st International Conference on Computer Applications and Quantitative Methods in Archaeology (CAA). Oxford: Archaeopress. pp. 451–456.

Doneus M., N. Doneus, C. Briese and G. Verhoeven. 2015. Airborne laser scanning and Mediterranean environments—Croatian case studies, In *The Island Research Projects. Croatian Archaeological Society, Zagreb*. pp. 147–165.

Eisenbeiss H., M. Sauerbier, L. Zhang and A. Gruen. 2005. Mit dem Modellhelikopter u¨ ber Pinchango Alto. *Geomatik Schweiz* 9: 510–515.

Eisenbeiss H. and L. Zhang. 2006. Comparison of DSMs generated from mini UAV imagery and terrestrial laser scanner in a cultural heritage application. *International Archives of Photogrammetry, Remote Sensing and Spatial Information Sciences* XXXVI-5: 90–96.

Eisenbeiss H. 2009. UAV photogrammetry. Dissertation ETH No.18515, Institute of Geodesy and Photogrammetry, ETH Zurich, Switzerland. p. 105.

FAO. 2006. *Global Forest Resources Assessment 2005*. FAO Forestry Paper 147. Rome: FAO Rome.

Fiorillo F., F. Remondino, S. Barba, A. Santoriello, C.B. De Vita and A. Casellato. 2013. 3D digitization and mapping of heritage monuments and comparison with historical drawings. *ISPRS Annals of the Photogrammetry, Remote Sensing and Spatial Information Sciences* Volume II-5/W1: 133–138. *XXIV International CIPA Symposium*, 2–6 September 2013, Strasbourg, France.

Jones R.J.A. and R. Evans. 1975. Soil and crop marks in the recognition of archaeological sites by air photography, In Wilson D.R. (ed.), *Aerial Reconnaissance for Archaeology*: 1–11. London: The Council for British Archaeology.

Lambers K., H. Eisenbeiss, M. Sauerbier, D. Kupferschmidt, T. Gaisecker, S. Sotoodeh and T. Hanusch. 2007. Combining photogrammetry and laser scanning for the recording and modeling of the late intermediate period site of Pinchango Alto, Palpa, Peru. *Journal of Archaeological Science* 34(10): 1702–1712.

Lang M., T. Behrens, K. Schmidt, D. Svoboda and C. Schmidt. 2016. A fully integrated UAV system for semi-automated archaeological prospection, In Campana S., R. Scopigno, G. Carpentiero and M. Cirillo (eds.). *CAA 2015 Keep the Revolution Going. Proceedings of the 43rd CAA*, 30 March–2 April 2015, Siena-Italy. Oxford: Archaeopress Publishing: 989–996. Available in open-access ISBN 9781784913380, e-PDF: https://www.academia.edu/23671738/CAA2015_KEEP_THE_REVOLUTION_GOING_Proceedings_of_the_43rd_Annual_Conference_on_Computer_Applications_and_Quantitative_Methods_in_Archaeology

Lo Brutto M., P. Meli, F. Ceccaroni and M. Casella. 2013. Studio delle potenzialità delle piattaforme UAV nel campo del rilievo dei Beni Culturali, In *Atti della XVII Conferenza Nazionale ASITA*, Riva del Garda, 5–7 novembre 2013.

Musson C., R. Palmer, and S. Campana. 2005. *In volo nel passato. In Aerofotografia e cartografia archeologica*. Florence: All'Insegna del Giglio.

Nex F. and F. Remondino. 2013. UAV for 3D mapping applications. a review. *Applied Geomatics* doi:10.1007/s12518-013-0120.

Nocerino, E., F. Menna, F. Remondino and R. Saleri. 2013. Accuracy and block deformation analysis in automatic UAV and terrestrial photogrammetry. Lesson learnt. *Proceedings of the 24th International CIPA Symposium*, 2–6 September, Strasbourg, France. *ISPRS Annals of the Photogrammetry, Remote Sensing and Spatial Information Sciences* II (5/W1): 203–208.

Oczipka M., J. Bemman, H. Piezonka, J. Munkabayar, B. Ahrens, M. Achtelik and F. Lehmann. 2009. Small drones for geo-archaeology in the steppes: locating and documenting the archaeological heritage of the Orkhon Valley in Mongolia. *Remote Sensing for Environmental Monitoring, GIS Applications, and Geology* 7478: 747806–1.

Opitz R. 2016. Airborne laserscanning in archaeology: maturing methods and democratizing applications, In Forte M., S. Campana, *Archaeology in the Age of Sensing*, New York: Springer.

Opitz R. and D.C. Cowley. 2012. *Interpreting Archaeological Topography. Airborne Laser Scanning, 3D Data and Ground Observation*. Oxford: Oxbow books.

Piani P. 2013. La strumentazione UAV nel rilievo e nella modellazione tridimensionale di un sito archeologico, In *Archeomatica, Marzo*: 6–10.

Piccarreta F. and G. Ceraudo. 2000. *Manuale di aereofotografia archeologica. Metodologia, tecniche, applicazioni*. Bari: Edipuglia.

Pueschel H., M. Sauerbier and H. Eisenbeiss. 2008. A 3D model of Castle Landenberg (CH) from combined photogrammetric processing of terrestrial and UAV-based images. *The International Archives of the Photogrammetry, Remote Sensing and Spatial Information Sciences* XXXVII. Part B6b, Beijing, China: 93–98.

Remondino F. and S. El-Hakim. 2006. Image-based 3D modelling: a review. *The Photogrammetric Record* 21(115): 269–291.

Remondino F., A. Gruen, J. Von Schwerin, H. Eisenbeiss, A. Rizzi, M. Sauerbier and H. Richards-Rissetto. 2009. *Multi-sensors 3D documentation of the Maya site of Copan*. Proceedings of 22nd CIPA Symposium, Kyoto, Japan, on CD-ROM.

Remondino F., L. Barazzetti, F. Nex, M. Scaioni and D. Sarazzi. 2011. UAV photogrammetry for mapping and 3D modelling—current status and future perspectives, In *International Archives of the Photogrammetry, Remote Sensing and Spatial Information Sciences,* ISPRS Zurich 2011 Workshop (Zurich, 14–16 settembre 2011), XXXVIII-1/C22, pp. 25–31.

Rinaudo F., F. Chiabrando, A. Lingua and A. Spanò. 2012. Archaeological site monitoring: UAV photogrammetry could be an answer. *International Archives of Photogrammetry, Remote Sensing and Spatial Information Sciences*, Melbourne, Australia, 39: 5.

Sauerbier M. and H. Eisenbeiss. 2010. UAVs for the documentation of archaeological excavations. The International Archives of the Photogrammetry, *Remote Sensing and Spatial Information Sciences* 38(5): 526–531.

Sconzo, P. 2014. *The Tübingen Eastern Habur Project Archaeological Survey in the Dohuk Region of Iraqi Kurdistan*. https://www.academia.edu/9547402/The_T%C3%BCbingen_Eastern_%E1%B8%AAabur_Project_Archaeological_Survey_in_the_Dohuk_Region_of_Iraqi_Kurdistan (accessed April 18, 2016).

Seitz C. and H. Altenbach. 2011. Project Archeye—the quadrocopter as the archaelogist's eye, In *International Archives of the Photogrammetry, Remote Sensing and Spatial Information Sciences*, ISPRS Zurich 2011 Workshop (Zurich, 14–16 settembre 2011), XXXVIII-1/C22, pp. 297–302.

Shaw R. and A. Corns. 2008. Recording archaeological excavation using terrestrial laser scanning and low cost bal- loon based photogrammetry. In *Proceedings of CAA 2008. Computer Applications and Quantitative Methods in Archaeology*, Budapest, Hungary.

Sonnemann T.F., E. H. Malatesta and C.L. Hofman. 2016. Applying UAS photogrammetry to analyze spatial patterns of indigenous settlement sites in the northern Dominican Republic, In Near-infrared aerial crop mark archaeology: From its historical use to current digital implementations Forte M., S. Campana, (eds.). *Archaeology in the Age of Sensing*, New York: Springer.

Sordini M., F. Brogi and S. Campana. 2016. 3D recording of archaeological excavation: the case study of Santa Marta, Tuscany, Italy, In Campana S., R. Scopigno, G. Carpentiero and M. Cirillo (eds.). *CAA 2015 Keep the Revolution Going. Proceedings of the 43rd CAA*, 30 March–2 April 2015, Siena-Italy. Oxford: Archaeopress Publishing, pp. 383–391 Available in open-access ISBN 9781784913380, e-PDF: https://www.academia.edu/23671738/CAA2015_KEEP_THE_REVOLUTION_GOING_Proceedings_of_the_43rd_Annual_Conference_on_Computer_Applications_and_Quantitative_Methods_in_Archaeology

Tokmakidis K. and D. Skarlatos. 2000. Mapping Excavations and Archaeological Sites Using Close Range Photos. In *Proceedings of the ISPRS Commission V Sympo- sium, WG V/4, Corfu, Greece*. International Society for Photogrammetry and Remote Sensing, 459–462.

Verhoeven G. 2008. Imaging the invisible using modified digital still cameras for straightforward and low-cost archaeological near-infrared photography. *Journal of Archaeological Science* 35: 3087–3100. doi:10.1016/j.jas.2008.06.012.

Verhoeven G. 2009a. Providing an archaeological bird's-eye view—an overall picture of ground—based means to execute low-altitude aerial photography (LAAP) in Archeology. *Archaeological Prospection* 16: 233–249. doi:10.1002/arp.354.

Verhoeven G. 2009b. Beyond conventional boundaries. New technologies, methodologies, and procedures for the benefit of aerial archaeological data acquisition and analysis. PhD thesis, Ghent: Ghent University.

Verhoeven G. 2012. Near-infrared aerial crop mark Archaeology: from its historical use to current digital implementations. *Journal of Archaeological Method and Theory* 19: 132–160. doi:10.1007/s10816-011-9104-5.

Verhoeven G. and J. Loenders 2006. Looking through black-tinted glasses—a remotely controlled infrared eye in the sky, In Campana S. and M. Forte (eds.), *From Space to Place, 2nd International Conference on Remote Sensing in Archaeology. Proceedings of the 2nd International Workshop*, CNR, Rome, Italy, December 4–7. Oxford: Archaeopress.

Verhoeven G., J. Loenders, F. Vermeulen and R. Docter, 2009. Helikite Aerial Photography (HAP)—a versatile means of unmanned, radio-controlled, low-altitude aerial archaeology. *Archaeological Prospection*. doi:10.1002/arp.353.

Verhoeven G. and K.D. Schmitt. 2010. An attempt to push back frontiers—digital near-ultraviolet aerial archaeology. *Journal of Archaeological Science* 37: 833–845.

Vermeulen F., S. Hay, G. Verhoeven. 2006. Potentia: an integrated survey of a roman colony on the Adriatic coast. *Papers of the British School at Rome* 74: 203–236. doi:10.1017/S0068246200003263.

WEB REFERENCES

Allsopp Helikites, Hampshire, SP6 1BD, England (UK) http://www.allsopp.co.uk/

International Association Unmanned Vehicle System (UVS) http://www.uvs-international.org/

MAIA - Multispectral camera http://www.salengineering.it/public/it/p/camera_multispettrale_maia.asp

MeshLab https://www.spectralcam.com

MicaSense RedEdge http://www.micasense.com/rededge

Microgeo spa (Firenze) http://www.microgeo.it/

FLIR Vue Pro http://www.flir.com/suas/content/?id=70728

Gruen A. 2014. Combined 3D modeling from UAV aerial images and Mobile Mapping laser-scan point clouds. From Imagery to map: Digital Photogrammetric Technologies, *14th International Scientific and Technical Conference*. https://www.youtube.com/watch?v=LurNobNVlEM

RIEGL Laser Measurement Systems (Vienna): unmanned laser scanning http://www.riegl.com/products/unmanned-scanning/

Tetracam ADC Lite http://www.tetracam.com/Products-Micro_MCA.htm

Drone-based radar sensor http://www.radarteam.se

Drone-based magnetic sensor http://www.gemsys.ca/uav-magnetometers/

14 Unmanned Aerial System (UAS) Applications in the Built Environment
Towards Automated Building Inspection Procedures Using Drones

Tarek Rakha
Georgia Institute of Technology

Alice Gorodetsky
Syracuse University

CONTENTS

INTRODUCTION

Humans live in ageing built environments. Maintaining the energy efficiency of such infrastructure and building stock is integral in moving towards an environmentally sensitive and sustainable future in the age of climate change. Forty percent of US homes were built before 1970 [1], and the building sector accounts for 40% of CO_2 emissions in the USA [2]. Therefore, to address the inefficiency of older infrastructure, building energy retrofitting practices typically identify, diagnose, and design solutions that address issues of building usage, systems, and envelope [3]. In Metropolitan Boston, MA, for example, heat loss through window surfaces, cracks, chimneys, and soffits were each present in more than 70% of 135 houses surveyed with infrared (IR) technology. Heat transfer and air leaks through cracks and ducts were identified as the reason behind about 40% of energy lost in a residential building [4].

To identify problematic issues specific to the building envelope, energy auditors typically use tools such as blower door tests to detect infiltration/exfiltration regions, as well as thermal bridges [3].

Technologically advanced tools such as IR cameras and Unmanned Aerial Systems/Vehicles (UASs/UAVs) enable professionals to analyse such issues rapidly and accurately while reducing operational costs and minimising safety risks, and when paired with video recording, photography, or multi-spectral imaging, drones can safely, economically, and efficiently carry out a broad variety of surveying services [5]. UASs provide experts across industries with a unique aerial perspective. This viewpoint allows easy access to remote or inaccessible areas without compromising the safety of the pilot [6], and this combination of technology has and will continue to permeate a wide range of varying fields as its applications, innovations, and capabilities are discovered and improved upon [7].

The primary barrier to energy-efficient retrofitting is the uncertainty over return on investment, an uncertainty that can be addressed when focusing on the building envelope with relatively quick and cheap visualisation of thermal anomalies [8,9]. Thermal patterns gathered with IR cameras attached to drones can be converted into automatically generated 3D CAD models using 3D photogrammetry software. These models, as well as the imagery gathered, are visual tools that aid the professional assessment of energy production and conservation in buildings, as well as communication between parties involved [10]. Although the use of drones and thermal cameras for surveying is an innovation that has benefited a variety of unrelated industries [5], a gap currently exists in the literature relevant to the built environment. In the few investigations that have combined UAS-based surveying with IR imagery and 3D modelling to audit energy use in buildings, the procedure and methodology used is typically not described in a scientific approach that can be standardised. This chapter's goal is it to report on a comprehensive review of contemporary developments in UAS technologies specific to such building performance inspection applications and presents a novel proof-of-concept study for thermography-based building auditing using UAS. The manuscript, therefore, addresses this aim through the following objectives: (i) developing a timeline for military-specific and commercial applications of drones, (ii) reviewing literature focusing on-site investigations, (iii) examining building analysis techniques using UAS, and (iv) investigating drone flight planning procedures for building performance inspection. The results specifically focus on applications of thermography and 3D CAD model generation, and a case study is presented for the methods on the Syracuse University campus in the USA. This chapter concludes with a discussion that contextualises past investigation directions, current research challenges, and a framework for future developments.

HISTORICAL BACKGROUND

MILITARY ORIGINS AND COMMERCIAL DEVELOPMENTS

The genesis of drones dates back to Italy in 1849 during the first Italian War of Independence when the Austrian Empire devised a system of unmanned hot air balloons that dropped bombs on Venice. Later, during the American Civil War and the Spanish American War, hot air balloons and kites were used to gather and telegraph reconnaissance [11]. Military needs were the predominant driving force behind developing UAS technology until the 21st century. The only commercial developments were iterations in radio-controlled toy technology [7]. Military practices, on the other hand, were quickly developing largely due to innovations in manned aviation technology and practices.

As a response to the growing tensions between the US and the Soviet Union during the Cold War, the US Government began a highly classified UAS research program which ran under the code name 'Red Wagon'. During the same year, the Defense Advanced Research Projects Agency

(DARPA) launched the first global satellite navigation system in the air which would later become the basis of all Global Positioning System (GPS) technology developments [12].

In 2006, Frank Wang with the help of his friends, colleagues, and teachers, established DJI Technology from his dorm room in the Hong Kong University of Science and Technology. Since its inception, DJI has steadily grown in influence around the world and is currently regarded as a leading commercial drone company [13]. DJI's initiation marks the start of the personal and commercial drone trend as it is the first company to produce drones that are easily accessible to the public. In 2012, US Congress passed the Federal Aviation Administration (FAA) Modernization Act to require the FAA to integrate small drones into airspace by 2015. The FAA missed this deadline which was expected especially after Congress released a public statement acknowledging the unlikeliness of the FAA's success in organising this transition [14]. Soon after, online retailer Amazon announced plans to deliver products with a drone taskforce [15]. At this point in history, the term 'drone' had become a common household topic as the public gradually warmed to this pacified product of military progress.

Figure 14.1 summarises the previously detailed timeline. The observation is now confirmed that as experts in various fields come together to develop this emerging practice, differing processes and methodologies arise [5]. No one field or industry has discussed a comprehensive review of drone surveying techniques. In addition, since each industry has specific needs or goals when using this combination of technology, established procedures are rarely applied outside of their original field. This literature review seeks to gather, compare, contrast, and assess current and emerging practices.

THERMOGRAPHY

People have been locating general anomalies by assessing temperature variance even before the ubiquitous use of electricity. Ancient Egyptians documented a practice wherein they would move their hands across the surface of a human body to scan and monitor changes in temperature [29]. Historians have also wondered what would have been the fate of the Titanic if the engineers had installed Bellingham's thermopile and mirror system to detect icebergs by means of temperature variance aboard the ship [7]. IR energy is absorbed and emitted by almost all materials and has a direct influence on the temperature of an object [30], and the efficacy of IR technology has been steadily growing since the Second World War [7]. Aside from military applications, IR technology has been adapted as a useful diagnostic tool as well as a cost-reducing monitoring system [31]. The main benefit of IR visualisation is the stark contrast and almost immediate awareness of conditions that deviate from the norm [30,32]. The most common indicator of depredation in any physical phenomenon is temperature change, and thermal imaging tools can detect that through depiction of radiance, temporal, and surface properties simultaneously [7]. As technology advances, the resolutions of new sensors improve in tandem [33,34], and as IR sensors became significantly smaller in size and lighter in weight, they were further integrated with drones for surveying purposes, including building surveys.

UAS-based surveying enables expedited auditing at minimal human effort and cost and is not limited by accessibility aside from drone obstacles which, naturally, are not encountered as often as human accessibility obstacles [35]. The scales, accuracies, and scopes of building energy audits are increased when the exterior façade is feasibly accessible with thermal cameras. Building audits conducted with thermal imaging can be executed using two varying methodologies: (i) active thermography, which begins with an external stimulus such as a rise in heat in order to observe hidden defects made clear by a contrast between typical and extreme temperatures, and (ii) passive thermography, which observes thermal patterns in a built environment's normal state with no manmade stimuli. To use active thermography to its fullest potential, it was recommended that auditors have preexisting knowledge of possible defects and focus on relatively smaller areas. Therefore, passive building thermography is more often utilised for audits of entire buildings with little to no preexisting knowledge of defects [36].

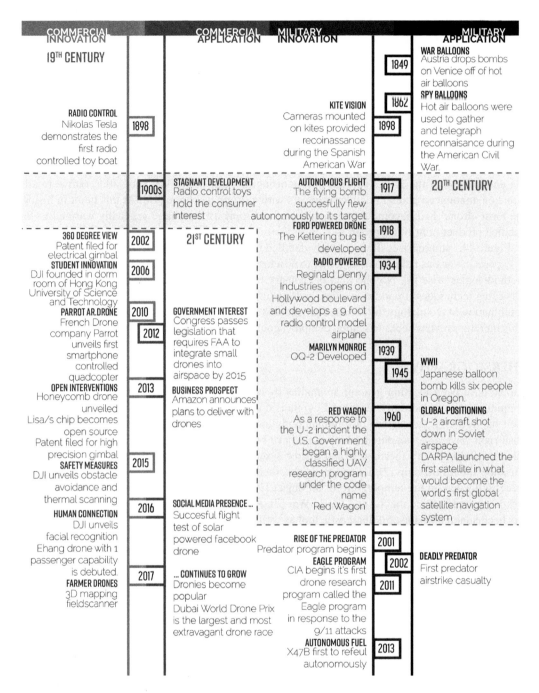

FIGURE 14.1   A historical timeline of UAS technology developments in the 19th, 20th, and 21st centuries [16–28].

RESEARCH METHODOLOGY

The review initiated with a literature survey via multiple online scholarly sources. The reviewed works comprise 92 sources, the majority of which were written by authors associated with higher education institutions. Literature was gathered via a keyword-based search. The sources were selected based on their ability to detail the applications and current abilities of UAS and thermal

IR technology in improving existing building audit practices. This necessitated scholarly sources that detail the design of flight paths in relation to buildings, large expanses of landscape, and/or geoclusters. Additionally, sources that detail methods of photogrammetry, IR-based thermal pattern identification, and 3D model generation from point clouds illustrate rapidly expanding capabilities for IR-based surveying.

As the review specifically focuses on stages of an energy audit using UAS equipped with thermal cameras, an experimental setup is proposed as part of the work in each of the reviewed topics. For pre-, during-, and post-flight analyses, investigation methods are developed, calibrated, and tested on the campus of Syracuse University as part of the findings for the literature review.

RESEARCH FINDINGS

The search for sources began as a broad investigation to understand the varied uses of thermography, benefits of UAS, and general procedures for applications of both technologies. Twenty nine percent of the sources were mostly concerned with the process and reasoning behind building audits. Fifteen percent looked at current UAS trends. Thirteen percent presented historical background and development of UAS technology. Ten percent focused on 3D and Building Information Modelling (BIM), 9% investigated applications towards engineering, and 4% explored historical preservation. Four percent detailed photogrammetry methods and 3% looked at applications towards agriculture. The remaining 9% or less investigated multiple disciplines including volcanology, cartography, computer science, construction, hydrology, and sociology (Figure 14.2).

Based on the compiled sources, Figure 14.3 illustrates a decision tree which categorises the current workflow of documented UAS-based IR surveying practices. The workflow is divided into three main categories, namely, site acquisition, flight path planning, and post-analysis. This chapter links building thermography methods with UAS by dissecting the process of the audit itself. The methodology is split into three steps: pre-flight drone path planning; in-flight IR thermography; and post-flight image processing, segmentation, and automatic 3D model generation. The main issues observed throughout the literature are drone distances from surfaces and flight paths, use of thermography to diagnose issues, photogrammetry, and generation of 3D models in order to better inform both the development of a comprehensive methodology as well as to create a review of current practices in UAS-based IR building energy audits.

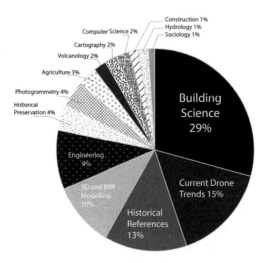

FIGURE 14.2 Investigated literature by discipline for 92 publications.

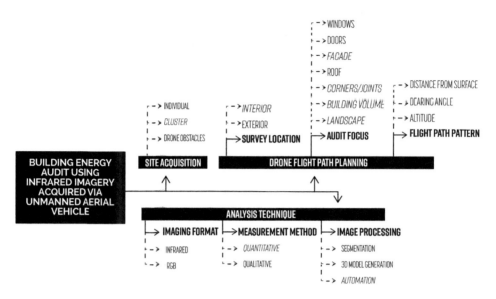

FIGURE 14.3   Categorical decision tree for the workflow of UAS literature investigations.

DRONE FLIGHT – PATH

One of the unavoidable questions encountered while designing a UAS-based process is the flight path or pattern. Planning a drone flight necessitates an awareness of multiple factors such as distance from target, altitude, speed, overlap, and pattern [37]. The accuracy of the drone flight as well as avoidance of obstacles is heavily reliant on the accuracy of the GPS and Inertial Navigation System (INS) [38]. Tahar has tested the accuracy of GPS onboard in combination with specific image registration algorithms versus ground control points (GCPs) informed by Google Earth Pro. The results found that the root mean square error was less for flights that utilised the onboard GPS than flights that used GCPs although the difference is equivalent to about ±2 m [39]. Volkmann and Barnes also specify using Google Earth to inform their flight plan, although they point out that it is only an acceptable tool for on-site reconnaissance and pre-flight surveying because the imagery is often not current enough and does not depict drone-specific obstacles [40]. Other common drone flight obstacles include battery life, a limited resource, and legal regulations of air space [40]. Figure 14.4 combines recommended distances provided by multiple sources across varying fields. While each has its benefits, the optimal distance relies heavily on industry specific needs. Throughout the literature, it became clear that although each industry has found a method that works for them, none have identified an ideal distance for building energy audits.

Not all sources include the reasoning behind their flight plans, and although recently released software innovations have established relative industry standards, the wide range of recommended flight paths necessitates further conclusive research. Investigated literature clarifies that UAS-based auditing requires an established flight plan that either targets regions of interest or audits the building and its surfaces in their entirety. Before addressing the building in question, it is necessary to note the surrounding natural environment and ensure that landscaping elements will not interfere with drone flight and that no relevant façade details are obscured. Often more than one audit is necessary to accurately capture the entirety of a building and often flight planners must determine the next best view (NBV) based on preliminary scans or surveys [41]. Blaer and Allen developed a voxel-based occupancy procedure for detecting areas of the building that were inadequately captured [42]. Corsi recommends auditing large surfaces to ensure accurate mean value and detection of local anomalies [7]. Son et al. propose determining regions of interest pre-flight based on construction drawings and knowledge of thermal patterns. This method minimises the flight time

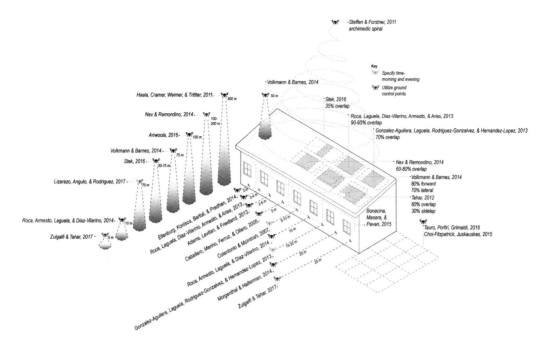

FIGURE 14.4 Drone inspection distances as reported in the literature survey of façade and building inspection, energy performance inspection, and BIM and mapping

by targeting predetermined areas but might allow room for human error and missed thermal leaks [43]. Martinez-de Dios et al. detail a flight plan wherein the inspection of a building is divided into three tasks: stabilisation, detection, and tracking. In this plan, the UAS hovers at predetermined inspection points and captures a series of images. Commonly available flight planning software has popularised the term 'waypoints' to refer to virtual reference points based on GPS coordinates that enable the UAS to autonomously fly and perform actions (record, photograph, hover, etc.) at each location. Collecting an abundance of images is useful in filtering out false positives, and the distortions between consecutive images show the translations and rotations of the drone [44,45]. For example, Haala et al. mainly use a system of waypoints which can be loaded before or during the flight [46]. Siebert and Teizer used an Extensible Markup Language (XML)-based file structure with a wireless data upload link from a ground control station to input waypoints and create a flight path with the aid of external software [47].

Multiple researchers approach the issue of flight planning mathematically, by taking into consideration the limitations of camera technology and battery life to create formulas that inform drone distance from the target, speed of flight, and path geometry [48,49]. Emelianov et al. use mathematical principles to optimise the flight path of the UAS and scientific process to test the feasibility of auditing a geocluster of 300 buildings with the use of a quadrotor. Their findings prove that with produced algorithms, the quadrotor could scan the facades of 300 buildings in approximately 6 hours given an unlimited battery life, although this hypothesis has only been tested at significantly smaller scales [50]. Laguela et al. developed algorithms to correct radial distortion and skewed scale. They achieved mixed results as neither method is optimal in every axis, and both methods result in one axis being skewed or inaccurate [51].

Although mathematical principles aid in the correction and optimisation of UAS-based auditing, various researchers choose to approach the question of flight path with a more geometric approach. The most common solution has been to fly the drone in either vertical or horizontal strips in a zig-zag pattern across the area of interest. This method is frequently referred to as the strip method [38,51–54]. Eschmann and Wundsam elaborate further on the strip method by testing the differences between utilising vertical strips or horizontal strips. Their study finds that vertical strips result in unfavourable movement of the lens which decreases the clarity and quality of data gathered. Horizontal strips are proven to be ideal especially when paired with a low flight speed [53]. In addition to being a proponent of the strip method, Steffen and Forstner present research which advocates for the use of Archimedic spirals to optimise flight plans [38]. Multiple researchers plan their flights according to a grid overlaid on the predetermined site boundaries, so that the drone flies to gather data of each grid unit [55,56].

The strip method frequently involves specified overlaps to ensure not only a robust and thorough collection of data but also an easy transition into photogrammetry, 3D model generation, and/or analysis. To survey archaeological ruins, Stek recommends a >35% overlap [57]. However, there is a great range found that includes 60% [39], >70% [58]; 60%–80% [59], 70% lateral and 80% forward [40], and 90%–95% [60]. For surveys of large or urban areas, a height of 100–200 m allows for high (unspecified) overlap and high-resolution data capturing [59]. Haala et al. further tested their methodology by surveying an agricultural site at 300 m above ground in an overlapping strip pattern with 90% forward overlap and 70% side overlap [46].

It is also critical to consider the altitude at which the UAS flies, as well as the distance it maintains from the target surface. Some researchers choose set distances such as 75 m above ground and 50 m above the tops of buildings [40] or a flight altitude of 50 m [54]. To survey archaeological ruins, Stek selected an altitude of 30–75 m above ground level to cover the full site [57]. Other distances vary dramatically, including a distance of 20 m away from target surface and an altitude maintained by the barometric height control [61]. Three to four metres away from the target surface as prescribed by a Kinect range sensor [60], 2–5 m while assessing damage in post-disaster areas [62], 5 m away [63], and less than 1 m while looking for small-scale cracks in concrete and masonry [64]. Others might simply choose a set altitude of 70 m and do not describe deviating from that set distance [52] or a distance of 25 m from the building and 5 m in the air [65]. A full account of all distances is represented in Figure 14.4. Special topic research, such as the detection of sinkholes by means of UAS-based thermal camera, was simulated with man-made sinkholes of varying depths and test flights at 50 m above ground. The UAS was successful in capturing thermal photos which identified the sinkholes of all depths [66].

The timing of the flight is important to note as well. Although the combination of image-based modelling and UAS-based image reduces occlusion by shadow, daylight and solar radiations still affect the emissivity of materials which can create false positives or result in exaggerated readings [47]. Gonzalez-Aguilera et al. [58] detail a unique process for acquiring thermal imagery where they choose to only take images before sunrise and after sunset to avoid false positives due to direct radiation. Colantonio and McIntosh explain that exterior inspections are best conducted at night to minimise solar heat gain effects on the building envelope cladding. Inspections that focus on mechanical or electrical systems, on the other hand, should be conducted during the daytime while under maximum load [67]. Lizarazo et al. conducted test flights between 9:00 am and 11:00 am in order to ensure a similar solar illumination throughout the data collected [52]. With the wide range of UAS applications possible, optimal flight planning varies depending on the goal of the flight. Flights geared towards data and image collection for the purposes of photogrammetry or 3D model generation often necessitate extra steps and considerations as opposed to flights meant for the sole purpose of auditing or visualising energy flows. For the purpose of aiding existing photogrammetry or 3D model generation software, some sources detail the use of GCPs. They describe marked physical locations around or on the surveyed area (such as coloured markers in agricultural surveys or targets taped on the surface of the roof for a roof audit) that act as physical manifestations of the virtual tie-in points. Although their flight was limited to one façade and one pass around the building, Zulgalfi and Tahar are one such team that used GCPs with the intention of aiding the photogrammetry software in identifying tie-in points [65]. Other researchers specify GCPs so as to easily place the location of the photo while auditing volcanoes [68,69] or use GCPs to achieve spatial accuracy in their flight path (although current software capabilities address limitations in these methods) [52,59]. Feifei et al. use a UAS with four attached cameras to ensure high-precision aerial triangulation for accurate results. The four combined wide-angle cameras increase the base–height ratio as well as expand the angle of view in lateral and longitudinal directions. This system requires less GCPs than typically recommended [70].

DRONE FLIGHT – THERMOGRAPHY

Building thermography for the purpose of anomaly detection can be carried out in a range of methods that include but are not limited to aerial surveys, automated fly-past surveys, street pass-by surveys, perimeter walk-around surveys, walk-through surveys, repeat surveys, and time-lapse surveys [36]. Innovative approaches have been detailed for many of these methods and include use of various combinations of vehicles (whether unpiloted or robotic) paired with cameras of various capabilities, including research on building energy audits conducted by a terrestrial robot laden with IR, photo, and Light Detection and Ranging (LIDAR) technology [41]. Research teams proposed various methods, an example of which is by Hoegner and Stilla's use of vans as surveying vehicles with IR imaging equipment to document and analyse thermal patterns in urban environments [71]. Lerma et al. use a combination of four different cameras to record data on World Heritage Monuments including a reflector-less total station, a terrestrial laser range scanning sensor, a digital photo camera, and a thermal camera [72]. Lopez-Fernandez et al. have utilised a UAS with IR camera to analyse solar radiation on roofs in order to determine the optimal placement of solar panels [73]. The use of UAS for identifying ideal locations for solar and PV panels has been previously explored by comparing the use of UAS and photogrammetry techniques versus LIDAR imaging [8]. Utne, Brurok, and Rodseth used thermal imaging and non-intrusive condition monitoring to maintain safe conditions of static process equipment without necessitating human entry to machinery vessels. This improves operational and occupational safety, as well as reducing operational costs and the need for maintenance [74].

Thermal cameras capture the IR energy emitted by the surface layer which is heavily influenced by the emissivity of the material as well as environmental conditions, building orientation, and camera settings [36,75,76]. This poses interpretation challenges to auditors who are unfamiliar with the basics of thermal emissivity and according to Fox et al. and Gonclaves et al. pose that this is typically the most challenging aspect of working with thermal imaging. Proper camera settings are also integral to the success of the operation, and informed decisions must be made as to the appropriate thermal and spatial resolution as well as the temperature range [77]. Barreira and Freitas conducted sensitivity studies to gauge how thermal cameras capture data with varying emissivity conditions. The tests conclude that cameras set to the improper emissivity will capture unreliable data [75]. The non-destructive and non-contact nature of IR thermography is especially of use to projects that deal with historical buildings or works of art [33,75,78] or to areas that have experienced catastrophic disasters [69]. Grinzato details an instance where a historic fresco, which seemingly survived an earthquake, was discovered to be full of cracks and weak points when captured with IR cameras. This same approach can easily be adapted for the assessment of structural integrity in the built environment [33].

The use of IR thermography in buildings is especially useful for visualising the thermal patterns within the building envelope as well as the movement through and across various building materials [33]. The non-contact nature of IR visualisation is less likely to be inaccurate than any contact measurement method readily available because it does not affect thermal equilibrium. By using airborne IR cameras, auditors can capture a large surface with minimal effort. This is thought to be the only way to achieve a correct mean value and measure local anomalies [7]. To ensure accurate results, Borrmann et al. recommend a temperature difference of about $10°$ K between the interior of the building and the exterior ambient air temperature. The ideal conditions for IR auditing require stable weather conditions over a long period of time [79]. In addition, extensive exposure to solar radiation is not ideal as it may oversaturate and obscure smaller instances of leaks. For this reason, ideal weather conditions are partly cloudy morning hours during the winter months [41]. To ensure an easy transition to the next step, photogrammetry, and 3D model generation, Ham and Golparvar-Fard reinforce the necessity for captured thermal imager to be ordered, calibrated, and geo-tagged [80].

DRONE FLIGHT – POST-PROCESSING (3D PHOTOGRAMMETRY AND CAD MODELLING)

Post-processing is as integral to the operation as the drone flight itself to create a comprehensive and complete image of the audit results [81]. Regarding this practice, current building energy modelling and simulation tools face multiple barriers. Aside from accuracy issues, they are time-consuming and labour-intensive due to the manual modelling and calibration processes. Additionally, simulation software cannot determine construction defects or degradation [80]. Post-processing begins with the awareness of five different conditions, namely, altitude, quality, timing, spectrum, and overlap [82]. Photogrammetry techniques can either utilise direct geo-referencing, GCPs, manual tie-in points, or a combination thereof [34]. Geo-referencing is most easily achieved with time-stamped GPS data recorded during flight [83]. It is typical to use photogrammetry techniques to create a large and complete image of a singular façade [81]. Specifically, regarding building energy inspection, the most common methodology for post-processing uses threshold techniques and software to segment or enhance image saturation to further emphasise the region of interest [5,45]. Eschmann and Wundsam developed a technique centred around the removal and immediate reapplication of a Gaussian blur to gathered RGB images to highlight and emphasise cracks in the building envelope [53]. Yahyanejad and Rinner designed a robust registration methodology of visual and thermal images that begins with extraction of features in the individual images followed by matching corresponding feature points and identifying inliers between them. They use computer software and mathematical algorithms to compute the transformations for aligning individual images [82]. Kung et al. begin their photogrammetry process with aerial triangulation algorithms based on binary local key points whose output is a geo-referenced orthomosaic. This method does not necessitate GCPs and mainly depends on the onboard GPS and geotags provided by the UAS. They found that eliminating the measurements of the GCPs results in lower accuracy, yet for difficult terrains, this sacrifice is often needed [81].

3D modelling techniques can be organised into two sets of goals: groups of buildings (geoclusters) or singular buildings [70]. Although extensive research has been conducted, Borrmann et al. bring attention to the fact that, to the best of their knowledge at the time of publication, no truly autonomous system for 3D model generation of building geometry using thermal imaging has been recorded in a scholarly article [41]. LIDAR is argued to be superior for this task [50,84]. Most literature focusing on photogrammetry experimented with point clouds and have yielded commendable results though none with as much fine detail as with LIDAR technology. Lizarazo et al. use RGB photos to create 3D geometry first because the spatial resolution tends to be more optimised than the IR photos. In addition, they developed algorithms and applied Wallis filters to successfully improve the accuracy of the final product [52]. Multiple researchers tested a methodology where RGB photos were used to create an initial 3D model where the IR images would later be overlaid as they found that 3D model generation software tends to be more successful with RGB photos [65,73]. Volkmann and Barnes utilised a technique known as 'multi-view stereopsis' that uses specific software to align the photos and create a sparse point cloud. The point cloud is then subject to linear transformation parameters and error analysis, which optimise the sparse point cloud to create a dense point cloud, which is translated into the final model [40]. And finally, Hackel et al. developed a systematic process for detecting contours in large-scale outdoor point clouds. It is divided into a two-step process: first a contour score for each individual point is extracted using a binary classifier, and the next step automatically selects the optimal set of contours from the candidates. According to the authors, this method out-performs the canny-style edge detection technique. The innovation in this method is detecting contours before surface reconstruction so that they drive the segmentation and fill in patches in the generated surface [85].

CASE STUDY

Reviewed literature varies and sometimes overlaps in recommended inspection methodologies. A case study was therefore developed as an application that combines and tests reviewed literature outcomes. The case study is investigated through this chapter's objectives, by detailing pre-flight path design, during-flight data gathering, and post-flight analysis. The inspected building was chosen as a generic residential structure on the Syracuse University campus. All flight parameters were varied, including flight path design, image capturing density, overlap, and distances from buildings and external environmental conditions for surveys. The aim was to develop empirical setups for building inspection using UAS equipped with thermal cameras.

PRE-FLIGHT PATH DESIGN

Field tests were conducted with a DJI Inspire 1 drone, paired with a FLIR Zenmuse XT thermal camera. The accompanying DJI app was used during flight to monitor the thermal data. The flight path was predetermined and automated using the Litchi and DroneDeploy apps for roof images. The images were processed and analysed using the FLIR Tools program. We empirically found that using both the strip pattern and the Archimedic spiral, as detailed earlier, have proven to be effective methods. The strip pattern with at least a 70% overlap is suitable for gathering data to audit or visualise energy use in buildings. An elliptical flight path with as much as 95% overlap is appropriate for photogrammetry and 3D model generation. A distance of 12 m away from the target surface with changing bay widths of 2–3 m is fit for capturing images every 1.5 m along the path. For the generation of 3D models, flying multiple elliptical flights at varying altitudes can be ideal. The most effective altitudes have been found to be 18, 22, and 27 m or twice the height of the building and twice the size of the building footprint. The flights should increase in altitude and size by roughly 1.25× per flight. The ellipses should overlap at least 70% and the frequency with which pictures are captured should result in a 90%–95% overlap in captured data. Although this seems redundant, current 3D model generation software, like Pix4D, works best with a repetitive surplus of images. Figure 14.5 represents the empirical parameters of flight path planning.

DURING-FLIGHT DATA GATHERING

For thermography purposes using UAS, our experimentation confirms that better conditions for thermal imaging are cloudy morning or evening hours with stable temperatures and no precipitation. At the time of writing this chapter, a difference between indoor and outdoor temperatures of at least 10° K is recommended when an FLIR camera is being used [86]. This may not be necessarily the case for more sensitive state-of-the-art IR cameras and also may not be feasible in regards to commercial year-round UAS applications. Appropriate emissivity values should be inputted to the thermal camera before flight, calibrated during flight when necessary, or modified post flight during analysis to ensure accurate readings. INS and GPS systems should be correctly calibrated to accurately geo-tag the images. Figure 14.6 demonstrates examples of compromises of an example building envelope, identified with thermal imaging.

    a. Windowpane and frame infiltration/exfiltration potential.
    b. Window frame side failure, creating infiltration/exfiltration.
    c. Door infiltration, with significant change in floor temperature.
    d. Thermal bridging by nails that compromise the roof.
    e. Thermal leakage as a result of envelope installation malfunction.

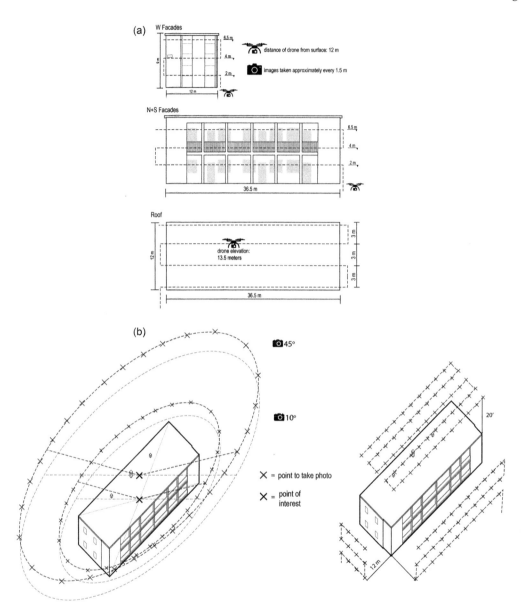

FIGURE 14.5 Flight path parameters for building inspection.

RESULTS: POST-FLIGHT 3D MODELLING

Photos taken during our test drone flight were used in the program Pix4D to generate a 3D point cloud from the 2D images. Multiple models may be recommended based on the number of photos taken. Based on empirical trials and observations, approximately 1,000–1,300 photos are recommended as a suitable number of photos for one simulation, due to current software limitations that may be resolved in the future. Multiple models may be merged upon completion, and this allows the operator to eliminate photos that may not calibrate properly. The program extracts pixels from 2D images by triangulation and locates individual pixels from photos within a 3D point cloud model. During this process, images of inferior quality and images that did not capture the subject will be rejected by the program. Upon generation of the model, we observed that the façade that produced the most detailed model is consistently located on the southern face of the inspected building.

FIGURE 14.6 Building envelope issues identified with a thermal camera.

Using Pix4D modeler, the program runs a 15-point process that results in a report output noting the efficiency of the photos used in generating the model, image overlap, and location, among other variables. This output notes weaknesses of the image retrieval process and should influence future flights. The model can be processed in other 3D modelling and CAD software such as Rhino3D. When opened in a .FBX file format, the 3D model will maintain render and texture capabilities. The model is compatible and can be used in rendering and 3D printing (Figure 14.7).

The data collected in the inspection flight was used to compare three software platforms: Pix4D, DroneDeploy, and AgiSoft Photoscan. The comparison reveals that there are variations in terms of the processing workflow, model production time, and the quality of the output. Table 14.1 details parameters pertaining to model production, where if a low quality and faster model is needed, AgiSoft Photoscan may be useful, while Pix4D produced better quality model with the most quality control comparably, but it needs the most time as well. There is a reliance on cloud computing in DroneDeploy, with a limited number of images, which makes the production of 3D models limited when geometric accuracy is needed. However, at the time of writing this chapter, DroneDeploy produced the highest quality model in terms of mesh smoothness and accuracy as compared to the

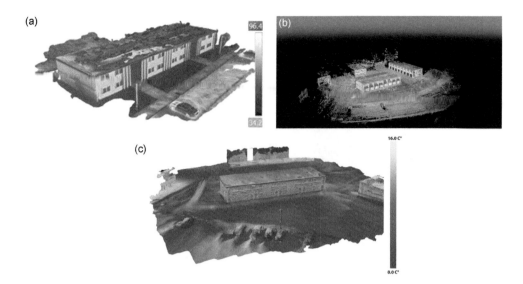

FIGURE 14.7   IR and RGB models generated using 3D photogrammetry using Pix4D and Rhino3D CAD modelling software.

TABLE 14.1

Comparison between 3D Photogrammetry Software (Case Study Is between Brackets)

	AgiSoft Photoscan	Pix4D	DroneDeploy
Standalone computing	X	X	-
Cloud-based computing	-	X	X
Maximum number of images	Unlimited (1,338 IR)	Unlimited (1,338 IR/139 RGB)	500 images
Merging RGB and IR images	-	X	-
Model quality (w/o merging RBG and IR)	Low	Mid	High
Pixilation as compared to images	High	Mid	Low
Geometric output as compared to images	Low	High	Mid
Accuracy as compared to images	Low	Mid	High
Interactive assessment of model	Standalone (limited)	Cloud and standalone	Cloud-based
Time to process model	Low (4 h)	High (18 h)	Mid (8 h)
Control of output model quality	Little to no control	High control	Some control

thermal images used as inputs. Pix4D produced a more reliable model in terms of building angles, because it combined both RGB and IR images, yet, thermal anomalies detected through IR imaging, such as water puddling, were more evident on the roof of the DroneDeploy model. Figure 14.8 showcases differences in the quality of the produced model for the case study.

DISCUSSION

Drones have been adapted to improve upon traditionally manual and human-reliant tasks across industries. To assess the current professional opinion towards drones, Mauriello et al. conducted a study wherein ten professional building auditors were presented with scenarios where UASs and thermal imaging aided in the building inspection process. The test subjects were then asked to respond critically to the scenarios based on their experience working in the field of building audit. The results were positive, as all auditors tested agreed that UASs and thermal imaging provided

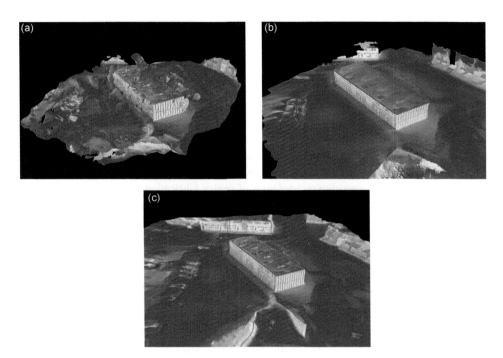

FIGURE 14.8   IR model comparison between (a) AgiSoft Photoscan, (b) Pix4D, and (c) DroneDeploy.

benefits and improved workflow. Additionally, auditors believed thermal imaging would also prove to be a valuable tool in communicating with clients [9]. Opfer and Shields conducted a similar study, where they informally surveyed 31 contractors who are already using UASs in their scope of work. Contractors surveyed explained that traditionally acquired aerial photography is expensive and typically not included in the project budget whereas UAS aerial photography is cheaper and more suited for job progress photography [87]. However, if IR imaging is to be used in such processes, developing thermography expertise is needed in order to correctly analyse and make meaningful interpretations from such images. This would include understanding environmental conditions of acquiring data, distinguishing critical differences in direct and indirect readings, as well as differences between qualitative and quantitative thermography [88].

The future of UAS-based applications for professional building surveying rests in experimentation-based research and subsequent standardisation. While improving technology typically makes standardisation and automation significantly more feasible, the limitations, considerations, and applications have the potential to change dramatically as well. In order to ensure accuracy and efficiency, more comprehensive methodology tests would be beneficial to not just the field of building surveying but any field that utilises UAS with thermal imaging capabilities. Standardised procedures or equations can be further developed for designing flight paths for optimal distances from the target surface, enhanced flight speed for clarity, and geometry that covers the entire target efficiently. Furthermore, post-processing procedures would benefit from similar research in order to further uncover uses for thermal imaging in the inspection and design of buildings. As more technology and software is released, more of the post-processing work will become automated, requiring less human input and participation and expectedly yielding more accurate results. More comparison between various methods is necessary to scientifically justify resulting standardisations. Flight procedures can then be designed for full automation, with minimum to no human interaction during building inspection activities. Global energy audit sectors could operate remotely, by sending UAS remotely for complete, on the fly, building energy inspection, and modelling. This is estimated to cut audit time and increase reporting accuracy, which would consequently result in reliable and unbiased recommendations for deep retrofit strategies. The challenges to be addressed in a fully automated UAS-based building inspection workflow include (i) limitations in GPS positioning and relevant communication between drones and remote operating systems, (ii) battery power limitations, with land and charge potential, (iii) obstacle interference with UAS and possible crashing, (iv) required input for drone navigation in unfamiliar or obstacle-heavy spaces, (v) challenging to reach building components due to the size and navigation of the UAS, (vi) UAS operation vibrations interfering with data acquisition, and (vii) signal acquisition risks as well as signal interpretation challenges.

Microsystem technologies used in both new markets and consolidated markets create a trend of technology transfer between commercial and military applications. Future trends for UAS and airborne building inspections rely on emerging technologies of sensor fabrication for mass production and the demand created by civil applications [7]. With the first emergence of UAS in the commercial market, the technology and flight preparation tools were insufficient to conduct building energy audits, yet within the last few years of technological advances, UASs have become better tools [83]. Incremental technological advances aimed at automation, cost reduction, weight reduction, payload increases, stability enhancement, and connectivity assurance cultivate an interest for UASs for professional building inspection use [5]. Emerging technology like micro drones expand accessibility and increase accuracy in an unexpected scale [89], while drones with increased payload capabilities and automated obstacle avoidance are being deployed for package delivery [15]. These innovations are aimed at increasing the independence, accuracy, and safety with which a UAS navigates a flight path which ultimately dictates the ease of use for the pilot as well as the quality of the end product of an airborne building inspection.

Greater GPS accuracy is still needed to expedite 3D model and flight path generation [38]. However, the miniaturisation of technology necessary for UAS flight such as GPS as well as the emergence of affordable platforms is expected to vitalise the industry towards an improved methodology [34,40]. As multi-spectral imaging capabilities are also expanded upon, and quality of multi-spectral photography increases, smaller details will be captured and analysed with greater accuracy. Within the scope of automated 3D model generation, the possibilities of generating four-dimensional energy performance models which factor time into the 3D model are anticipated [80]. This volume of research into applications and further innovations has resulted in a strong community which is fully invested in the progression of a new field affectionately dubbed 'droneography' [40].

Innovations in the technology continue to grow as both consumer and military demand increases. Global expenditure on commercial UASs in 2014 totalled over $700 million with DJI accounting for the majority of sales. It is estimated that UAS sales will increase to $1 billion by 2018 [90]. This substantial increase links to how the technology will be growing in perception as well. Technoethics, an interdisciplinary approach to assessing technological impacts on societal morals and ethics, have the potential to be the driving force behind the integration of UAS into commercial and public airspace. To address public safety and privacy issues, Luppicini and So gather and review literature that outlines the main technoethical stigmas surrounding drone use which are safety, legality, privacy, airspace use, informational integrity, and commercial concerns. The overarching issue is that the significance of drone use is greatly underrepresented in common media, and therefore, the general public maintains its general distrust towards UASs [91].

As the general public is more exposed to UAS technologies, and with the progression of time, more research will explore UAS used across industries and fields. More minute areas of life and work will become either automated or expedited with the presence of UAS, and the social stigma around unmanned flight may continue to develop alongside increased consumer awareness. The adaptation of this technology in standard building audit companies and organisations will necessitate a baseline understanding of thermography as well as a standardised procedure. Necessary UAS regulations are currently under development around the world to specify where, when, how, and why drone technology can be used and allow for more legally sound exploration of potential uses [92].

Limited opposition to this methodology should not be neglected. Vavilov claims that the use of mathematical equations and existing heat flow patterns is more accurate than IR imaging, although the investigation concludes that the difference is between $1°$ and $3°$ K. The research argues that exterior surveying is insufficient for comprehensive analysis of building energy use and waste, interior inspections with handheld thermal cameras or traditional tests are described to be time proof and dependable [93]. Other researchers confirm this remark since areas of thermal leaks are more clearly observed from the interior as opposed to the exterior [36].

CONCLUSION

This review detailed current procedures and methodologies of UAS-based thermal imaging practices. An experiment was conducted to empirically assess reviewed work, and a UAS-based building inspection method was presented, tested, and results were stated. Currently ageing infrastructure and building stock necessitate energy retrofitting action, and advancements in the methods with which thermal issues are identified will enable more action. In the age of climate change, the use of UASs and IR imaging has proven to be a significant improvement on traditional auditing methods and techniques [7]. The increased accessibility, efficiency, and safety present a unique opportunity to expedite the improvement and retrofitting of ageing and energy inefficient building stock and infrastructure. Existing software and mathematical concepts present a variety of options for post-processing, analysis, and visual representation with reduced manual workflow, as a step closer towards fully automated building performance inspections using drones. Future research should build on the presented workflows to develop a standardised approach for building energy audits. This should include references to existing technological capabilities and further parameterisation of the process to become more global through replicated experiments that validate the presented work.

ACKNOWLEDGEMENTS

This chapter is based on a peer-reviewed journal paper: Rakha, T., & Gorodetsky, A. (2018). Review of Unmanned Aerial System (UAS) applications in the built environment: Towards automated building inspection procedures using drones. Automation in Construction, 93, 252-264. It is based on work funded in part by Gryphon Sensors, Syracuse University's Office of Research (Grant #SP-29403-2) and the Campus as a Lab for Sustainability program at Syracuse University. The author would like to thank Mr. Ian Joyce, the Center for Advanced Systems Engineering (CASE) and the Syracuse Center of Excellence at Syracuse University for drone flight, data gathering and faculty development support. The author is especially grateful for the contributions of Alice Gorodetsky in developing this chapter and is also grateful for the student work Amanda Liberty and Rasan Taher provided to support this manuscript.

REFERENCES

1. U.S. Census Bureau, [Online]. Retrieved from: https://www.census.gov/programs- surveys/ahs/data/interactive/ahstablecreator.html#?s_areas=a00000&s_year=n2015&s_tableName=Table1&s_byGroup1=a4&s_byGroup2=a1&s_filterGroup1= t1&s_filterGroup2=g1, (2015), [Accessed 06 02 2017].
2. U.S. Energy Information Administration, How much energy is consumed in U.S. residential and commercial buildings? [Online]. Retrieved from: https://www.eia.gov/tools/faqs/faq.php?id=86&t=1, (2017), [Accessed 14 05 2017].
3. U.S. Department of Energy, Thermographic inspections, [Online]. Retrieved from: https://energy.gov/energysaver/thermographic-inspections, (2012), [Accessed 18 05 2017].
4. E.C. Shao, Detecting sources of heat loss in residential buildings from infrared imaging, Doctoral Dissertation, Massachusetts Institute of Technology, Cambridge, MA, (2011), http://hdl.handle.net/1721.1/68921 [Accessed 24 08 2017].
5. A. Kylili, P.A. Fokaides, P. Christou, S.A. Kalogirou, Infrared thermography (IRT) applications for building diagnostics: a review, *Appl. Energy* 134 (2014) 531–549, doi:10.1016/j.apenergy.2014.08.005 [Accessed 22 05 2017].

6. L.E. Mavromatidis, D.J. L, R. Saleri, J.C. Batsale, First experiments for the diagnosis and thermo-physical sampling using impulse IR thermography from unmanned aerial vehicle (UAV), Quantitative InfraRed Thermography Conference, Bordeaux, (2014), doi:10.21611/qirt.2014.213 [Accessed 04 06 2017].

7. C. Corsi, History highlights and future trends of infrared sensors, *J. Mod. Opt.* 57 (18) (2010) 1663–1686, doi:10.1080/09500341003693011 [Accessed 07 05 2017].

8. S. Schuffert, T. Voegtle, N. Tate, A. Ramirez, Quality assessment of roof planes extracted from height data for solar energy systems by the EAGLE platform, *Remote Sens.* 7 (12) (2015) 17016–17034, doi:10.3390/rs71215866 [Accessed 03 05 2017].

9. R. Ariwoola, Use of drone and infrared camera for a campus building envelope study, Electronic theses and Dissertations. East Tennessee University in Johnson City, Tennessee State. Paper 3018, (2016), http://dc.etsu.edu/etd/3018 [Accessed 30 07 2017].

10. M.L. Mauriello, L. Norooz, J.E. Froehlich, Understanding the role of thermography in energy auditing: current practices and the potential for automated solutions, *Proceedings of the 33rd Annual Association for Computing Machinery Conference on Human Factors in Computing Systems*, Seoul, (2015), [ISBN: 978-1-4503-3145-6. Accessed 18 05 2017].

11. R. Naughton, Remote piloted aerial vehicles, [Online]. Retrieved from: http://www.ctie.monash.edu.au/hargrave/rpav_home.html/, (2003), [Accessed 24 08 2017].

12. E. Howell, What is DARPA? [Online]. Retrieved from: https://www.space.com/29273-what-is-darpa.html, (30 04 2015), [Accessed 24 08 2017].

13. R. Mac, Bow to your billionaire drone overlord: Frank Wang's quest to put DJI robots into the sky, [Online]. Retrieved from: https://www.forbes.com/sites/ryanmac/2015/05/06/dji-drones-frank-wang-china-billionaire/, (06 05 2015), [Accessed 24 08 2017].

14. C. Dillow, A brief history of drones, [Online]. Retrieved from: http://fortune.com/2014/10/09/a-brief-history-of-drones/, (09 10 2014), [Accessed 24 08 2017].

15. G. Wallace, Amazon says drone deliveries are the future, [Online]. Retrieved from: http://money.cnn.com/2013/12/01/technology/amazon-drone-delivery/index.html, (02 12 2013), [Accessed 24 08 2017].

16. I. Shaw, History of U.S. drones, [Online]. Retrieved from: https://understandingempire.wordpress.com/2-0-a-brief-history-of-u-s-drones/, (2014), [Accessed 24 08 2017].

17. A. Tarantola, This flying bomb failure was America's WWI cruise missile, [Online]. Retrieved from: http://gizmodo.com/this-flying-bomb-failure-was-americas-wwi-cruise-missi-1184824802, (04 09 2013), [Accessed 24 08 2017].

18. J. Stamp, Unmanned drones have been around since world war I, [Online]. Retrieved from: http://www.smithsonianmag.com/arts-culture/unmanned-drones-have-been-around-since-world-war-i-16055939/#ZOkewSDbAgEoRHhA.99, (12 02 2013), [Accessed 24 08 2017].

19. J.W. Saveriano, Superman, Marilyn, robot planes, and UAVs, [Online]. Retrieved from: https://sandacom.wordpress.com/2010/03/26/superman-marilyn-and-robot-planes-uavs-part-2/, (26 03 2010), [Accessed 24 08 2017].

20. J. Rizzo, Japan's secret WWII weapon: balloon bombs, [Online]. Retrieved from: http://news.nationalgeographic.Com/news/2013/05/130527-map-video-balloon-bomb-wwii-japanese-air-current-jet-stream/, (27 05 2013), [Accessed 24 08 2017].

21. A. Orlov, The U-2 program: A Russian officer remembers, [Online]. Retrieved from: https://www.cia.gov/library/center-for-the-study-of-intelligence/csi-publications/csi-studies/studies/winter98_99/art02.html, (2007), [Accessed 24 08 2017].

22. E. Helmore, US retires predator drones after 15 years that changed the 'war on terror', [Online]. Retrieved from: https://www.theguardian.com/world/2017/mar/13/predator-drone-retire-reaper-us-military-obama, (2017), [Accessed 24 08 2017].

23. D. Prindle, DJI's new obstacle avoidance tech aims to make drones crash proof, [Online]. Retrieved from: https://www.digitaltrends.com/cool-tech/dji-obstacle-avoidance-matrice-100-guidance/, (09 06 2015), [Accessed 24 08 2017].

24. B. Popper, DJI's revolutionary phantom 4 drone can dodge obstacles and track humans, [Online]. Retrieved from: https://www.theverge.com/2016/3/1/11134130/dji-phantom-4-drone-autonomous-avoidance-tracking-price-video, (01 03 2016), [Accessed 24 08 2017].

25. A. Dalton, Engadget, [Online]. Retrieved from: https://www.engadget.com/2016/07/21/facebooks-solar-powered-drone-makes-first-full-test-flight/, (21 07 2016), [Accessed 24 08 2017].

26. E. Codel, Drones doing the darndest things, [Online]. Retrieved from: https://iq.intel.com/mind-blowing-drone-innovations/, (20 04 2016), [Accessed 24 08 2017].

27. Lockheed Martin Aeronautics Company, Lockheed martin supports U.S. navy and Northrop Grumman in X-47B UCAS-D successful first flight, [Online]. Cision PR Newswire https://www.prnewswire.com/news-releases/lockheed-martin-supports-us-navy-and-northrop-grumman-in-x 47b ucas-d-successful-first flight-115496549.html, (07 February 2011), [Accessed 30 07 2017].

28. L. Mallonee, Move over, selfies. 'Dronies' are where it's at, [Online]. Retrieved from: https://www.wired.com/2015/09/move-selfies-dronies/, (2015), [Accessed 18 08 2017].

29. A. Bar-Sela, The history of temperature recording from antiquity to the present, In: M. Abernathy, S. Uematsu (Eds.), *Medical Thermology*, American Academy of Thermology, Washington, DC, (1986), (ISBN: 10:0961490500).

30. M.R. Clark, D.M. McCann, M.C. Forde, Application of infrared thermography to the non-destructive testing of concrete and masonry bridges, *NDT E Int.* 36 (4) (2003) 265–275, doi:10.1016/S0963-8695(02)00060-9 [Accessed 28 08 2017].

31. R.J.P. Ferreira, A. Teixeira de Almeida, C.A.V. Cavalcante, A multi-criteria decision model to determine inspection intervals of condition monitoring based on delay time analysis, *Reliab. Eng. Syst. Saf.* 94 (5) (2008) 905–912, doi:10.1016/j.ress.2008.10.001 [Accessed 02 06 2017].

32. F. Lizak, M. Kolcun, Improving reliability and decreasing losses of electrical system with infrared thermography, *Acta Electrotechnica et Informatica* 8 (1) (2008) 60–63, ISSN 1335–8243. [Accessed 20 08 2017].

33. E. Grinzato, IR thermography applied to the cultural heritage conservation, *18th World Conference on Nondestructive Testing*, Durban, (2012), http://citeseerx.ist.psu.edu/viewdoc/download?doi=10.1.1.638.1122&rep=rep1&type=pdf [Accessed 27 08 2017].

34. M. Hollaus, High resolution aerial images from UAV for forest applications, http://www.newfor.net/wp-content/uploads/2015/02/DL14-Newfor-SoA-UAV.pdf, (2014), [Accessed 29 06 2017].

35. L. Ma, M. Li, L. Tong, Y. Wang, L. Cheng, Using unmanned aerial vehicle for remote sensing application, *21st International Conference on Geoinformatics*, Kaifeng, (2014), doi:10.1109/Geoinformatics.2013.6626078 [Accessed 18 07 2017].

36. M. Fox, D. Coley, S. Goodhew, P. de Wilde, Thermography methodologies for de-tecting energy related building defects, *Renew. Sust. Energ. Rev.* 40 (2014) 296–310, doi:10.1016/j.rser.2014.07.188 [Accessed 08 08 2017].

37. K. Kakaes, F. Greenwood, M. Lippincott, S. Dosemagen, P. Meier, S. Wich, Drones and aerial observation: new technologies for property rights, human rights, and global development, New America, https://www.newamerica.org/international-security/events/drones-and-aerial-observation/, (2015), [Accessed 07 09 2017].

38. R. Steffen, W. Forstner, On visual real time mapping for unmanned aerial vehicles, *Int. Arch. Photogramm. Remote. Sens. Spat. Inf. Sci.* XXXVII (B3a) (2011) 57–62 [Accessed 12 08 2017].

39. K.N. Tahar, Aerial terrain mapping using unmanned aerial vehicle approach, *Int. Arch. Photogramm. Remote. Sens. Spat. Inf. Sci.* XXXIX (B7) (2012) 493–498, doi:10.5194/isprsarchives-XXXIX-B7-493-2012 [Accessed 18 07 2017].

40. W. Volkmann, G. Barnes, Virtual surveying: mapping and modeling cadastral boundaries using Unmanned Aerial Systems (UAS), *Proceedings of the Federation Internationale des Geometres (FIG) Congress: Engaging the Challenges—Enhancing the Relevance*, Kuala Lumpur, (2014), http://www.fig.net/resources/proceedings/fig_proceedings/fig2014/ppt/TS09A/TS09A_barnes_volkmann_7300_ppt.pdf [Accessed 17 06 2017].

41. D. Borrmann, A. Nuchter, M. Dakulovic, I. Maurovic, I. Petrovic, D. Osmankovic, J. Velagic, A mobile robot based system for fully automated thermal 3D mapping, *Adv. Eng. Inform.* 28 (4) (2014) 425–440, doi:10.1016/j.aei.2014.06.002 [Accessed 11 08 2017].

42. P.S. Blaer, P.K. Allen, View planning and automated data acquisition for three-dimensional modeling of complex sites, *J. Field Rob.* 26 (11–12) (2009) 865–891, doi:10.1002/rob.20318 [Accessed 18 05 2017].

43. H. Son, S. Lee, C. Kim, Automated 3D model reconstruction to support energy-efficiency, *Procedia Eng.* 145 (2016) 571–578, doi:10.1016/j.proeng.2016.04.046 [Accessed 20 08 2017].

44. J.R. Martinez-de Dios, A. Ollero, *Automatic Detection of Windows Thermal Heat Losses in Buildings Using UAVs*, World Automation Congress, Budapest, (2006), doi:10.1109/WAC.2006.375998 [Accessed 20 08 2017].

45. J.R. Martinez-de Dios, A. Ollero, J. Ferruz, Infrared inspection of buildings using autonomous helicopters, *4th International Federation of Automatic Control (IFAC) Symposium on Mechatronic Systems*, Loughborough, (2006), doi:10.3182/20060912-3-DE-2911.00105 [Accessed 20 08 2017].

46. N. Haala, M. Cramer, F. Weimer, M. Trittler, Performance test on UAV-based photogrammetric data collection, *International Society for Photogrammetry and Remote Sensing Zurich 2011 Workshop*, Zurich, (2011), doi:10.5194/isprsarchives-XXXVIII-1-C22-7-2011 [Accessed 24 08 2017].

47. S. Siebert, J. Teizer, Mobile 3D mapping for surveying earthwork projects using an Unmanned Aerial Vehicle (UAV) system, *Autom. Constr.* 41 (19 02 2014) 1–14, doi:10.1016/j.autcon.2014.01.004 [Accessed 14 08 2017].

48. L. Lopez-Fernandez, S. Laguela, J. Fernandez, D. Gonzalez-Aguilera, Automatic evaluation of photovoltaic power stations from high-density RGB-T 3D point clouds, *Remote Sens.* 9 (6) (2017) 631, doi:10.3390/rs9060631 [Accessed 26 07 2017].

49. W.H. Maes, A. Huete, K. Steppe, Optimizing the processing of UAV-based thermal imagery, *Remote Sens.* 9 (5) (2017) 476–493, doi:10.3390/rs9050476 [Accessed 28 06 2017].

50. S. Emelianov, A. Bulgakow, D. Sayfeddine, Aerial laser inspection of buildings fa-cades using quadrotor, *Procedia Eng.* 85 (2014) 140–146, doi:10.1016/j.proeng.2014.10.538. *Part of special issue: Selected papers from Creative Construction Conference*, Prague, 2014 [Accessed 23 08 2017].

51. S. Laguela, L. Diaz-Vilarino, D. Roca, J. Armesto, Aerial oblique thermographic imagery for the generation of building 3D models to complement geographic in-formation systems, *12th International Conference on Quantitative Infrared Thermography (QIRT)*, Bordeaux, (2015), doi:10.21611/qirt.2014.041 [Accessed 04 08 2017].

52. I. Lizarazo, V. Angulo, J. Rodriguez, Automatic mapping of land surface elevation changes from UAV-based imagery, *Int. J. Remote Sens.* 38 (8–10) (2017) 2603–2622, doi:10.1080/01431161.2016.1278313 [Accessed 25 06 2017].

53. C. Eschmann, T. Wundsam, Web-based georeferenced 3D inspection and monitoring of bridges with unmanned aircraft systems, *J. Surv. Eng.* 143 (3) (2017) 04017003-1–04017003-10, doi:10.1061/(ASCE)SU.1943-5428.0000221 [Accessed 02 06 2017].

54. S. Laguela, L. Diaz-Vilarino, D. Roca, H. Lorenzo, Aerial thermography from low-cost UAV for the generation of thermographic digital terrain models, *Opto-Electron. Rev.* 23 (1) (2015) 76–82, doi:10.1515/oere-2015-0006 [Accessed 20 06 2017].

55. A. Choi-Fitzpatrick, T. Juskauskas, Up in the air: applying the Jacobs Crowd for-mula to drone imagery, *Procedia Eng.* 107 (2015) 273–281, doi:10.1016/j.proeng.2015.06.082 [Accessed 20 08 2017].

56. F. Tauro, M. Porfiri, S. Grimaldi, Surface flow measurements from drones, *J. Hydrol.* 540 (2016) 240–245, doi:10.1016/j.jhydrol.2016.06.012 [Accessed 22 07 2017].

57. T.D. Stek, Drones over Mediterranean landscapes. The potential of small UAV's (drones) for site detection and heritage management in archaeological survey projects: a case study from le Pianelle in the Tappino Valley, Molise (Italy), *J. Cult. Herit.* 22 (2016) 1066–1071, doi:10.1016/j.culher.2016.06.006 [Accessed 18 07 2017].

58. D. Gonzalez-Aguilera, S. Laguela, P. Rodriguez-Gonzalvez, D. Hernandez-Lopez, Image-based thermographic modeling for assessing energy efficiency of building facades, *Energ. Buildings* 65 (2013) 29–36, doi:10.1016/j.enbuild.2013.05.040 [Accessed 01 07 2017].

59. F. Nex, F. Remondino, UAV for 3D mapping applications: a review, *Appl. Geomatics* 6 (1) (2014) 1–15, doi:10.1007/s12518-013-0120-x [Accessed 06 06 2017].

60. D. Roca, S. Laguela, L. Diaz-Vilarino, J. Armesto, P. Arias, Low-cost aerial unit for outdoor inspection of building facades, *Autom. Constr.* 36 (2013) 128–135, doi:10.1016/j.autcon.2013.08.020 [Accessed 20 08 2017].

61. G. Morgenthal, N. Hallermann, Quality assessment of unmanned aerial vehicle (UAV) based visual inspection of structures, *Adv. Struct. Eng.* 17 (3) (2014) 289–302. [Accessed 10 06 2017].

62. S.M. Adams, M.L. Levitan, C.J. Friedland, High resolution imagery collection utilizing unmanned aerial vehicles (UAVs) for post-disaster studies, *Adv. Hurricane Eng. Learn. Past* (2013) 777–793, doi:10.1061/9780784412626.067 [Accessed 12 08 2017].

63. F. Caballero, L. Merino, J. Ferruz, A. Ollero, A visual odometer without 3D re-construction for aerial vehicles. Applications to building inspection, *International Conference on Robotics and Automation*, Barcelona, (2005), doi:10.1109/ROBOT.2005.1570841 [Accessed 09 08 2017].

64. A. Ellenberg, A. Kontsos, I. Bartoli, A. Pradhan, Masonry crack detection application of an unmanned aerial vehicle, *International Conference on Computing in Civil and Building Engineering*, Orlando, (2014), doi:10.1061/9780784413616.222 [Accessed 20 07 2017].

65. M.N. Zulgafli, K.N. Tahar, Three dimensional curve hall reconstruction using semi-automatic UAV, *Asian Res. Pub. Net. J. Eng. Appl. Sci. (ARPN-JEAS)* 12 (10) (2017) 3228–3232, ISSN 1819–6608. [Accessed 03 06 2017].

66. E.J. Lee, S.Y. Shin, B.C. Ko, C. Chang, Early sinkhole detection using a drone based thermal camera and image processing, *Infrared Phys. Technol.* 78 (2016) 223–232, doi:10.1016/j.infrared.2016.08.009 [Accessed 13 07 2017].

67. A. Colantonio, G. McIntosh, The differences between large buildings and residential infrared thermographic inspections is like night and day, *11th Canadian Conference on Building Science and Technology*, Banff, (2007), http://citeseerx.ist.psu.edu/viewdoc/download?doi=10.1.1.543.1644&rep=1 ep1&type=pdf [Accessed 25 06 2017].

68. A.P.M. Tarigan, D. Suwardhi, M.N. Fajri, F. Fahmi, Mapping a volcano hazard area of Mount Sinabung using drone: preliminary results, *1st Annual Applied Science and Engineering Conference*, Pendidikan, (2017), doi:10.1088/1757-899X/180/1/012277 [Accessed 18 06 2017].

69. M.C. Harvey, J.V. Rowland, K.M. Luketina, Drone with thermal infrared camera provides high resolution georeferenced imagery of the Waikite geothermal area, New Zealand, *J. Volcanol. Geotherm. Res.* 325 (2016) 61–69, doi:10.1016/j.jvolgeores.2016.06.014 [Accessed 14 08 2017].

70. X. Feifei, L. Zongjian, G. Dezhu, L. Hua, Study on construction of 3D building based on UAV images, *Remote. Sens. Spat. Inf. Sci.* XXXIX (B1) (2012) 469–473, doi:10.5194/isprsarchives-XXXIX-B1-469-2012 [Accessed 30 07 2017].

71. L. Hoegner, U. Stilla, Texture extraction for building models from IR sequences of urban areas, *Urban Remote Sensing Joint Event*, Munich, (2007), doi:10.1109/URS.2007.371812 [Accessed 20 08 2017].

72. J.L. Lerma, M. Cabrelles, T.S. Akasheh, N.A. Haddad, Documentation of weathered architectural heritage with visible, near infrared, thermal and laser scanning data, *Int. J. Heritage Digital Era* 1 (2) (2012) 251–275, [Accessed 30 07 2017].

73. L. Lopez-Fernandez, S. Laguela, I. Picon, D. Gonzalez-Aguilera, Large scale automatic analysis and classification of roof surfaces for the installation of solar panels using a multi-sensor aerial platform, *Remote Sens.* 7 (9) (2015) 11226–11248, doi:10.3390/rs70911226 [Accessed 30 07 2017].

74. I.B. Utne, T. Brurok, H. Rodseth, A structured approach to improved condition monitoring, *J. Loss Prev. Process Ind.* 25 (3) (2011) 478–488, doi:10.1016/j.jlp.2011.12.004 [Accessed 04 08 2017].

75. E. Barreira, V.P. Freitas, Evaluation of building materials using infrared thermography, *Constr. Build. Mater.* 21 (1) (2007) 218–224, doi:10.1016/j.conbuildmat.2005.06.049 [Accessed 24 06 2017].

76. N.P. Avdelidis, A. Moropoulou, Emissivity considerations in building thermography, *Energ. Buildings* 35 (7) (2003) 663–667, doi:10.1016/S0378-7788(02)00210-4 [Accessed 29 07 2017].

77. M.D. Gonçalves, P. Gendron, T. Colantonio, Commissioning of exterior building envelopes of large buildings for air leakage and resultant moisture accumulation using infrared thermography and other diagnostic tools, *Thermal Performance of the Exterior Envelope of Whole Buildings X International Conference*, Clearwater Beach, (2007), https://www.cebq.org/documents/RD-0123-AThermalSolutions2007-FINALPAPERMGr1_000.pdf [Accessed 07 07 2017].

78. V. Bonora, G. Tucci, V. Vaccaro, 3D data fusion and multi-resolution approach for a new survey aimes to a complete model of Rucellai's Chapel by Leon Battista Alberti in Florence, *The International Committee for Architectural Photogrammetry (CIPA) 2005 XX International Symposium*, Torino, (2005), OCLC: 420816504. [Accessed 12 08 2017].

79. N.D. Opfer, D.R. Shields, Unmanned aerial vehicle applications and issues for construction, *121st American Society of Engineering Education (ASEE) Annual Conference and Exposition*, Indianapolis, (2014), https://peer.asee.org/23235, ISSN 2153-5965 [Accessed 05 06 2017].

80. Y. Ham, M. Golparvar-Fard, An automated vision-based method for rapid 3D energy performance modeling of existing buildings using thermal and digital imagery, *Adv. Eng. Inform.* 27 (3) (2013) 395–409, doi:10.1016/j.aei.2013.03.005 [Accessed 28 08 2017].

81. O. Kung, C. Strecha, A. Beyeler, J.C. Zufferey, D. Floreano, P. Fua, F. Gervaix, The accuracy of automatic photogrammetry techniques on ultra-light UAV imagery, *Unmanned Aerial Vehicle in Geomatics*, Zurich, (2011), doi:10.5194/isprsarchives-XXXVIII-1-C22-125-2011 [Accessed 04 07 2017].

82. S. Yahyanejad, B. Rinner, A fast and mobile system for registration of low-altitude visual and thermal aerial images using multiple small-scale UAVs, *ISPRS J. Photogramm. Remote Sens.* 104 (2014) 189–202, doi:10.1016/j.isprsjprs.2014.07.015 [Accessed 16 06 2017].

83. P. Rodriguez-Gonzalvez, D. Gonzalez-Aguilera, G. Lopez-Jimenez, I. Picon-Cabrera, Image-based modeling of built environment from an Unmanned Aerial System, *Autom. Constr.* 48 (2014) 44–52, doi:10.1016/j.autcon.2014.08.010 [Accessed 17 07 2017].

84. D. Roca, J. Armesto, S. Laguela, L. Diaz-Vilarino, LIDAR-equipped UAV for building information modelling, *The International Archives of the Photogrammetry, Remote Sensing and Spatial Information Sciences, Volume XL-5, International Society for Photogrammetry and Remote Sensing (ISPRS) Technical Commission V Symposium*, 23–25 June 2014, Riva del Garda, Italy, (2014), doi:10.5194/isprsarchives-XL-5-523-2014 [Accessed 04 05 2017].

85. T. Hackel, J.D. Wegner, K. Schindler, Contour detection in unstructured 3D point clouds, *The Institute of Electrical and Electronics Engineers Conference on Computer Vision and Pattern Recognition*, Las Vegas, (2016), doi:10.1109/CVPR.2016.178 [Accessed 08 06 2017].

86. FLIR Systems, Thermal imaging guidebook for building and renewable energy applications, http://www.flirmedia.com/MMC/THG/Brochures/T820325/T820325_EN.pdf, (2013) [Accessed 04 06 2017].

87. Z. Qamar, Zero gravity, fully cute: New Japanese drone in space, [Online]. Retrieved from: http://www.cnn.com/2017/07/18/tech/cute-japanese-space-drone/index.html, (18 07 2017), [Accessed 08 02 2017].

88. M. Vollmer, M.Ã. Klaus-Peter, *Infrared Thermal Imaging: Fundamentals, Research and Applications*, John Wiley & Sons, Hoboken, NJ (2017) ISBN: 78-3-527-64155-0 [Accessed 05 06 2017].

89. K. Finley, World's smallest drone autopilot system goes open source, [Online]. Retrieved from: https://www.wired.com/2013/08/drone-autopilot/, (2013), [Accessed 08 01 2017].

90. B. Rao, A.G. Gopi, R. Maione, The societal impact of commercial drones, *Technol. Soc.* 45 (2016) 83–90, doi:10.1016/j.techsoc.2016.02.009 [Accessed 29 07 2017].

91. R. Luppicini, A. So, A technoethical review of commercial drone use in the context of governance, ethics, and privacy, *Technol. Soc.* 46 (2016) 109–119, doi:10.1016/j.techsoc.2016.03.003 [Accessed 30 07 2017].

92. F. Remondino, L. Barazzetti, F. Nex, M. Scaioni, D. Sarazzi, UAV photogrammetry for mapping and 3D modeling—current status and future perspectives, *International Archives of the Photogrammetry, Remote Sensing and Spatial Information Sciences, Volume XXXVIII-1/C22, 201, International Society for Photogrammetry and Remote Sensing (ISPRS) Zurich 2011 Workshop*, Zurich, Switzerland, (14–16 September 2011), doi:10.5194/isprsarchives-XXXVIII-1-C22-25-2011 [Accessed 08 08 2017].

93. V.P. Vavilov, How accurate is the IR thermographic evaluation of heat losses from buildings? *Quan. Infra. Thermo. J.* 7 (2) (2010) 255–258, doi:10.1080/17686733.2010.9737142 [Accessed 06 07 2017].

15 The Application of UAVs to Inform Coastal Area Management

Euan J. Provost
Southern Cross University

Paul A. Butcher
NSW Fisheries

Andrew P. Colefax
Southern Cross University

Melinda A. Coleman
NSW Fisheries

Belinda G. Curley
NSW Fisheries

Brendan P. Kelaher
Southern Cross University

CONTENTS

INTRODUCTION

Unmanned Aerial Vehicles (UAVs) are being used increasingly for cost-effective environmental monitoring to collect data that is beneficial for coastal management (Ouellette and Getinet, 2016). Recent improvements in affordability and accessibility of UAV systems have driven the development of a range of applications that benefit coastal management, such as wildlife monitoring (e.g. Kelaher et al., 2019), marine habitat mapping (e.g. Joyce et al., 2018), fisheries compliance (e.g. Bloom et al., 2019), topographic mapping (e.g. Gonçalves and Henriques, 2015), including modelling wave runup (e.g. Casella et al., 2014) and assessing storm damage (e.g. Turner et al., 2016) and erosion (e.g. Clark, 2017). Compared to conventional manned aircraft, UAVs often provide financial, logistic, and safety benefits (Colefax et al., 2018). Compared to satellite data, UAVs can provide higher resolution imagery and flexibility in frequency of data collection, as well as generate data in conditions where satellites are not particularly effective (e.g. periods of extensive cloud cover, Joyce et al., 2018). Currently, UAVs bridge the gap between large-area low-resolution remote sensing techniques and small-scale labour-intensive field sampling techniques, allowing for efficient data collection in fine detail across relatively large areas. As the capabilities, benefits and cost-efficiencies of UAVs evolve with continual technological improvements and reforms to legislation, their application for coastal management will expand rapidly.

Small consumer-grade UAVs (<7 kg; see Carmody, 2019) are currently affordable, compact, and quiet, making them ideal for cost-effective monitoring with limited disturbance to wildlife and people (Linchant et al., 2015). Additionally, the noise effects of small quadcopter UAVs on submerged animals are minimal below the immediate surface of the water (Christiansen et al., 2016). Thus, UAVs have been used extensively in coastal areas for wildlife monitoring of a diverse range of species, including whales (Aniceto et al., 2018), seals (Krause et al., 2017), dugongs (Hodgson et al., 2013), dolphins, and manatees (Ramos et al., 2018), sharks and rays (Kiszka et al., 2016; Raoult et al., 2018, Colefax et al., 2019, sea turtles and crocodiles (Bevan et al., 2018), coastal birds, including penguins (Goebel et al., 2015) and numerous wetland species (Han et al., 2017). Broader UAV-based surveys recording entire assemblages of large marine fauna off sandy beaches have also been undertaken, allowing for the comparison of marine communities both spatially and temporally (Kelaher et al., 2019). Innovative biological sampling of marine wildlife using UAVs includes the non-invasive collection of DNA from whale blow, which allows for research vessels to be kept at safe distances and negating the need for invasive techniques (Geoghegan et al., 2018; Russell et al., 2017).

The potential for UAVs to undertake discrete and cost-effective monitoring and surveillance may be applied to improve fisheries and marine-protected area management (Jiménez López and Mulero-Pázmány, 2019; Nyman, 2019). The detection of illegal or abandoned fishing gear using UAVs has been trialled successfully (Bloom et al., 2019) with ~75% of unmarked submerged crab traps detected in an estuarine environment. An increase to the frequency of monitoring and surveillance within marine-protected areas may also be achieved using UAVs, with a potential reducetion in illegal fishing and the associated impacts (Toonen and Bush, 2018). UAVs have been used to support fisheries stock assessments, such as providing a means for rapid biomass estimates of edible jelly fish (Raoult and Gaston, 2018) and quantifying salmon nests during spawning events (Groves et al., 2016).

The collection of data on the usage of coastal areas by people is another beneficial application of UAVs. Human use data is used to inform management agencies when developing strategies to enhance the long-term ecological and economic sustainability of exploited coastal habitats (King and McGregor, 2012). For example, beaches are important assets for coastal regions which attract people who undertake recreational activities such as swimming, surfing, and fishing. Intensive recreational use of beach habitats can lead to degraded habitat quality and amenity, reducing their economic, ecological, and recreational value (Lucrezi et al., 2016). Active beach management is essential to avoid unsustainable levels of usage, by implementing strategies that ensure beach habitats are adequately protected (i.e. limiting access, dune protection and revegetation programs, water quality monitoring, and run-off control; Chen and Teng, 2016; Papageorgiou, 2016). These strategies are most effectively designed and implemented through an understanding of the human requirements (i.e. usage and values), in addition to the physical and biological beach characteristics (i.e. morphology and environmental influences; Domínguez-Tejo et al., 2016; James, 2000; Semeoshenkova et al., 2017).

The recreational use of beaches is conventionally assessed using shore-based surveys (fixed video cameras and photography; Huamantinco Cisneros et al., 2016; Ibarra, 2011), manned aerial surveys (trained observers and/or digital imagery; Blackweir and Beckley, 2004), manual counting by researchers on beaches (King and McGregor, 2012), and community surveys (Dwight et al., 2007; Zhang et al., 2015). Relative to the strengths and weakness of these techniques, UAVs may represent a useful step forward for improved assessments of recreational beach use that can effectively underpin coastal management decisions. The following case study explores this supposition and demonstrates how UAVs were used to effectively collect human use data to compare four popular beaches on the east coast of Australia.

CASE STUDY: RECREATIONAL BEACH USE ON THE NORTH COAST OF NEW SOUTH WALES, AUSTRALIA

Australia has a diverse beach culture with high-quality beaches that drive regional, national, and international tourism (Maguire et al., 2011). Beach tourism is especially important for small regional towns, or 'holiday towns', that rely on tourism to boost local economies. During school holiday periods, a surge in beach attendance can pressure beach management strategies and the integrity of the beaches habitats (Hadwen et al., 2011). Beaches are often managed by local authorities and lifeguard services provided by volunteer Surf Life Saving clubs. For effective management, the provision of adequate resources allows for appropriate levels of services, which reduces risks to habitats and to beach users. However, this requires a detailed understanding of the number of people using beaches and the types of activities undertaken, as well as an appropriate regulatory framework to implement and execute effective beach management strategies. Despite the importance of quantitative human use on beaches for supporting beach management strategies, there is a paucity of this type of data in many parts of the world, limiting effective beach management, undermining the quality of the beach going experience, and impacting the biota associated with these natural habitats.

For the north coast of New South Wales (NSW), an opportunity to evaluate the effectiveness of UAVs to collect recreational beach use data became available as part of shark monitoring flights. Here, very small quadcopters (~1.4 kg, see Carmody, 2019; Colefax et al., 2019) were used to survey recreational beach use at four popular beaches during the austral summer (27 December 2016–29 January 2017), winter (30 June 2017–17 July 2017), and spring (23 September 2017–08 October 2017) school holiday periods. From north to south, the beaches surveyed were: Byron Bay (flights; *winter n* = 16, *spring* = 16, GPS = 28.641751°S–153.619049°E), Lennox Head (flights; *summer* = 29, *winter* = 16, *spring* = 14, GPS = 28.795967°S–153.594969°E), Ballina (flights; *summer* = 19, *winter* = 18, *spring* = 13, GPS = 28.867810°S–153.592742°E), Evans Head (flights; *summer n* = 20, *winter* = 18, *spring* 14, GPS = 29.110761°S–153.434296°E) (Figure 15.1).

At each beach, flights were undertaken daily by commercially licensed UAV pilots at 10.30 am. Each pilot flew a DJI Phantom 4 quadcopter with circular polarising lenses over the water at 60 m altitude and at a speed of 8 m/s. For each location, the beach and the water sections 1 km north and 1 km south of the local Surf Life Saving club, were sampled, resulting in a 2 km transect. During flights, the UAV's camera was facing towards the beach to capture beach users from the dunes to water interface. The aircraft then moved further seaward and returned flying a parallel flight path to capture water users from the back of the breakers to water interface (see Figure 15.1 for examples of the video footage collected, which was recorded in UHD resolution $3,840 \times 2,160$ at 25 frames per second). For each flight, the wind speed, wind direction, air temperature, humidity, air pressure, cloud cover, and Beaufort Sea state at each location was recorded, and water temperature and surf rating were obtained from Coastalwatch Networks (https://www.coastalwatch.com). Flights were not undertaken if the weather was unsuitable for flying (raining or wind velocities exceeding 33 km/h).

Videos were reviewed post collection, and individual beach users were counted and classified into one of the following user categories: sunbathing (those sitting, laying, standing still, and beach games), walking (including running and dog walking), swimming, surfing (including Stand Up Paddleboarding (SUP) and bodyboarding), and fishing. Permutational analyses of variance (PERMANOVAs, Anderson et al., 2008) were used to test whether participation in beach user categories, and all user groups combined, differed among the beaches sampled, sampling

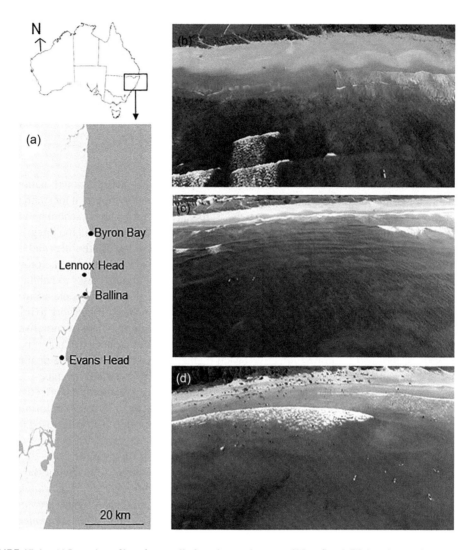

FIGURE 15.1   (a) Location of beaches studied on the north coast of New South Wales, Australia. An example of the footage collected with UAVs of recreational beach use: (b) monitoring of beach users at Lennox Head; (c) monitoring of water users at Lennox Head; and (d) the patrolled surf lifesaving area at Evans Head.

periods, and weekdays versus weekends/public holidays. Generalised linear models (GLMs) were used to test hypotheses about the influence of environmental variables on participation in the main user categories.

RESULTS

During the study, 36,618 beach users were documented from 193 flights. Rain or strong wind forced the cancellation of 28 of the 221 planned flights. Overall participation in sunbathing was 42.5%, swimming 22.0%, walking 22.1%, surfing 13.0%, and fishing 0.4% of attendees (Figure 15.2). During the summer holiday period, participation in sunbathing was 38.7%, swimming 33.1%, walking 18.7%, surfing 9.2%, and fishing 0.3% of attendees. During the winter holiday period, participation in sunbathing was 30.4%, swimming 7.9%, walking 39.7%, surfing 21.1%, and fishing 0.9%

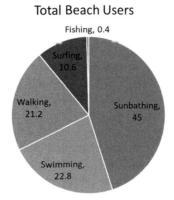

Total Beach Users

FIGURE 15.2 Comparison of beach users engaged in swimming, sunbathing, walking, surfing, and fishing, over the sampling periods. Figures display average users observed per flight (±SE).

of attendees. During the spring holiday period, participation in sunbathing was 51.6%, swimming 19.8%, walking 16.2%, surfing 12.2%, and fishing 0.2% of attendees (Table 15.1). Overall, the study found during warmer periods, there was an increased participation in sunbathing and swimming; however, participation in walking, surfing and fishing did not vary substantially throughout sampled periods (Figure 15.3).

TABLE 15.1
Average Daily Participation in Recreational Beach Categories Collected during the Case Study ± Standard Error

Season	Sunbathing	Swimming	Walking	Surfing	Fishing	Total Users	Days Sampled
Summer	73.9 ± 151.3	63.3 ± 65.5	35.7 ± 60.8	17.6 ± 44.4	0.6 ± 0.3	191.1 ± 303.4	68
Winter	34.9 ± 82.5	9.1 ± 22.5	45.7 ± 64.1	24.3 ± 46.8	1.0 ± 3.1	115.1 ± 208.5	68
Spring	143.0 ± 243.4	54.9 ± 86.0	44.9 ± 78.2	33.8 ± 70.4	0.6 ± 1.1	277.2 ± 303.4	57

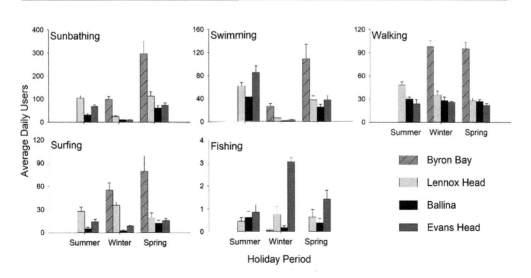

FIGURE 15.3 Percentage of total beach users documented.

Patterns of Use among Beaches

Multi-variate PERMANOVA of all user groups combined found a significant interaction between beach and sample period (PERMANOVA, pF = 3.88, $P \leq 0.001$, Figure 15.3). Post hoc pairwise test revealed beach attendance differed significantly between the summer and winter holiday period (Pairwise test, $P < 0.01$), and that Byron Bay was significantly different from the other beaches during the spring holiday period (Pairwise test, $P < 0.01$, Figure 15.3).

Participation in swimming significantly differed among beaches (PERMANOVA, pF = 27.38, $P \leq 0.001$, Figure 15.3), with significantly greater numbers of swimmers at Byron Bay compared to the other beaches (pairwise test, $P < 0.001$). Swimming participation at Lennox Head and Evans Head did not significantly differ (pairwise test, $P > 0.05$), but this was significantly greater than at Ballina (pairwise test, $P < 0.001$). Significant interactive effects of sample period and weekend resulted from increased swimming on weekdays compared to weekends during the winter and spring holiday periods (PERMANOVA, pF = 15.26, $P \leq 0.01$).

Participation in sunbathing significantly differed among sampling periods (PERMANOVA, pF = 36.07, $P \leq 0.01$) and beaches sampled (PERMANOVA, pF = 42.49, $P \leq 0.001$). There was a significant reduction in participation in sunbathing during the winter period compared to the summer and spring holiday periods (pairwise test, $P < 0.05$). Participation in sunbathing was similar at Evans Head and Ballina (pairwise test, $P > 0.05$), while significantly more sunbathing occurred at Lennox Head, and significantly more again at Byron Bay (pairwise test, $P < 0.001$).

Participation in walking varied significantly among beach (PERMANOVA, pF = 49.33, $P \leq 0.001$), and an interaction between the sample period and weekend versus weekdays was also observed (PERMANOVA, pF = 10.85, $P \leq 0.05$). There was significantly higher participation in walking at Byron Bay than the other beaches (pairwise test, $P < 0.001$). The interaction between period and weekend was a result of increased participation in walking on weekends during the summer and winter holiday periods (pairwise test, $P < 0.05$).

Participation in surfing was shown to have a significant interaction between the period and beach sampled (PERMANOVA, pF = 36.11, $P \leq 0.05$). Significantly more surfing was undertaken at Lennox Head and Evans Head than at Ballina during the summer holiday period (pairwise test, $P < 0.005$). During the winter sampling period, participation at Lennox Head and Byron Bay was significantly greater than at Evans Head or Ballina (pairwise test, $P < 0.001$). During the spring sampling period, the number of surfers at Byron Bay was significantly greater than all other beaches (pairwise test, $P < 0.05$). During the spring sampling period there was, in fact, 95% more surfers at Byron Bay than the other beaches combined.

Although participation in fishing was small, there was significantly more participation on weekends (PERMANOVA, pF = 36.11, $P \leq 0.05$). Additionally, a significant interaction among beaches and period was also found (PERMANOVA, pF = 3.86, $P \leq 0.01$), due to most of the fishing being undertaken at Evans Head during winter and spring holiday periods.

Relationships between Patterns of Use and Environmental Variables

Participation in sunbathing was significantly influenced by wind direction (greater participation during onshore wind), air temperature (more when warmer), humidity (less when higher humidity), and water temperature (less when water was colder) (GLM, $P < 0.05$). The number of people swimming was significantly influenced by wind direction (greater participation during onshore winds), air temperature (more when warmer), and humidity (less when humidity was higher) (GLM, $P < 0.01$). Cloud cover was the only environmental variable to be significantly associated with walking (GLM, $P < 0.001$), with participants displaying a preference for days that were finer (less clouds). Surfing was negatively influenced by wind speed, air temperature, and sea state (preference for low wind, colder temperatures, and lower sea states with reduced wind chop) (GLM, $P < 0.05$). No environmental variables were found to be significantly related to participation in fishing (GLM, $P > 0.05$ for all tests). Despite similarities in general patterns of usage among seasons, the patterns of beach attendance and the proportional participation in particular activities at each beach were relatively unique (e.g. surfing popular at Lennox Head and fishing popular in Evans Head) (GLM, $P > 0.001$).

DISCUSSION

Quantification of recreational beach use by collecting high-resolution video with UAVs, was undertaken successfully across four beaches. It was possible to simultaneously survey multiple beaches due to the low cost and flexibility in deployment of UAV systems. Our results clearly highlighted the variation in beach usage patterns among popular beaches. While UAV-based surveys reported well-known patterns of beach use, such as increased attendance during warmer seasons, they also revealed less obvious patterns, for example, similar levels of beach use on weekends and weekdays during holiday periods. The detailed human use information also provided a strong basis for delivering effective evidence-based management strategies to minimise environmental impacts and maximise beach user safety, for example, UAVs detected that most swimmers at Byron Bay were outside of the designated safe swimming zone with lifeguard monitoring (~260% more swimmers outside these areas) suggesting that more services may be required. Overall, these results suggest that while there is potential for general beach management practices at a regional scale, site-specific strategies may be more effective for optimising the management of recreational beach use. This case study demonstrates the capacity for UAVs to undertake cost-effective quantification of recreational beach usage to assist coastal management. For additional information on this case study, see Provost et al. (2019).

UAVs are easily transportable, rapidly deployed, cost-effective, and do not require permanent infrastructure (e.g. airports or helipads), allowing for use in locations that may not be suitable for other techniques. UAVs allow for the collection of human use data in new areas or areas not considered worth the cost of manned aircraft. Additionally, the use of ground control points or RTK (real-time kinematic) GPS allows for geographic information system and advanced mapping applications (Agüera-Vega et al., 2017), which may be implemented to quantify impacts from environmental factors and human use activities over time (Turner et al., 2016). Pre-determined flight paths and autonomous operation may further reduce cost and increase the accuracy of surveys. As human use is an important consideration for the management of natural areas, the increased capacity to collect this type of data will improve management strategies and outcomes intended to prevent unsustainable use and habitat degradation.

In addition to monitoring human use, UAVs can provide a means of data collection for informing other aspects of coastal area management, such as wildlife monitoring. The benefits of using a UAV-based approach include cost-effective data collection, greater precision (from high-resolution data), and in some cases, better outcomes when compared to conventional methods. For example, UAVs were found to be between 43% and 96% more accurate when compared to a standard field counting technique to survey bird colonies (Hodgson et al., 2018), and photographs captured by UAVs provided highly accurate estimates of leopard seal mass, removing the need for animal captures (Krause et al., 2017). There may be additional situations where a UAV-based approach may be a more suitable method for data collection; however, additional well-designed comparative studies are needed. Current challenges for successful UAV techniques include technological (i.e. battery capacity) and regulatory (i.e. operations must be within line of sight) limitations. As these restrictions are resolved, the utility and range of beneficial applications will increase.

CONCLUSION

The high-resolution data collected by UAVs provided enough detail to determine differences in human usage among beaches within regions. The results suggest that site-specific beach management strategies may be required to avoid unsustainable use and environmental degradation. The results can also inform beach management authorities where services and infrastructure may be required to improve beach user experiences. As well as human use monitoring, UAVs can be used for additional beneficial tasks in coastal areas (e.g. search and rescue, wildlife surveys, quantifying storm impacts), and the cost-effectiveness of this sampling platform provides managers with the flexibility to gather a broader range of data valuable for informing management decisions. Overall, UAVs can be a valuable addition to the coastal manager's monitoring toolbox, and as the range of applications increase, so too will the associated value of this aerial sampling platform.

REFERENCES

Agüera-Vega, F., Carvajal-Ramírez, F. and Martínez-Carricondo, P. (2017). Assessment of photogrammetric mapping accuracy based on variation ground control points number using unmanned aerial vehicle. *Measurement*, 98, 221–227. doi:10.1016/j.measurement.2016.12.002

Anderson, M., Gorley, R. and Clarke, K. (2008). *PERMANOVA+ For PRIMER: Guide to Software and Statistical Methods*. Plymouth, UK: PRIMER-E Ltd.

Aniceto, A. S., Biuw, M., Lindstrøm, U., Solbø, S. A., Broms, F. and Carroll, J. (2018). Monitoring marine mammals using unmanned aerial vehicles: Quantifying detection certainty. *Ecosphere*, 9(3). doi:10.1002/ecs2.2122

Bevan, E., Whiting, S., Tucker, T., Guinea, M., Raith, A. and Douglas, R. (2018). Measuring behavioral responses of sea turtles, saltwater crocodiles, and crested terns to drone disturbance to define ethical operating thresholds. *PLoS ONE*, 13(3), 1–17. doi:10.1371/journal.pone.0194460

Blackweir, D. G. and Beckley, L. E. (2004). Beach usage patterns along the Perth metropolitan coastline during shark surveillance flights in summer 2003/04, (December 2003). School of Environmental Science, Murdoch University.

Bloom, D., Butcher, P., Colefax, A., Provost, E., Cullis, B. and Kelaher, B. (2019) Drones detect illegal and derelict crab traps in a shallow water estuary. *Fisheries Management and Ecology*. doi:10.1111/fme.12350

Carmody, S. (2019). Part 101 (Unmanned Aircraft and Rockets) Manual of Standards 2019, 101(April).

Casella, E., Rovere, A., Pedroncini, A., Mucerino, L., Casella, M., Alberto, L., Vacchi, M., Ferrari, M. and Firpo, M. (2014). Study of wave runup using numerical models and low-altitude aerial photogrammetry: A tool for coastal management. *Estuarine, Coastal and Shelf Science*, 149, 160–167. doi:10.1016/j.ecss.2014.08.012

Chen, C. L. and Teng, N. (2016). Management priorities and carrying capacity at a high-use beach from tourists' perspectives: A way towards sustainable beach tourism. *Marine Policy*, 74(June), 213–219. doi:10.1016/j.marpol.2016.09.030

Christiansen, F., Rojano-Doñate, L., Madsen, P. T. and Bejder, L. (2016). Noise levels of multi-rotor unmanned aerial vehicles with implications for potential underwater impacts on marine mammals. *Frontiers in Marine Science*, 3(December), 1–9. doi:10.3389/fmars.2016.00277

Clark, A. (2017). Small unmanned aerial systems comparative analysis for the application to coastal erosion monitoring. *GeoResJ*, 13, 175–185. doi:10.1016/j.grj.2017.05.001

Colefax, A. P., Butcher, P. A. and Kelaher, B. P. (2018). The potential for unmanned aerial vehicles (UAVs) to conduct marine fauna surveys in place of manned aircraft. *ICES Journal of Marine Science*, 75(1), 1–8. doi:10.1093/icesjms/fsx100

Colefax, A. P., Butcher, P. A., Pagendam, D. E. and Kelaher, B. P. (2019). Reliability of marine faunal detections in drone-based monitoring. *Ocean and Coastal Management*, 174(October 2018), 108–115. doi:10.1016/j.ocecoaman.2019.03.008

Domínguez-Tejo, E., Metternicht, G., Johnston, E. and Hedge, L. (2016). Marine spatial planning advancing the ecosystem-based approach to coastal zone management: A review. *Marine Policy*, 72, 115–130. doi:10.1016/j.marpol.2016.06.023

Dwight, R. H., Brinks, M. V., SharavanaKumar, G. and Semenza, J. C. (2007). Beach attendance and bathing rates for southern California beaches. *Ocean and Coastal Management*, 50(10), 847–858. doi:10.1016/j.ocecoaman.2007.04.002

Geoghegan, J. L., Pirotta, V., Harvey, E., Smith, A., Buchmann, J. P., Ostrowski, M., Eden, J.S., Harcourt, R. and Holmes, E. C. (2018). Virological sampling of inaccessible wildlife with drones. *Viruses*, 10(6), 1–7. doi:10.3390/v10060300

Goebel, M. E., Perryman, W. L., Hinke, J. T., Krause, D. J., Hann, N. A., Gardner, S. and LeRoi, D. J. (2015). A small unmanned aerial system for estimating abundance and size of Antarctic predators. *Polar Biology*, 38(5), 619–630. doi:10.1007/s00300-014-1625-4

Gonçalves, J. A. and Henriques, R. (2015). UAV photogrammetry for topographic monitoring of coastal areas. *ISPRS Journal of Photogrammetry and Remote Sensing*, 104, 101–111. doi:10.1016/j.isprsjprs.2015.02.009

Groves, P. A., Alcorn, B., Wiest, M. M., Maselko, J. M. and Connor, W, P, (2016) Testing unmanned aircraft systems for salmon spawning surveys. *Facets*, 1(1), 187–204. doi:10.1139/facets-2016-0019

Hadwen, W. L., Arthington, A. H., Boon, P. I., Taylor, B. and Fellows, C. S. (2011). Do climatic or institutional factors drive seasonal patterns of tourism visitation to protected areas across diverse climate zones in eastern Australia? *Tourism Geographies*, (May 2019), 13(2), 187–208. doi:10.1080/1461668 8.2011.569568

Han, Y. G., Yoo, S. H. and Kwon, O. (2017). Possibility of applying unmanned aerial vehicle (UAV) and mapping software for the monitoring of waterbirds and their habitats. *Journal of Ecology and Environment*, 41(1), 1–7. doi:10.1186/s41610-017-0040-5

Hodgson, A., Kelly, N. and Peel, D. (2013). Unmanned aerial vehicles (UAVs) for surveying marine fauna: A dugong case study. *PLoS ONE*, 8(11), 1–15. doi:10.1371/journal.pone.0079556

Hodgson, J. C., Mott, R., Baylis, S. M., Pham, T. T., Wotherspoon, S., Kilpatrick, A. D., Segaran, R. R., Reid, I., Terauds A. and Koh, L. P. (2018). Drones count wildlife more accurately and precisely than humans. *Methods in Ecology and Evolution*, 9(5), 1160–1167. doi:10.1111/2041-210X.12974

Huamantinco Cisneros, M. A., Revollo Sarmiento, N. V., Delrieux, C. A., Piccolo, M. C. and Perillo, G. M. E. (2016). Beach carrying capacity assessment through image processing tools for coastal management. *Ocean and Coastal Management*, 130, 138–147. doi:10.1016/j.ocecoaman.2016.06.010

Ibarra, E. M. (2011). The use of webcam images to determine tourist-climate aptitude: Favourable weather types for sun and beach tourism on the Alicante coast (Spain). *International Journal of Biometeorology*. doi:10.1007/s00484-010-0347-8

James, R. J. (2000). From beaches to beach environments: Linking the ecology, human-use and management of beaches in Australia. *Ocean and Coastal Management*, 43(6), 495–514. doi:10.1016/S0964-5691(00)00040-5

Jiménez López, J. and Mulero Pázmány, M. (2019). Drones for conservation in protected areas: Present and future. *Drones*, 3(1), 10. doi:10.3390/drones3010010

Joyce, K. E. A., Duce, S., Leahy, S. M., Leon, J. and Maier, S. W. (2018). Principles and practice of acquiring drone-based image data in marine environments. *Marine and Freshwater Research*, 70(7), 952–963.

Kelaher, B. P., Colefax, A. P., Tagliafico, A., Bishop, M. J., Giles, A. and Butcher, P. A. (2019). Assessing variation in assemblages of large marine fauna off ocean beaches using drones. *Marine and Freshwater Research*. doi:10.1071/mf18375

King, P. and McGregor, A. (2012). Who's counting: An analysis of beach attendance estimates and methodologies in southern California. *Ocean and Coastal Management*, 58, 17–25. doi:10.1016/j.ocecoaman.2011.12.005

Kiszka, J. J., Mourier, J., Gastrich, K. and Heithaus, M. R. (2016). Using unmanned aerial vehicles (UAVs) to investigate shark and ray densities in a shallow coral lagoon. *Marine Ecology Progress Series*, 560(November), 237–242. doi:10.3354/meps11945

Krause, D. J., Hinke, J. T., Perryman, W. L., Goebel, M. E. and LeRoi, D. J. (2017). An accurate and adaptable photogrammetric approach for estimating the mass and body condition of pinnipeds using an unmanned aerial system. *PLoS ONE*, 12(11), 1–20. doi:10.1371/journal.pone.0187465

Linchant, J., Lisein, J., Semeki, J., Lejeune, P. and Vermeulen, C. (2015). Are unmanned aircraft systems (UASs) the future of wildlife monitoring? A review of accomplishments and challenges. *Mammal Review*, 45(4), 239–252. doi:10.1111/mam.12046

Lucrezi, S., Saayman, M. and Van der Merwe, P. (2016). An assessment tool for sandy beaches: A case study for integrating beach description, human dimension, and economic factors to identify priority management issues. *Ocean and Coastal Management*, 121, 1–22. doi:10.1016/j.ocecoaman.2015.12.003

Maguire, G. S., Miller, K. K., Weston, M. A. and Young, K. (2011). Being beside the seaside: Beach use and preferences among coastal residents of south-eastern Australia. *Ocean and Coastal Management*, 54(10), 781–788. doi:10.1016/j.ocecoaman.2011.07.012

Nyman, E. (2019). Techno-optimism and ocean governance: New trends in maritime monitoring. *Marine Policy*, 99(September 2018), 30–33. doi:10.1016/j.marpol.2018.10.027

Ouellette, W. and Getinet, W. (2016). Remote sensing for marine spatial planning and integrated coastal areas management: Achievements, challenges, opportunities and future prospects. *Remote Sensing Applications: Society and Environment*, 4, 138–157. doi:10.1016/j.rsase.2016.07.003

Papageorgiou, M. (2016). Ocean and coastal management coastal and marine tourism: A challenging factor in marine spatial planning. *Ocean and Coastal Management*, 129, 44–48. doi:10.1016/j.ocecoaman.2016.05.006

Provost, E. J., Butcher, P. A., Colefax, A. P., Coleman, M. A., Curley, B. G. and Kelaher B. P. (2019). Using drones to quantify beach users across a range of environmental conditions. *Journal of Coastal Conservation*. doi:10.1007/s11852-019-00694-y

Ramos, E. A., Maloney, B.M., Magnasco, M. O. and Reiss, D. (2018). Bottlenose dolphins and Antillean mana-tees respond to small multi-rotor unmanned aerial systems. *Frontiers in Marine Science*, 5(September), 316. doi:10.3389/fmars.2018.00316

Raoult, V. and Gaston, T. F. (2018). Rapid biomass and size-frequency estimates of edible jellyfish populations using drones. *Fisheries Research*, 207(July), 160–164. doi:10.1016/j.fishres.2018.06.010

Raoult, V., Tosetto, L. and Williamson, J. (2018). Drone-based high-resolution tracking of aquatic vertebrates. *Drones*, 2(4), 37. doi:10.3390/drones2040037

Russell, D., Pirotta, V., Ostrowski, M., Harcourt, R., Jonsen, I. D., Grech, A. and Smith, A. (2017). An economical custom-built drone for assessing whale health. *Frontiers in Marine Science*, 4(December), 1–12. doi:10.3389/fmars.2017.00425

Semeoshenkova, V., Newton, A., Contin, A. and Greggio, N. (2017). Development and application of an Integrated Beach Quality Index (BQI). *Ocean and Coastal Management*, 143, 74–86. doi:10.1016/j.ocecoaman.2016.08.013

Toonen, H. M. and Bush, S. R. (2018). The digital frontiers of fisheries governance: Fish attraction devices, drones and satellites. *Journal of Environmental Policy and Planning*, 0(0), 1–13. doi:10.1080/1523908X.2018.1461084

Turner, I. L., Harley, M. D. and Drummond, C. D. (2016). UAVs for coastal surveying. *Coastal Engineering*, 114, 19–24. doi:10.1016/j.coastaleng.2016.03.011

Zhang, F., Wang, X. H., Nunes, P. A. L. D. and Ma, C. (2015). The recreational value of gold coast beaches, Australia: An application of the travel cost method. *Ecosystem Services*, 11, 106–114. doi:10.1016/j.ecoser.2014.09.001

16 From Land to Sea
Monitoring the Underwater Environment with Drone Technology

David R. Green
UCEMM – University of Aberdeen

Billy J. Gregory
DroneLite

CONTENTS

INTRODUCTION

In the last 10 years, UAV or UAS technology has evolved very quickly to provide a number of low-cost platforms to carry numerous miniaturised sensors (e.g. RGB, NDVI, thermal, LIDAR) that can be used to gather aerial photography and sensor imagery for numerous land-based monitoring, mapping, and modelling environmental applications, some of which have been described already in this volume.

There are now also many examples of airborne platforms that can be used to survey the coast including coastal and marine waters (Green et al., 2019). In addition, there are now a number of small, relatively low-cost waterborne platforms available that can be used to collect data and information about the underwater coastal and marine environment. Most recently, a number of small

271

affordable waterborne and underwater platforms and sensors have emerged with the capability to see through the water column to gather different types of data and imagery below the surface in both freshwater and saltwater environments. Already there are waterborne platforms available – mostly resembling small boats or catamarans – that can carry small sonar and multibeam sensors – to provide autonomous surveys of water depths and the seabed. Other tethered robotic remotely operated vehicles (ROVs) can be used to carry similar sensors as well as cameras and video. Low-cost versions of these typically carry RGB and video cameras as well as lights to illuminate the underwater environment. Accurate data concerning bathymetry, as well as the environmental conditions in shallow waters, can now be acquired using small specialist sensors such as multibeam and sonar integrated into waterborne drones. To operate effectively in the harsh maritime environment, this technology has been developed to withstand storm force winds and heavy rain and snow and salt spray, and as technology advances, so does the flight time available on drones, meaning more area can be covered in a quicker timeframe. Drone technology is therefore revolutionising bathymetric surveys (Matsuba and Sato, 2018; https://www.martek-marine.com/blog/are-drones-the-future-of-marine-surveying/; https://www.ecagroup.com/en/solutions/mini-uav-survey; https://www.blueyerobotics.com/?gclid=EAIaIQobChMIy9SVwPOY4AIVwojVCh1Z7AlgEAMYASAAEgJyPfD_BwE; https://www.maritime-survey.fr/maritime-surveillance-by-waterborne-drone.html; https://www.maritime-survey.fr/the-different-types-of-drone.html). Faster turnaround times and advanced data is offering more detailed nautical charts, and improving global maritime safety.

This chapter briefly explores some of the small and affordable waterborne and underwater platforms and sensors now coming onto the market and some of their potential uses, advantages, and limitations.

AIRBORNE PLATFORMS

There are a number of different types of airborne platforms carrying different types of sensors that can be used to acquire data and information about the water environment.

ABOVE WATER

A good example is the Riegl BathyCopter carrying a bathymetric LIDAR that can be used for clear and shallow water to generate bathymetric survey data. In effect, this is the water equivalent of the terrestrial Riegel LIDAR system but designed to gather data over a freshwater or saltwater body. This system is mainly designed for hydrographic surveying of inland waters (http://www.riegl.com/products/unmanned-scanning/bathycopter/). It uses the RIEGL BDF-1- a compact and lightweight bathymetric depth finder – comprising a tilt compensator, an Inertial Measurement Unit/Global Navigation Satellite System (IMU/GNSS) with antenna, a control unit, and up to two external digital cameras. BDF-1 can optionally be equipped with a miniVUX-1UAV LIDAR sensor and is ideally suited for generating profiles of inland water bodies.

ON WATER

There are also a number of other small survey-type platforms available that provide small-scale systems for gathering both surface and water column data. Although not strictly devices designed for water operation, there are now some aerial platforms available that can land on and take off from water and can carry sensors and cameras to monitor the water surface or below e.g. using a waterproof camera.

One such example that can take some underwater photography and video is the Aquacopter Bullfrog (www.aquacopters.com) (Figure 16.1). This is a small aerial UAV that has waterproof motors and components that prevents water from getting into the motors or the drone electronics, thereby allowing the platform to fully operate in freshwater and saltwater environments.

FIGURE 16.1   Aquacopter Bullfrog waterproof drone (www.aquacopters.com). (Photograph: Harvey Mann (Buzzflyer Ltd.)

FIGURE 16.2   Splashdrone waterproof drone (https://swellpro-uk.co.uk/).

Whilst various types of sensors can be mounted on this platform, the most common unit in operational use to date has been the GoPro camera series mounted in a standard waterproof GoPro camera housing. Another similar platform is the Splashdrone (https://swellpro-uk.co.uk/). Using a unique waterproof camera and gimbal, this aerial platform can land on water, float, and in a similar way capture underwater imagery and video in clear water (Figure 16.2).

SURFACE PLATFORMS AND SENSORS

A number of examples from different manufacturers have led to several different solutions.

SUPERBATHY

For example, SuperBathy by Heliceo (http://www.heliceo.com/en/produits-pour-geometres/superbathy-bathymetric-drone/) is an autonomous catamaran-type platform using both aerial and water-based propulsion systems to deal with water currents. This is typically used for bank and bridge survey and provides an opportunity to gather navigational data and imagery for freshwater and saltwater areas such as ports, marinas, seabed areas, rivers, creeks, lakes, ponds, and lagoons. It can also facilitate monitoring of siltation and marine sites generating volumes of DTMs (Digital Terrain Models).

CEE-USV

CEE-USV™ by CEE HydroSystems (http://www.ceehydrosystems.com/products/unmanned-survey-vessels/cee-usv/) is a small survey boat (Figure 16.3) incorporating a single beam echo sounder, GNSS positioning, live video, and on-board data management. Using a high bandwidth radio link to a dedicated CEE-LINK™ shore station provides an operator with access to precise bathymetric survey results in real time. CEE echo sounder technology on the CEE-USV™ maximises the potential for field operation. Using a network (Ethernet) data link accessible through a shore laptop internal WIFI or a dedicated long-range CEE-LINK™ shore radio module, data telemetry requires no user management. All survey data – position, bathymetry, and quality control information – are recorded on-board the platform ready for direct import into the survey software.

UNDERWATER DRONE TECHNOLOGY

Whilst most people are now relatively familiar with the small airborne platforms and camera sensors so widely available (called drones or Unmanned Aerial Platforms), whether multi-rotor, fixed-wing, or VTOL (vertical take-off and landing), many are not yet so aware of the recent developments in platforms and sensors that sit either on the surface of the water or that can dive beneath the surface of the water.

Both of these developments are really an extension of the capabilities of current robotic technologies that allow (i) different data and imagery and (ii) information about the water column and seabed, as well as boat hulls and the underwater side of e.g. quays to be collected that are not possible in any other way without the help of larger submersibles (e.g. AUVs (autonomous underwater vehicles), ROVs, or divers).

FIGURE 16.3   CEE-USV™ by CEE HydroSystems (http://www.ceehydrosystems.com/products/unmanned-survey-vessels/cee-usv/).

SUBMERSIBLE PLATFORMS: AN EXAMPLE (POWERVISION)

A low-cost example of a submersible platform is the PowerVision (https://www.powervision.me/eu/). At about the same price as a small airborne drone (£1,500), the PowerVision is a small tethered platform (Figure 16.4) that can float on the surface of the water (either freshwater or saltwater) as well as dive to depths of 100 m with the aid of four electric motor thrusters (two at the back, one on top, and one below).

Controlled wirelessly with a custom hand-held controller, a mobile phone, and a special app, the PowerVision can be submersed and manoeuvred in much the same way as an airborne platform albeit with slightly less controllability.

It also carries small fishfinder sonar with an app that displays a vertical profile of the water column beneath the sonar. Although primarily a 'fishing' aid, the video camera and ease with which the unit can be controlled makes the PowerVision potentially a very useful remote sensing unit that can reveal some interesting information about the underwater environment such as fish populations (Figure 16.5) and can even be used to survey the seabed (Figure 16.6) and the hulls of leisure craft (Figure 16.7).

SUBMERSIBLE PLATFORMS: AN EXAMPLE (GLADIUS MINI)

The Gladius Mini is a very similar portable tethered underwater drone to the PowerVision that can be operated in both saltwater and freshwater. It is distinguished by having five thrusters – rather than three of the PowerVision – giving the operator vertical and horizontal control of the

FIGURE 16.4 PowerVision underwater drone (https://www.powervision.me/eu/) (Photograph: David R. Green)

FIGURE 16.5 PowerVision (seabed). (Photograph: Billy J. Gregory)

FIGURE 16.6   PowerVision (fish populations). (Photograph: Billy J. Gregory)

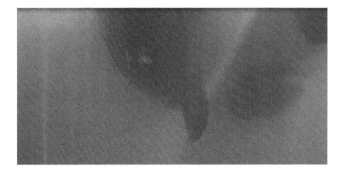

FIGURE 16.7   PowerVision (underside of leisure craft hull). (Photograph: Billy J. Gregory)

drone and sensor. The five-thruster design enables smooth precision movement and stabilisation. As with the other lower-cost examples of underwater platforms, it comes with a 4 K/12 MP ultra HD camera, headlights, and up to 2-hour battery runtime. Like the PowerVision, it is also pressure-rated to 100 m. This makes the platform highly suitable for general underwater operations involving photography, cinematography, as well as underwater exploration, inspection, and other commercial uses, and stable video footage. Control via an app allows one to take advantage of social media functionality such as livestreaming video footage to YouTube and Facebook (Figure 16.8).

FIGURE 16.8   Computer control and instrumentation. (https://www.chasing.com/gladius-mini.html)

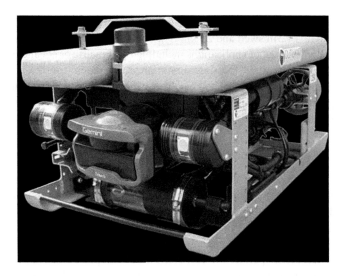

FIGURE 16.9 Outland ROV (http://www.outlandtech.com/) and Tritech Gemini multibeam sonar.

SUBMERSIBLE PLATFORMS: EXAMPLES (OUTLAND, SEAMOR, AND VIDEORAY)

These are still tethered platforms for control, with similar sensors but much heavier duty and more robust equipment for operation in harsher environments. Each one of these platforms – the Outland (http://www.outlandtech.com/), SeaMor (https://seamor.com/), and VideoRay (http://www.videoray.com/) – by comparison with those described above is a higher level solution than the two previous examples and by comparison would be considered true 'professional' ROVs, based on a robust frame construction with the addition of electric motors and other components including cameras, lights, and sensors such as multibeam and sonar, with operating depths of 300 m.

These platforms are constructed using materials such as acrylic, anodised aluminium, stainless steel, and Kevlar rather than the plastic of lower-cost examples. In addition, they include additional instrumentation such as a compass, depth sensor, have auto depth-holding capability, and can be augmented with a wide assortment of accessories, including grippers, positioning systems, scanning sonar, and other sensors.

Computer control and instrumentation are also available, along with many customisations to meet the needs of diverse underwater professions (Figure 16.9).

SOME EXAMPLE APPLICATIONS

SEWER INSPECTION

Drone-based remote sensing technology is gradually finding its way into many applications including sewer surveys (Green, 2016). Anglian Water along with Barhale in the UK, for example, are using drone technology to survey their sewers. The floating Multi-sensory Inspection Unit is being used to survey thousands of metres of sewer pipes underneath Cambridge and Grimsby in England. This unit, operated by Anglian Water's partners Draincare, is operated by floating the platform and sensor down the sewer pipes and using laser, sonar, and high-definition CCTV is then used to scan sewers. The output provides a detailed report of the corrosion and silt build-up in the sewer pipes and is used to plan ahead so as to ensure the sewer network is robust and free flowing. At present, the unit is the only one of its kind in the UK and is developed by the US firm Redzone Robotics (http://www.barhale.co.uk/news/floating-drone-technology-to-survey-sewers/).

The *RSV* (Remote Survey Vehicle) is a lightweight, remote-controlled watercraft – of semi-rigid structure – equipped with multiple sensors. The system integrates navigation, communication, and data acquisition tools installed in watertight modules, and the *RSV* can be used in a number of environments including the sea in up to a 1.5 m swell and 25 knots wind speed as well as in shallow water. This platform can be used for a wide range of applications including bathymetric or oceanographic cartography; analysis of sea depths for wind turbine location; surveillance and imaging of rock accumulation for port jetties; detection and imaging of reefs, wrecks, pipelines, shipwrecks, and monitoring; and surveillance of the laying of telephone cables or pipelines, monitoring, and imaging of port works, visualisation and images of ship hulls in a port, and the surveillance of marine protected areas.

UNDERWATER IMAGERY AND INSPECTION

Dronelite (www.dronelite.co.uk) has utilised the PowerVision tethered underwater drone to undertake a number of simple photographic and video records of the underwater environment. These include camera and video footage of e.g. marina pontoons, the seabed, fish, and the hulls of leisure craft (see Figures 16.5–16.7). Whilst the platform and sensors are very basic, the PowerVision is stable in operation, although navigation, and accurate locational and depth positioning are quite difficult and limited. In addition, operation in turbulent water can place a fair amount of strain on the umbilical tether. Retrieval can also be difficult if the battery power is low. Nevertheless, as a basic underwater drone, the PowerVision is a useful and capable platform considering its modest cost.

SUMMARY AND CONCLUSIONS

In a very short period of time, airborne UAVs or drones and sensors have evolved from toys to professional platforms of varying different types able to carry a wide range of photographic, video, and sensor equipment.

Many of the airborne platforms have also become waterproofed not only to protect them from damage if they have to *ditch* over water but also to fly in damp and wet conditions or to be used in river, lake, coastal, and marine environments for a number of applications which expose the drones to inclement conditions and environments that would normally lead to corrosion or damage.

With time, this robotic technology has found its way into the water environment either floating on the surface or able to go beneath the waves. With miniaturisation, there are now a range of small waterborne platforms – in effect the size of small model boats – together with a number of small submersibles that – whilst constrained by the need to remain tethered to help send signals through the water for control and navigation – have nevertheless opened up a low-cost solution under the water – in either freshwater or saltwater – to allow for the collection of underwater photographs and video. In addition, these platforms usually carry a high-intensity light to illuminate the underwater environment. Some have additional technology on-board such as fishfinders that provide a slice through the water column.

Whilst there are more sophisticated and expensive solutions available, the fact that these examples have become affordable means that there are many exciting developments now possible that reveal the potential to explore the underwater environment in rivers or along the coast that open up opportunities to collect information about the seabed, the coastal ecology, fish and other organisms in the water, together with other information about the state of sea defences, seawalls, harbour structures, buoys, and also boats.

REFERENCES

Green, D.R., 2016. Drones a useful survey tool. *Water and Sewerage Journal*. Issue 96. p. 27.
Green, D.R., Hagon, J.J., Gómez., C., and Gregory, B.J., 2019. Chapter 21: Using low-cost UAVs for environmental monitoring, mapping, and modelling: Examples from the Coastal Zone. In Krishnamurthy, R.R., Jonathan, M.P., Srinivasalu, S., and Glaeser, B., (Eds.). *Coastal Management: Global Challenges and Innovations*. Amsterdam: Elsevier. p. 546.
Matsuba, Y., and Sato, S., 2018. Nearshore bathymetry estimation using UAV. *Coastal Engineering Journal*. Vol. 60. pp. 51–59.

ONLINE SOURCES

https://www.martek-marine.com/blog/are-drones-the-future-of-marine-surveying/
https://www.ecagroup.com/en/solutions/mini-uav-survey
http://www.ceehydrosystems.com/products/unmanned-survey-vessels/cee-usv/
https://www.blueyerobotics.com/?gclid=EAIaIQobChMIy9SVwPOY4AIVwojVCh1Z7AlgEAMYASAAEgJ
     yPfD_BwE
https://www.maritime-survey.fr/maritime-surveillance-by-waterborne-drone.html
https://www.maritime-survey.fr/the-different-types-of-drone.html

17 A Question of UAS Ground Control
Frequency and Distribution

Jason J. Hagon
GeoDrone Survey Ltd and UCEMM

CONTENTS

INTRODUCTION

THE UAS SECTOR AND TOPOGRAPHICAL SURVEYING

The civilian Unmanned Aerial System (UAS) industry has experienced significant growth within the last decade, primarily driven by an increase in the number of UAS manufactures and the relative falling costs of miniature electronic components (Rock et al., 2011). UAS with Global Navigation Satellite System (GNSS) units, Inertial Measurement Units (IMUs), and high-resolution cameras, capable of capturing 20 MPEG images are now available for less than £1,000. Consequently, there has been a surge in popularity for use of these systems within the Geosciences, particularly topographic mapping and environment reconstruction using Structure-from-Motion (SfM [SFM]) software (Gerke and Przybilla, 2016).

The advantages of surveying with UAS include the speed at which large areas can be covered – in comparison with terrestrial surveying techniques – resulting in significant economic savings. Ramirez and Hargraves (2016) estimate that overall efficiency savings can be as large as 66% when directly comparing a traditional survey crew with a UAS for corridor surveying applications. Furthermore, safety concerns associated with working in potentially dangerous environments including quarries can be partly mitigated by using UAS. Finally, the spatial resolution at which modern UAS can collect data is unrivalled by current satellites or manned aerial photography, meaning this technology occupies a unique section in the market between technologies like terrestrial laser scanning and traditional manned aerial photogrammetry.

New developments in photogrammetric SfM (SFM) software have led to claims by software manufactures that UAS imagery can be used to produce survey grade results, leading to dozens of investigations dedicated to ascertaining if these off-the-shelf, low-cost, UAS can produce consistent and high-quality results. Barry and Coakley (2015) concluded that UAS are now capable of 1:200 scale topographical surveys, a conclusion that many others have come to (see Nakano et al., 2016; Yastikli et al., 2013). The process of creating accurate topographical surveys and orthophotos requires a multistage process: (i) camera calibration, involving principle point positioning, focal length, radial distortion, and tangential distortion calculations; (ii) aerial triangulation; (iii) bundle block adjustment; and (iv) DSM and orthophoto generation. Much investigation has focused on understanding and improving these techniques; however, equally important are the UAS flight path, data acquisition parameters (Ground Sampling Distance (GSD), image overlap, and the use of ground control points (GCPs) (Figure 17.1).

GCPs are points where the coordinates are known and are widely used in photogrammetric processing (Pix4D, 2017). Often, GCPs are specifically designed aerial targets used to correct and verify SfM (SFM) project accuracy. GCPs are commonly needed due to most UAS being equipped with single-frequency GNSS receivers, typically capable of achieving 2 m accuracies – not the centimetre accuracy needed for surveying applications. Often, Real-Time Kinematic (RTK) GNSS systems are used to measure the GCPs – due to their superior accuracy, typically <5 cm. This information is then imported during the aerial triangulation stage to improve and verify the accuracy of the final outputs.

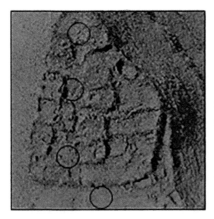 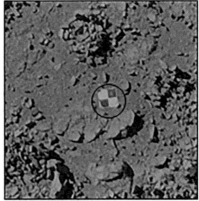

FIGURE 17.1 Aerial survey targets used as GCPs taken from a UAV at 80 m AGL on the left. The same image cropped on the right.

AIMS AND OBJECTIVES

Relatively few studies have comprehensively investigated; (i) how many GCPs are needed in a survey site? and (ii) where should the GCPs be placed within the area being surveyed? These questions are particularly relevant due to the amount of time and effort required to collect and record survey grade points at multiple locations throughout the area of interest.

This chapter aims to investigate how the number of GCPs and their spatial distribution effects the overall accuracy throughout three different surveyed environments in Scotland and England, using three RGB sensors all processed in Pix4D Mapper Pro.

REVIEW OF PREVIOUS WORKS

UAS-BASED STRUCTURE-FROM-MOTION PHOTOGRAMMETRY

UAS-based image acquisition can be subject to higher inclination angles, poor consistency in image overlap, and high distortion levels in comparison with traditional manned aircraft acquired aerial images, complicating the photogrammetric process (Rock et al., 2011). Software manufactures have recently aimed to overcome these problems with a variety of complicated semi-automatic, semi-global algorithms allowing consumer-grade UAS to produce high-quality results. Consequently, the market is filled with many UAS-specific software packages ranging in level of functionality, user interface, and price.

Figure 17.2 shows a characteristic UAS workflow. The GCP data – under additional parameters – directly influences the image triangulation process; hence, there is an obvious importance of ensuring that the GCP data is fit for purpose (Figure 17.3).

The placement and recording of GCP locations are comparatively time-consuming when considering other steps in an entire UAS field to finish workflow. Therefore, it is of great advantage to place an optimum number of GCPs to ensure that the area is surveyed to an appropriate accuracy, while also ensuring that no time is wasted unnecessarily. The consequences of placing a minimal number of GCPs can lead to image block deformation and overall inaccurate results (Gerke and Przybilla, 2016), while the extra economic and time cost of placing a large number of GCPs can lead to a site survey becoming economically unviable. Overall, placing an optimal amount of GCPs is of mutual benefit for the surveyor/UAS operator and the client.

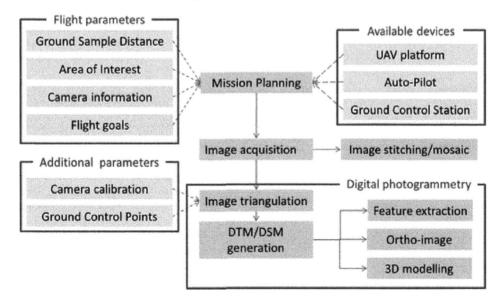

FIGURE 17.2 A typical UAS image acquisition and processing workflow. (Nex and Remondio, 2014.)

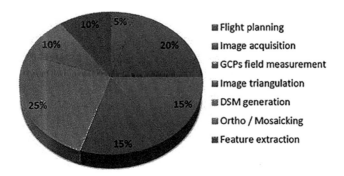

FIGURE 17.3 Approximate time effort in a typical UAS-based photogrammetric workflow. (Nex and Remondio, 2014.)

GCP AND SfM (SFM) PHOTOGRAMMETRY RESEARCH

Several articles investigating the effects of the placement and distribution of GCPs have been conducted within the last few years. Many of the conclusions validate logical hypotheses including that a higher number of GCPs considered during aero triangulation does improve the quality of the final products (Rock et al., 2011); however, others have concluded that for corridor surveys, little change in the error with the reduction of the number of GCP *was observed* (Prajwal et al., 2016). In Table 17.1, a summary of relevant works can be viewed.

TABLE 17.1
A Tabulated Summary of Related Works

Paper Title (Year)	UAS Type (Sensor)	Area Covered	GCP Usage	Methodology	Conclusion
"Sensitivity Analysis of UAV Photogrammetry for Creating Digital Elevation Models" (2011)	MAVinci Sirius 1 (6 Mpeg Canon 300D with Sigma 28 mm/1.8 Ex lens)	Quarry 84.2 ha	1,042+ LIDAR aerial survey reference data.	Reduce the number of GCPs used when generating the DEM to determine if RMSE is changed.	A higher number of GCPs considered during aero triangulation has shown to improve the overall accuracy of the DEM. If a high-quality DEM is needed, the terrain has to be prepared with the time-consuming placement of GCPs.
"Optimal Number of Ground Control Points for UAV Based Corridor Mapping" (2016)	Skylark Osprey (Sony PowerShot SX230hs)	Agricultural land corridor 600×60 m^2	9 GCPs	Reduce the number of GCPs to see if a reduction influences the RMSE of the processed data.	For the 600-m corridor, there was little change detected when reducing the number of GCPs (min used 3).
"Accuracy Analysis of Photogrammetric UAV Blocks: Influence of Onboard RTK-GNSS and Cross Flight Patterns" (2016)	MAVinci Sirius Pro (Lumix GX1-Pancake 14 mm-PRO)	Quarry site $1,100 \times 600$ m^2	17 GCPs	Reduce the number of GCPs to see the influence on RMSE. Implement a gridded flight plan to see if this increases the quality of the results.	Block deformation is a typical problem in UAV image blocks, especially when the block is not supported by well-distributed GCPs. Different flight directions contribute to a more accurate image block.

(Continued)

TABLE 17.1 (*Continued*)
A Tabulated Summary of Related Works

Paper Title (Year)	UAS Type (Sensor)	Area Covered	GCP Usage	Methodology	Conclusion
"Testing Accuracy and Repeatability of UAV Blocks Oriented with GNSS-Supported Aerial Triangulation" (2017)	SenseFly eBee RTK (Sony Cyber-shot DSC WX220 CMOS Exmor R Sensor 18.2 Mpeg)	Built up area 500×400 m^2	14 CPs and 12 GCPs	Reduce the number of GCPs to see if they influence accuracy of the photogrammetric results from an RTK-enabled UAV.	UAS RTK systems can be relied on to supply accurate longitudinal and latitudinal positions; however, in elevation, biases of several centimetres may arise – these can be controlled by 1 GCP.

METHODOLOGY

WORKFLOW DIAGRAM

See Figure 17.4.

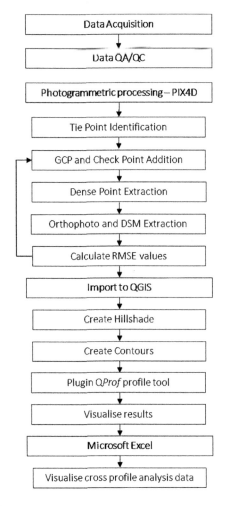

FIGURE 17.4 Project workflow.

DATA ACQUISITION AND SITE LOCATIONS

Data was acquired using a Phantom 3 Advanced, Inspire 1 V2.0, and a Falcon 8 (A7R) across two locations in Scotland and England. Leica Viva GS15 GNSS and Trimble 5000 R8 RTK GNSS units were used to record the positions of the aerial survey targets. Although different techniques including Post-Processing Kinematics and Virtual Reference Stations were used, the accuracy of the surveyed points was <5 cm. This accuracy was reflected during the ingestion of the data in Pix4D (Figures 17.5 and 17.6).

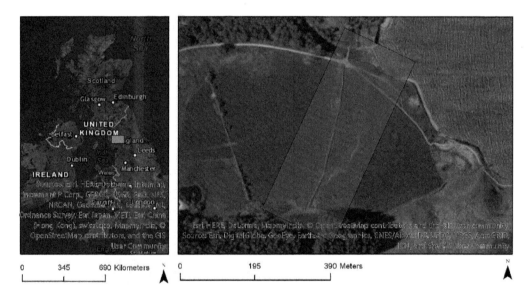

FIGURE 17.5   River Ehen – Ben Gill Channel site location and survey area.

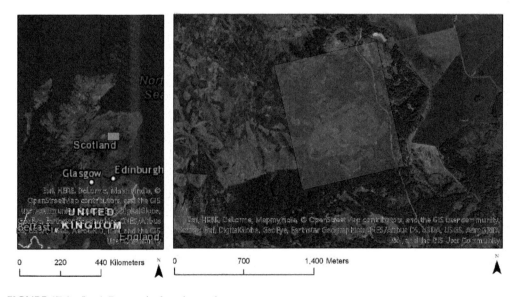

FIGURE 17.6   Loch Davan site location and survey area.

Loch Davan, Scotland – DJI Inspire 12.0
See Table 17.2.

Loch Davan, Scotland – Falcon 8
See Table 17.3.

Ennerdale River Channel, England – DJI Phantom 3 Advanced
See Table 17.4.

TABLE 17.2
UAS and Flight Parameters of Loch Davan Site

UAS	DJI Inspire 1 V2.0
Camera	Zenmuse X3 12 Mpeg
Area covered	84.58 ha
GSD	3.91 cm
Flight plan type	Single pass
Image front/side lap	80%/70%
Number of images used	720
Maximum no. of GCPs used	8

TABLE 17.3
UAS and Flight Parameters of Loch Davan Site

UAS	AscTec Trinity Falcon 8
Camera	Sony Alpha A7R 36.2 Mpeg
Area covered	13.09 ha
GSD	1.29 cm
Flight plan type	Single pass
Image Front/side lap	80%/70%
Number of images used	271
Maximum no. of GCPs used	8

TABLE 17.4
UAS and Flight Parameters of Loch Davan Site

UAS	DJI Phantom 3 Advanced
Camera	Sony EXMOR 12 Mpeg
Area covered	2.97 ha
GSD	1.02 cm
Flight plan type	Dual pass
Image front/side lap	75%/70%
Number of images used	516
Maximum no. of GCPs used	126

DATA QA/QC

Post data acquisition, the images were analysed to ensure that they were suitable for use. Images which appeared underexposed or overexposed, out of focus, or of poor sharpness were excluded before the image processing stage.

PHOTOGRAMMETRIC PROCESSING

Pix4D Mapper Pro version 3.1.2.3 was used to process the aerial images to render tie points, dense point clouds, orthophotos, and DSMs. GCPs and checkpoints (CPs) were added after the initial automatic tie points had been generated. A variety of different test scenarios were run where the frequency and spatial distribution of the GCPs were changed. The acquisition methodology of CPs is the same as with GCPs; however, CPs are not taken into consideration during the Automatic Aerial Triangulation (AAT) or Bundle Block Adjustment (BBA) but serve as a control against which the accuracy of the final outputs can be measured. The same CPs were used throughout each site to ensure that the only variable was the frequency and spatial distribution of the GCPs. In total, 14 test scenarios were run with a range of GCPs from 3 to 126 and a range of CPs from 2 to 63. Processing for the Loch Davan site was completed in WGS84 and the associated UTM zone, with heights referenced to EGM96. Processing for the Ben Gill River Channel was completed to OSGB36 with heights referenced to EGM96.

In all of the processing tests, none of the lens distortion parameters were pre-defined; rather, default settings were used to deduce focal length, principle points, and radial and tangential distortion values. Default image GNSS accuracy settings were also used in Pix4D Mapper Pro.

POST-PROCESSING ANALYSIS IN QGIS

The orthophotos and DSMs were imported into QGIS where a hillshade and contours were created for all sites. The RMSE information associated with each CP value was converted to .csv format and displayed as a shapefile with an average error from all x,y, and z axes for each CP proportionally visualised. A total RMSE for all CP values was also calculated to give an indication of the entire projects average RMSE.

To analyse the DSMs, the plugin *qProf* – which calculates height and slope profiles from DSMs – was used to complete cross-profile analysis along three pre-determined polylines at each site. This process was completed on two DSMs – one DSM created using the maximum number of GCPs available at each site, and one using the minimum (3). The values were imported into Microsoft Excel where the profiles were graphed for visual analysis. Additionally, the records from each profile were differenced to provide information on how much the cross-profile sections differed along the length of each pre-determined polyline.

RESULTS

LOCH DAVAN – DJI INSPIRE 1 V2.0 X3 CAMERA

See Figures 17.7–17.11.

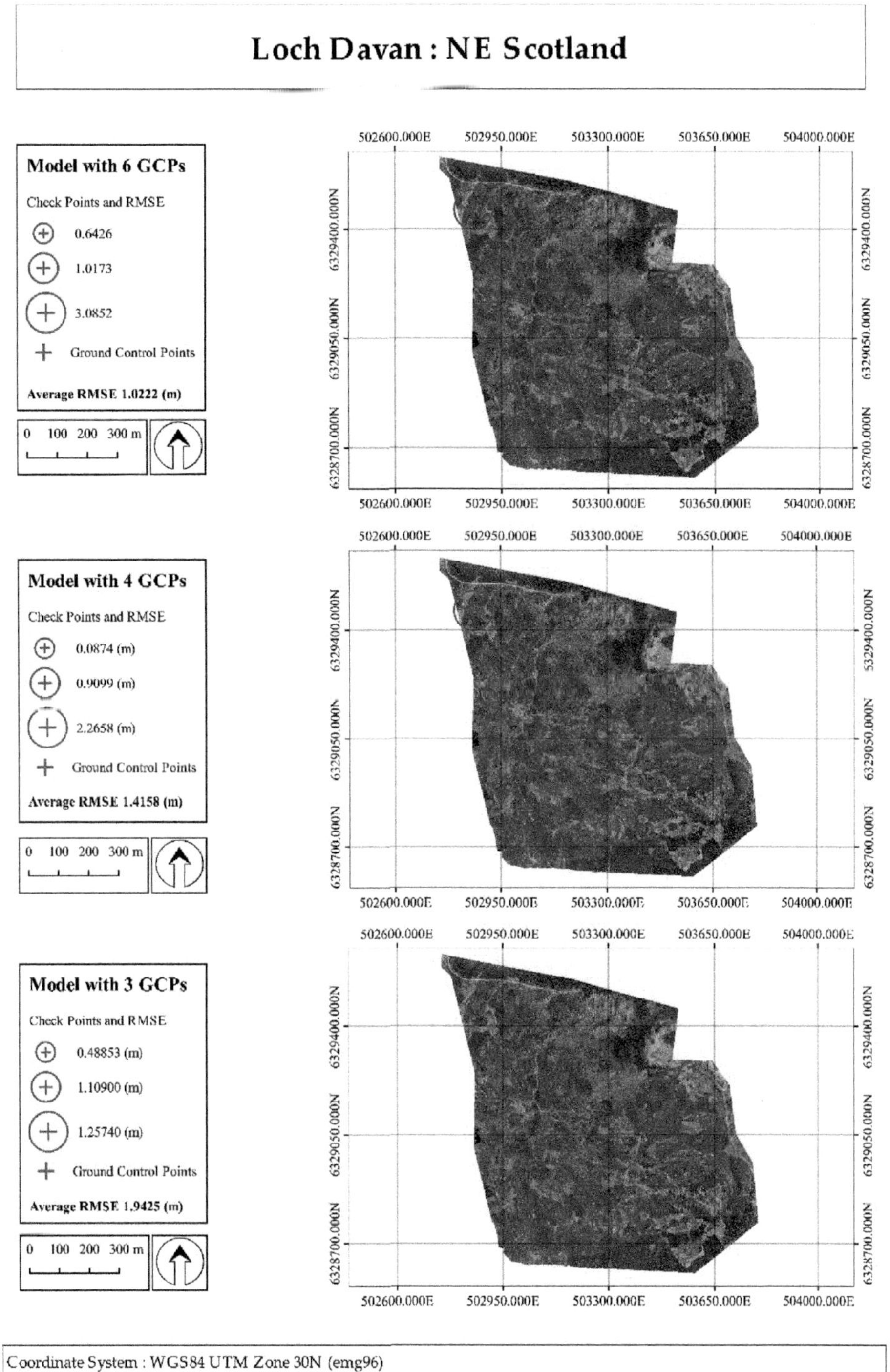

Loch Davan : NE Scotland

Model with 6 GCPs

Check Points and RMSE

⊕   0.6426

⊕   1.0173

⊕   3.0852

+   Ground Control Points

Average RMSE 1.0222 (m)

0   100   200   300 m

Model with 4 GCPs

Check Points and RMSE

⊕   0.0874 (m)

⊕   0.9999 (m)

⊕   2.2658 (m)

+   Ground Control Points

Average RMSE 1.4158 (m)

0   100   200   300 m

Model with 3 GCPs

Check Points and RMSE

⊕   0.48853 (m)

⊕   1.10900 (m)

⊕   1.25740 (m)

+   Ground Control Points

Average RMSE 1.9425 (m)

0   100   200   300 m

Coordinate System : WGS84 UTM Zone 30N (emg96)
Units : Metre
Note that the red circle is proportional to the RMSE caclulated as an average of x, y & z errors. The size of the circles are not comparable between maps.

FIGURE 17.7 Loch Davan – showing the proportional RMSE of the three CPs as the number of GCPs is reduced.

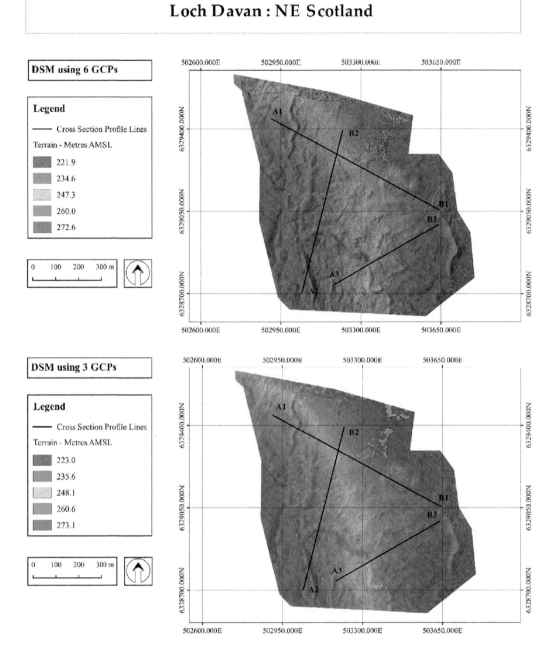

FIGURE 17.8 Loch Davan – Two DSMs with the location of the pre-determined polylines, used for the cross-section profile analysis shown.

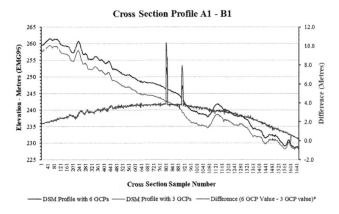

FIGURE 17.9 Results from the qProf tool for polyline A1–B1. The red line shows the results from the DSM rendered with 3 GCPs, while the black line shows the results of the DSM rendered with 6 GCPs. The y axis on the right and the blue line show the elevational difference.

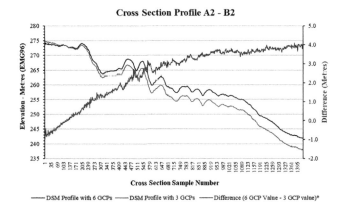

FIGURE 17.10 Results from the qProf tool for polyline A2–B2. The red line shows the results from the DSM rendered with 3 GCPs, while the black line shows the results of the DSM rendered with 6 GCPs. The y axis on the right and the blue line show the elevational difference.

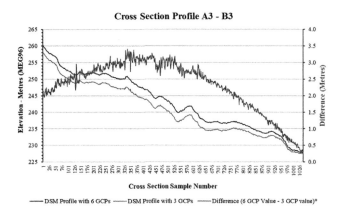

FIGURE 17.11 Results from the qProf tool for polyline A3–B3. The red line shows the results from the DSM rendered with 3 GCPs, while the black line shows the results of the DSM rendered with 6 GCPs. The y axis on the right and the blue line show the elevational difference.

Loch Davan – Falcon 8

See Figures 17.12–17.16.

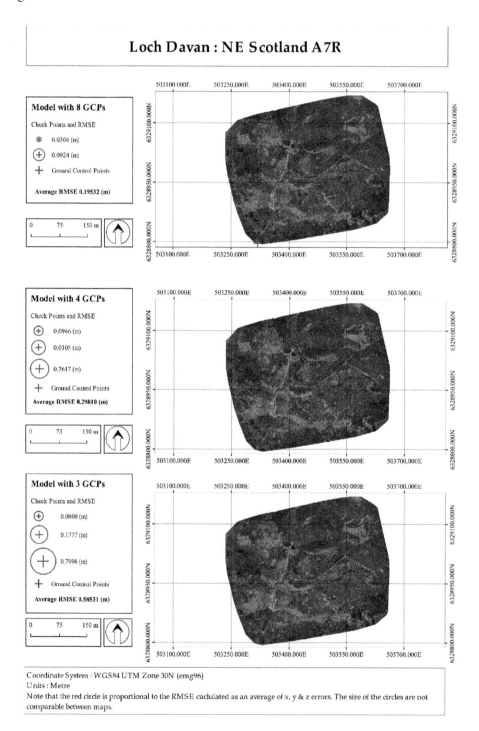

FIGURE 17.12 Loch Davan – showing the proportional RMSE of the three CPs as the number of GCPs is reduced.

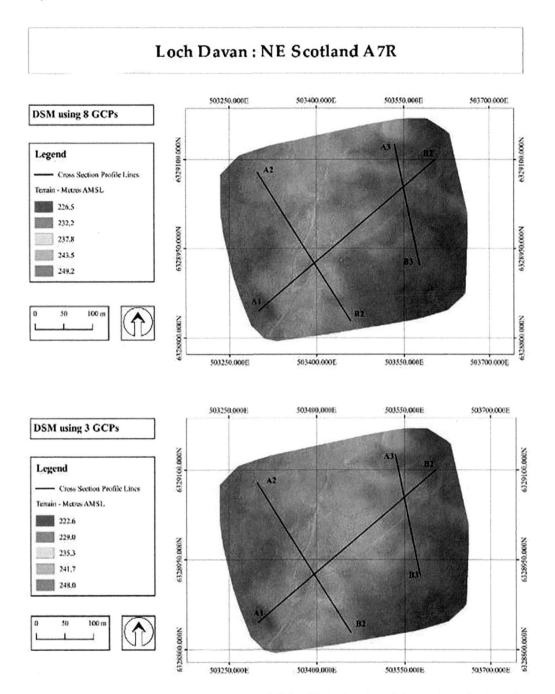

FIGURE 17.13 Loch Davan (A7R Project) – two DSMs with the location of the pre-determined polylines, used for the cross-section profile analysis shown.

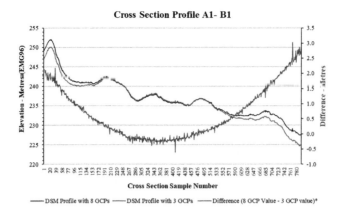

FIGURE 17.14 Results from the qProf tool for polyline A1–B1. The red line shows the results from the DSM rendered with 3 GCPs, while the black line shows the results of the DSM rendered with 8 GCPs. The y axis on the right and the blue line show the elevational difference.

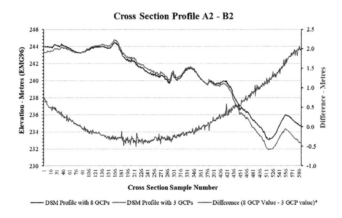

FIGURE 17.15 Results from the qProf tool for polyline A2–B2. The red line shows the results from the DSM rendered with 3 GCPs, while the black line shows the results of the DSM rendered with 8 GCPs. The y axis on the right and the blue line show the elevational difference.

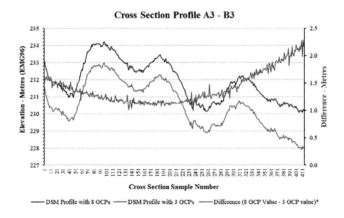

FIGURE 17.16 Results from the qProf tool for polyline A3–B3. The red line shows the results from the DSM rendered with 3 GCPs, while the black line shows the results of the DSM rendered with 8 GCPs. The y axis on the right and the blue line show the elevational difference.

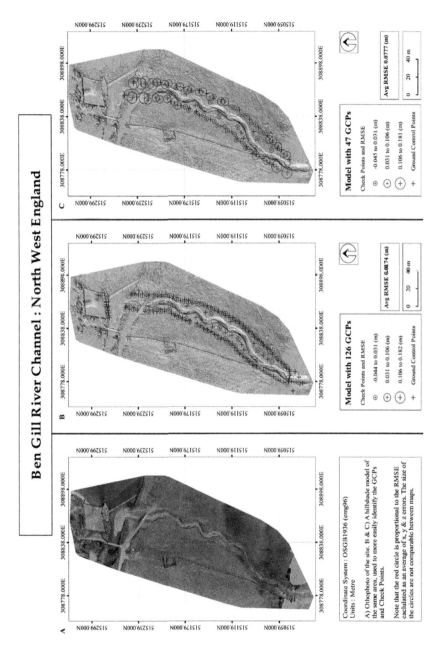

FIGURE 17.17   River Ehen – Ben Gill Channel showing the proportional RMSE of the 63 CPs as the number of GCPs is reduced.

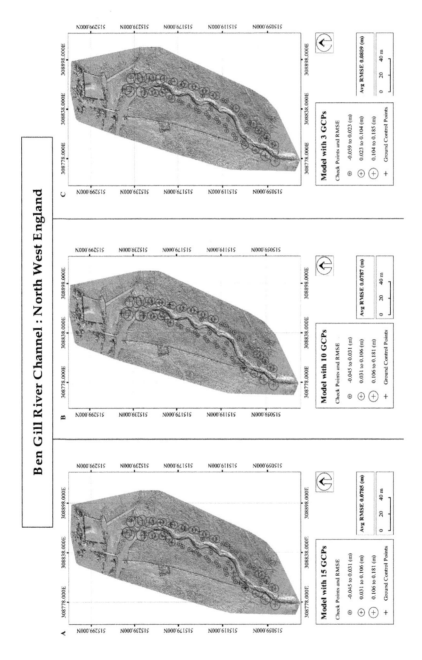

FIGURE 17.18 River Ehen – Ben Gill Channel showing the proportional RMSE of the 63 CPs as the number of GCPs is reduced.

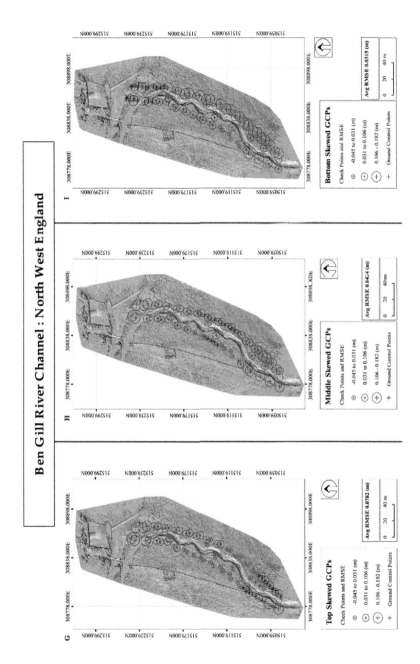

FIGURE 17.19 River Ehen – Ben Gill Channel showing the proportional RMSE of the 63 CPs as the spatial distribution of the GCPs is changed.

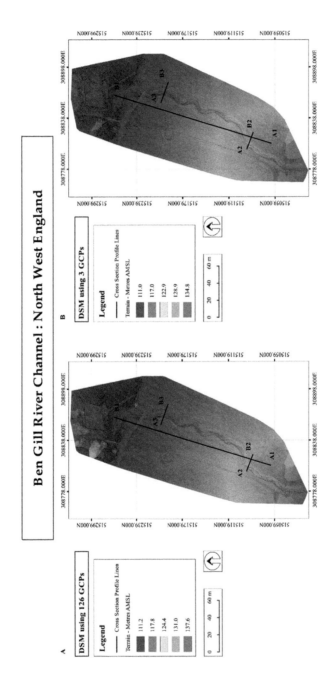

FIGURE 17.20 River Ehen – Ben Gill Channel – Two DSMs with the location of the pre-determined polylines shown.

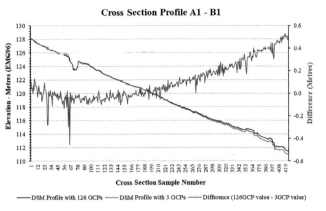

FIGURE 17.21 Results from the qProf tool for polyline A1–B1. The red line shows the results from the DSM rendered with 3 GCPs, while the black line shows the results of the DSM rendered with 126 GCPs. The y axis on the right and the blue line show the elevational difference.

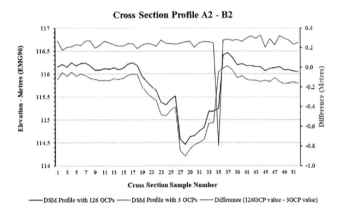

FIGURE 17.22 Results from the qProf tool for polyline A2–B2. The red line shows the results from the DSM rendered with 3 GCPs, while the black line shows the results of the DSM rendered with 126 GCPs. The y axis on the right and the blue line show the elevational difference.

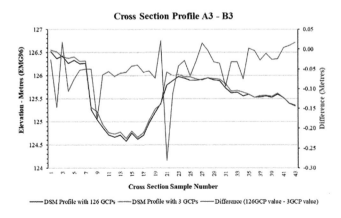

FIGURE 17.23 Results from the qProf tool for polyline A3–B3. The red line shows the results from the DSM rendered with 3 GCPs, while the black line shows the results of the DSM rendered with 126 GCPs. The y axis on the right and the blue line show the elevational difference.

DISCUSSION

LOCH DAVAN – INSPIRE 1 V2.0 X3

Overall, the RMSE for this site are higher than at other areas. This is likely due to several factors, including poor lighting conditions and a high UAS flight speed leading to linear rolling shutter distortion. However, using the DSM constrained by six GCPs as a baseline, it is possible to compare the effects of reducing the number of GCPs despite the initially large RMSE (Table 17.5).

Overall, there is a very strong correlation showing that as the number of GCPs is reduced there is an increase in RMSE. There is a much lower overall reduction in accuracy when reducing the number of GCPs from 6 to 3 in this instance. This is likely due to the spatial distribution of the three remaining GCPs being relatively good. It is notable that the error of CPs does increase when closely situated GCPs are removed.

Characteristic image block deformation can also be seen in profile A1–B1, as error increases towards the edges of the area of interest, where proximity to GCPs is reduced. Interestingly, there seems to be areas in the 3 GCP-controlled model which have been missed completely (small trees) which results in a difference of up to 10 m between the two DSMs.

Cross-section profile A2–B2 shows that the 3 GCP constrained model is on average below the 6 GCP model, reaches a point where there is no difference, followed by a positive skew, ending with a +4 m difference in elevation. Cross section A3–B3 shows that as CPs are further away from the GCPs, the difference increases, only to reduce again as the profile line nears a GCP used to control both DSMs.

Overall, it seems clear that spatial proximity from GCPs is a significant influencing factor in the RMSE of CPs.

Loch Davan – Falcon 8 (A7R)

The Sony Alpha A7R used in this test produced a GSD of 1.29 cm when flown at the same height as the DJI Inspire 1 V2.0. Consequently, the DSM resolution was greatly improved. Overall RMSEs were also significantly lower during this test. However, it is important to note that distance between the GCPs in this test was significantly lower (Table 17.6).

Supporting results from the previous tests, as the number of GCPs is reduced, the RMSE increases. Similarly to the results from the Inspire V2.0 test, as the GCP numbers are reduced and the spatial distribution becomes poor, the outlying CPs report much larger errors – the result is verified by the significant reduction in accuracy seen by a 52.62%–299.66% RMSE increase.

Cross-section profile A1–A2 shows image block deformation being propagated towards the edges of the reconstruction block when the spatial distribution of the GCPs is poor. This is mirrored by the results in profile A2–B2.

Profile A3–B3 is of interest as it shows the 3 GCP controlled DSM is consistently lower than the 8 GCP controlled DSM by approximately 1.1–2.1 m. The profile line runs north–south in an area that profile A1–B1 indicates has suffered from block deformation due to the poor distribution of GCPs. Therefore, it is likely that block deformation has occurred quite evenly towards the entire eastern section of the surveyed area, resulting in the consistent offset seen in profile A3–B3.

TABLE 17.5
How the RMSE of the CPs Increases as the Number of GCPs Is Reduced

GCPs Number Used	RMSE % Increase from Baseline	RMSE (m)
6 GCP	N/A	1.0222
4 GCP	+38.53	1.4158
3 GCP	+90.31	1.9425

TABLE 17.6
How the RMSE of the CPs Increases as the Number of GCPs Is Reduced

GCPs Number Used	RMSE % Increase from Baseline	RMSE (m)
8 GCP	N/A	0.19532
4 GCP	+52.62%	0.29810
3 GCP	+299.66%	0.58531

River Ehen: Ben Gill River Channel

The low flight altitude of this survey produced a GSD of 1.02 cm. Again, in general, as the number of GCPs decreases, the RMSE increases. A significant drop in accuracy is noticed from using 126 GCPs to 47 GCPs where the overall project accuracy decreases from 1.7 to 7 cm. However, in contrast to previous results, reducing the number of GCPs from 47 to 3 sees very little additional error introduced – only 0.3 cm. This is likely due to the relatively low elevation difference throughout the survey site, good distribution of remaining GCPs, and possibly due to the crossed flight path, all of which have been noted to positively influence accuracy (see: Benassi et al., 2017; Gerke and Przybilla, 2016; Prajwal et al., 2016) (Table 17.7).

The tests designed to identify spatial effects show that clustering of GCPs in the top, middle, and bottom of the site does not appear to reduce accuracy significantly. What is clear from Figure 17.20 is that the spatial proximity to the GCPs appears to influence the CP accuracy – it can be clearly seen that the RMSE values are reduced in the middle of the channel when the GCPs used to reference the model are in the same location.

With regards to the cross-section profiles, there are generally much lower differences attributed to the increased GSD. Profile A1–B1 shows very little difference between the two models with a range of −0.4 to +0.45 m – significantly less than the other tests. Additionally, it should be noted that the axis which most error is attributed to is by far the z axis. Significantly less variation is recorded in the x and y axes, even when reducing the number of GCPs to 3.

Profiles A2–B2 and A3–B3 also show very little difference; however, there is a small shift of the 3 GCP-controlled DSM in easting and northing which results in a sharp, but still relatively small, difference in cross-section elevation readings near the edge of the steep river channel.

Summary of Deductions

Considering all three test sites, it is clear that reducing the number of GCPs increases the RMSE in DSMs. However, the distribution of the GCPs is potentially more important – with notable exception of the river Ehen site – as failure to distribute the GCPs towards the edge of the survey area results in block deformation, something which is evident in many of the cross profiles.

TABLE 17.7
How the RMSE of the CPs Increases as the Number of GCPs Is Reduced

GCPs Number Used	RMSE % Increase from Baseline	RMSE (m)
126 GCP	N/A	0.0174
47 GCP	+446.55%	0.0777
15 GCP	+451.14%	0.0785
10 GCP	+452.29%	0.0787
3 GCP	+464.94%	0.0809
Top GCP	+449.42%	0.0782
Middle GCP	+243.67%	0.0424
Bottom GCP	+181.03%	0.0315

Importantly it is clear that GSD directly influences RMSE. These sets of tests showed that the higher the GSD, the lower the RMSE. Additionally, it is likely that the dual flight plan – and therefore high overlap – employed at the river Ehen site, contributed towards the minimisation of block deformation effects, along with good distribution of GCPs in most cases.

CONCLUSION

It is clear that the number of GCPs effects the RMSE of UAS-derived SfM (SFM) outputs; however, seemingly in these examples, there is no threshold at which below RMSE are always high. Furthermore, the spatial distribution of GCPs is particularly influential on RMSE. A poor distribution of GCPs can lead to block deformation being propagated in numerous directions throughout the survey areas, in some instances leading to ±10 m difference in elevation figures. Notably, reducing the number of GCPs has much less effect on the x and y accuracy of CPs; rather, most of the errors referred to are attributed to elevation differences in the z axis.

REFERENCES

Benassi, F., Dall'Asta, E., Diotri, F., Forlani, G., Morra di Cella, U., Roncella, R. and Santise, M. (2017) 'Testing accuracy and repeatability of UAV blocks oriented with GNSS-supported aerial triangulation', *Remote Sensing*, Vol. 9(172), pp. 1–23.

Barry, P. and Coakley. (2015) *'Accuracy of UAV photogrammetry Compared with network RTK GPS'*, *C-astral website*. [Online] Available at: http://www.c-astral.com/media/uploads/file/Bramor%20Accuracy%20compare_RTK_GPS.pdf [Accessed 28/04/2020].

Gerke, M. and Przybilla, H.J. (2016) 'Accuracy analysis of photogrammetric UAV image blocks: Influence of onboard RTK-GNSS and cross flight patterns', *Photogrammetrie – Geoinformation*, Vol. 2016(1), pp. 17–30.

Nex, F. and Remondio, F. (2014) 'UAV for 3D mapping applications: A review', *Applied Geomatics*, Vol. 6 (1), pp. 1–15.

Nakano, K., Suzuki, H., Tamino, T. and Chikatsu, H. (2016) 'On fundamental evaluation using uav imagery and 3D modelling software', *International Archives of the Photogrammetry, Remote Sensing and Spatial Information Sciences*, Vol. XLI-B5, pp. 93–97.

Pix4D. (2017) 'Using GCPs', *Pix4D Support Website*. [Online] Available at: https://support.pix4d.com/hc/en-us/articles/202558699-Using-GCPs#gsc.tab=0 [Accessed 25/04/2017].

Prajwal, M., Jain, R., Srinivasa, V. and Karthick, K. S. (2016) 'Optimal number of ground control points for a UAV based corridor survey', *International Journal of Innovative Research in Science Engineering and Technology*, Vol. 5(9), pp. 28–32.

Ramirez, P.D. and Hargraves, A. (2016) 'A new approach to survey using unmanned aerial vehicles (UAV's)', *Presented April 11th, 2016 at SPAR3D Conference*, The Woodlands, TX.

Rock, G., Rees, J. B. and Udelhoevn, T. (2011) 'Sensitivity analysis of UAV-Photogrammetry for creating digital elevation models', *International Archives of the Photogrammetry, Remote Sensing and Spatial Information Services*, Vol. XXXVIII-1/C22, 69–73.

Yastikli, N., Bagci, I. and Beser, C. (2013) 'The processing of image data collected by light UAV systems for GIS data capture and updating', *International Archives of the Photogrammetry, Remote Sensing and Spatial Information Sciences*, Vol. XL-7/W2, pp. 267–270.

18 Launch and Recovery System for Improved Fixed-Wing UAV Deployment in Complex Environments

Alastair Skitmore
Skyports Ltd

CONTENTS

INTRODUCTION

In the past two decades, Unmanned Aerial Vehicles (UAVs) have been increasingly relied upon to facilitate the collection of high-resolution spatial and temporal remote sensing data. UAVs can provide centimetre-accurate aerial data at a fraction of the cost of traditional manned aircraft and satellite-based methods (Colomina and Molina, 2014). The low-cost nature of small, lightweight UAVs, coupled with continual improvements in 'off-the-shelf' platforms, such as DJI's revolutionary Phantom series, has caused rapid proliferation of UAVs across a wide variety of different industries. UAVs have been utilised in a range of applications, including disaster management, search and rescue, precision agriculture, mapping, and environmental monitoring (Koh and Wich, 2012; Erdelj and Natalizio, 2016; Silvagni et al., 2016; Khan et al., 2017; Ju and Son, 2018). Multi-rotor UAVs are the current 'go-to' platforms for most tasks, due to their unique ability to hover, take-off, and land vertically, allowing them to be operated in even the most confined areas, and those which lack the infrastructure to support traditional manned aircraft operations (e.g. runways or helipads). At present, the use of fixed-wing UAVs is substantially lower than that of their multi-rotor counterparts, due to the considerable challenge of finding suitable take-off and landing zones in such environments. A recent survey by DroneDeploy, a cloud-based UAV mapping service, revealed that just 3% of its users were conducting mapping operations using fixed-wing platforms, compared to 97% using multi-rotors (Drone Deploy, 2017), presumably in part, due to the increased space required for fixed-wing take-offs and landings.

Fixed-wing UAVs offer a range of benefits compared to their multi-rotor counterparts, including greater range, endurance, and payload capacity (Nex and Remondino, 2013). Fixed-wing UAVs are ideally suited to long-range applications (Anderson and Gaston, 2013), where range and endurance requirements typically exceed the capabilities of current multi-rotor UAVs.

Current restrictions on 'Beyond Visual Line of Sight' (BVLOS) operations (in most countries) favour the use of multi-rotor UAVs, as the increased range and endurance capabilities of fixed-wing platforms cannot be properly utilised due to the regulations in place. Future regulations are expected to lessen restrictions on BVLOS flights. As such, the demand for long-range/endurance fixed-wing UAVs is expected to increase. Correspondingly, the onset of the Anthropocene is increasing the need for large-scale, repetitive monitoring of Earth's natural systems (Schimel et al., 2013), due to the profound changes being impacted upon them by human activities (e.g. melting ice sheets). Fixed-wing UAVs offer the ideal means to facilitate such large-scale monitoring studies. Therefore, expanding their capabilities to allow deployment in increasingly complex environments, such as dense rainforests and rugged glacial terrain, is of vital importance. As such, the proposed launch and recovery concept presented in this study is of particular relevance to the future of UAV-based environmental monitoring.

Recent advances in the military sector, driven by the expanding need to operate fixed-wing UAVs from the decks of ships, have yielded an array of specialist systems designed to launch and recover fixed-wing UAVs in confined areas and those without a runway. Unfortunately, such systems are generally prohibitively expensive and bulky. As such, there is a need for a cost-effective launch and recovery system to more easily permit the use of fixed-wing UAVs in complex environments to perform long-range and/or endurance tasks (e.g. rainforest or glacier monitoring). Hence, the aim of this study is to demonstrate the feasibility of using a low-cost multi-rotor UAV to launch and recover fixed-wing UAVs in mid-air, in order to reduce the amount of space required for fixed-wing UAV operations and, therefore, facilitate improved fixed-wing UAV deployment in complex environments.

This chapter details the design and implementation of an inexpensive fixed-wing UAV launch and recovery system, using a low-cost multi-rotor UAV as a versatile airlift and retrieval platform. A fixed-wing UAV is successfully airlifted to a pre-determined deployment altitude and launched, eliminating the need for a runway or costly rail launch system. The fixed-wing UAV is subsequently recovered by the multi-rotor in mid-air, using a net-based recovery approach. The finalised launch and recovery mechanisms are extremely inexpensive (~£40), easy to manufacture using widespread 3D printing methods, and are readily adaptable to a number of multi-rotor platforms, allowing researchers to make use of their existing systems. Crucially, the developed system completely alleviates the need for a flat, open area to operate a fixed-wing UAV. Instead, the UAV is launched and recovered at altitude, clear of any obstacles on the ground. The system therefore has particular relevance to UAV operations in complex environments, such as areas of rugged terrain or dense vegetation, where the lack of a suitable take-off and landing space has traditionally prevented the use of fixed-wing UAVs. Finally, since the fixed-wing UAV is launched and recovered using the multi-rotor airlifter, its design can be optimised specifically for cruise flight, rather than take-off and landing, improving its range and endurance performance.

REVIEW OF EXISTING FIXED-WING UAV LAUNCH AND RECOVERY TECHNIQUES

As aforementioned, fixed-wing UAVs require large areas of flat, unobstructed terrain to complete their take-off and landing manoeuvres. Take-offs traditionally occur through a conventional 'take-off roll' along a runway to generate lift over the wings (Austin, 2010) or by hand-launching. Landings are typically achieved using a shallow glideslope approach and flare to reduce horizontal and vertical velocity (Austin, 2010). The aircraft then 'rolls out' to a complete stop, assisted by the friction between the aircraft's wheels/fuselage and the landing surface (Novaković et al., 2016). Occasionally, active braking is deployed on larger fixed-wing aircraft, where friction alone is insufficient to bring the aircraft to a stop in the space available.

The space required for traditional take-off 'rolls' and glideslope landings is considerable and currently poses the greatest challenge to fixed-wing UAV operations in complex environments. Numerous techniques exist to reduce the space required for fixed-wing take-off and landing manoeuvres. This section will provide a brief overview and critique of the alternative take-off and landing methods in use today. Due to the scarcity of research relating to fixed-wing take-off and recovery methods, the majority of the review is based on commercial and military systems.

ALTERNATIVE LAUNCH METHODS

HAND-LAUNCHING

In the absence of a suitable runway surface, or for non-landing gear equipped UAVs, most lightweight fixed-wing UAVs can be launched by hand through the use of a simple upwind throw. Heavier fixed-wing UAVs (typically >3 kg) are increasingly difficult to hand-launch due to their size and weight. Furthermore, hand-launching can be hazardous due to the proximity of the aircraft's propellers to the person launching. Risk is further increased when hand-launching a 'pusher' style fixed-wing (with the propeller at the rear), as the propeller travels directly over the forearm/wrist of the person launching.

BUNGEE LAUNCH SYSTEMS

The most widespread alternative to the hand-launch for small fixed-wing UAVs is the bungee launch. Here, the fixed-wing is equipped with a fuselage hook that attaches to a tensioned bungee cable. The fixed-wing is placed on an inclined ramp, and a foot pedal or switch is used to release the bungee, transforming the stored potential energy of the bungee into kinetic energy, which accelerates the fixed-wing along the ramp and into the air (Novaković and Medar, 2013).

Bungee launch systems provide a reliable method of launching smaller fixed-wing platforms which are unsuitable for hand-launching. Their modular construction makes them ideally suited to packing down into a small form factor for use in remote field locations.

CATAPULT LAUNCH SYSTEMS

Catapult launch systems (pneumatic or hydraulic) are a derivative of the bungee launch principle and are particularly favoured by the military for launching large fixed-wing UAVs from ships. They operate on the same basic principle as the bungee launch but use compressed gas to accelerate the UAV along the ramp, rather than a tensioned bungee cable. Catapult systems offer gentler (less sudden) acceleration along the take-off ramp compared to a bungee system, which can be preferable when operating with sensitive sensors on-board the UAV (Novaković et al., 2014). Recent trends in system miniaturisation have yielded portable catapult launch systems, such as the collapsible UAV Factory pneumatic launch system. Unfortunately, such systems are still largely impractical for remote operating locations due to their requirement for an external power source.

A recent study by Bhatia and Aljadiri (2017) presented a novel catapult launch system which utilised electromagnets to accelerate the UAV along the ramp, rather than a bungee cable or compressed gas. While it is an interesting concept, the system is impractical for remote field site operations due to its extremely high power source requirements. The potential for electromagnetic interference with the UAV's systems is also a significant issue.

SPECIALIST LAUNCH SYSTEMS

In addition to bungee and catapult launch methods, several novel launch systems have been developed, primarily by the military for launching fixed-wing UAVs from ships. The 'Rocket-Assisted Take-Off' or 'RATO' system, uses a miniaturised rocket motor to accelerate the UAV forwards and upwards off the deck of the ship (Bettella et al., 2013). Another novel technique is the parasail launch system, developed by the US Coast Guard in the early 2000s. This method, designed for use on moving ships, uses a large parasail to hoist a fixed-wing UAV to a suitable deployment altitude, before releasing the UAV to conduct its mission (McDonnell, 2006).

Whilst the launch methods described in this section successfully alleviate the need for a runway, a large area of reasonably unobstructed space is still required in front of the launcher to allow the UAV sufficient time to gain airspeed and initiate a climb (Lin et al., 2012). As such, the aforementioned launch techniques are inadequate for deploying fixed-wing UAVs in complex environments, such as a rainforest or urban canyon, where only a small clearing may be available for launch and recovery.

ALTERNATIVE LANDING METHODS

Numerous approaches exist to reduce the footprint of fixed-wing UAV landing manoeuvres. This section will outline the main techniques in use today.

NET-BASED RECOVERY

The most widely used non-runway retrieval method is the net-based recovery. This approach makes use of a vertically oriented net erected in the landing zone to catch the incoming UAV, usually by a hook on the nose of the aircraft. The simplicity and effectiveness of the process provides an elegant solution to recovering fixed-wing UAVs without a runway. However, despite their effectiveness, net recoveries are accompanied by a range of drawbacks. Primarily, the potential for damage to the UAV and sensitive payloads caused by the massive deceleration experienced by the UAV when it hits the static recovery net. Additionally, the net-based approach is limited to pusher-style UAVs as traditional front-mounted propeller UAVs would damage the net and in all likelihood the motor and propeller too. Finally, erecting a suitable-sized net requires significant infrastructure and manpower, and an unobstructed approach path is still required for the incoming UAV. As such, the net-based recovery method is not well suited to remote and/or complex environments, where transportation of potentially bulky net infrastructure is impractical and/or there is no clear approach path for the incoming UAV through buildings, trees, or terrain.

PARACHUTE-BASED RECOVERY

Another widespread approach used to recover fixed-wing UAVs is through the use of on-board parachutes. Once the UAV enters the intended recovery zone, a parachute is deployed from a compartment within the UAV. The UAV then descends downwards at a controlled rate, suspended beneath the chute. A good example of successful integration of a parachute recovery system on-board a small commercial fixed-wing UAV is the AeroMapper Talon. Unfortunately, the simplicity of parachute recoveries comes at the cost of accuracy. Once deployed, the parachute and UAV are affected by wind, making it impossible to land the UAV at a specific point on the ground. Therefore, parachute recoveries are not well suited to complex environments, particularly those with tall and/or dense vegetation, as the precise landing site cannot be guaranteed. Furthermore, whilst parachutes are highly effective at reducing the horizontal speed of fixed-wing UAVs, the vertical speeds are often harder to control, resulting in rough landings which pose a damage risk to sensitive payloads (Cartwright, 2009) and will induce progressive wear and tear on the airframe.

WIRE-BASED RECOVERY

Recent advances in recovery technologies have seen net-based recovery systems being increasingly replaced by wire-based systems. Wire-based systems benefit from a smaller footprint compared to their net-based counterparts, a fact that has spurred military development of the systems for use on the decks of ships, where space is at a premium. The primary system in use today is Boeing's InSitu SkyHook, designed for use with their ScanEagle and RQ-21 UAVs (InSitu, 2015). The SkyHook consists of a vertically oriented cable attached to a pivoting arm. The approaching UAV impacts the cable along its wing and the cable becomes snagged by a hook at the wingtip.

The SkyHook and SideArm systems are ideally suited to the recovery of UAVs in military settings, where the space, manpower, and resources permit the assembly of the significant infrastructure associated with each system. Conversely, the systems are poorly suited for use in complex and/or remote environments, where resources including transportation and manpower are scarce. Ultimately, as with most of the recovery techniques discussed in this section, wire-based recovery methods are likely to damage aircraft and payloads due to high-impact speeds (e.g. DARPAtv, 2017) and require unobstructed approach paths for the incoming UAV, a prerequisite which may be difficult to satisfy in complex environments.

VTOL FIXED-WING UAVs

Recent advances in the commercial and military sectors have yielded a new breed of vertical take-off and landing (VTOL)-enabled fixed-wing UAV platforms. These so-called 'VTOL fixed-wings' combine the precision manoeuvrability of a multi-rotor, with the range and endurance of a traditional fixed-wing UAV. VTOL fixed-wing flight is typically achieved through the addition of four multi-rotor motors in an 'H' configuration to a standard fixed-wing platform.

While VTOL fixed-wing platforms have proven popular in both military and research settings, they suffer from several key drawbacks. The primary disadvantage is that the additional motors are only used during VTOL manoeuvres and remain idle while the UAV is in forward cruise flight (which accounts for the majority of the flight). Thus, the added VTOL powertrain (motors, Electronic Speed Controllers (ESCs), and propellers) contributes significant weight and drag to the UAV during cruise flight, reducing the UAV's range and endurance.

Tiltrotor VTOL fixed-wing platforms mitigate these effects to some degree, as most of the motors are tilted during transition between VTOL and forward flight, providing both the vertical thrust required for VTOL manoeuvres and the horizontal thrust required during cruise flight. However, typically at least one motor remains idle during cruise flight, increasing the weight and drag of the aircraft.

Furthermore, VTOL manoeuvres are power intensive and, therefore, deplete the UAV's flight batteries, reducing the potential range and endurance of the UAV. This reduction in the range and endurance of the UAV could eliminate the benefits of using a fixed-wing UAV for the desired task in the first place, rather than a multi-rotor.

PROPOSED LAUNCH AND RECOVERY SYSTEM

The launch and recovery system has been designed around a DJI F550 hexacopter being retrofitted to airlift a small flying wing. Initially, it was hoped that the system could be applied to a traditional V- or T-tail fixed-wing aircraft. However, problems arose when attempting to mount these airframes beneath the F550, due to the height of their vertical stabilisers. As such, the system has been designed with flying wings in mind, due to their largely flat profile.

LAUNCH MECHANISMS

Two different launch mechanisms were designed and implemented, to allow for a comparison and selection of the most effective launch technique. The primary mechanism involved an electromagnet, attached to a mounting plate beneath the F550, attracting a steel plate mounted on top of the fixed-wing, securing it firmly in place (Figure 18.1). To release the fixed-wing, power to the electromagnet would be removed, causing the magnetic field to dissipate and the fixed-wing to release and take flight. Laboratory testing of several electromagnets resulted in the selection of a 12 V DC electromagnet, rated for 10 kg of holding strength. Extensive testing revealed that despite the close proximity of the electromagnet to the flight electronics of both the F550 and fixed-wing, no unwanted interference was noted.

The secondary launch mechanism consisted of a pair of mechanical clamps, designed to attach to the F550's retractable landing legs. The mechanism used the vertical articulation of the landing legs to clamp and release the fixed wing. The clamping arms were designed to match the fore and aft curves of the fixed-wing to hold it securely in place until the landing gear was retracted (Figure 18.2).

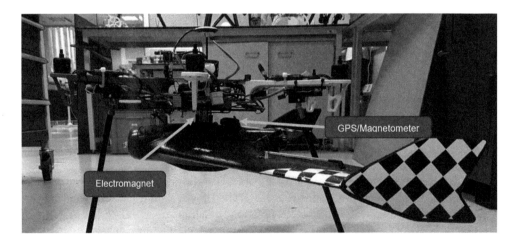

FIGURE 18.1 Electromagnet mechanism holding the fixed-wing beneath the F550. Note the close proximity between the electromagnet and GPS/magnetometer on the upper surface of the fixed-wing.

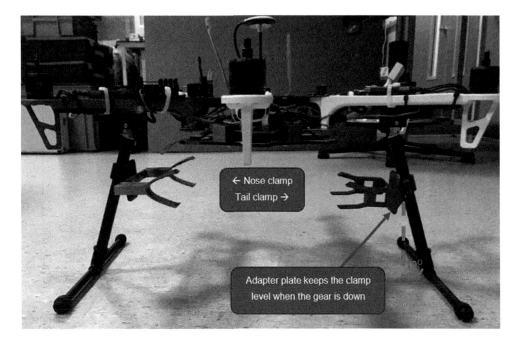

FIGURE 18.2   Mechanical clamping mechanism.

RECOVERY MECHANISM

Due to their proven effectiveness at recovering fixed-wing UAVs on the ground, an airborne net-based recovery system was designed. Traditional ground-based net systems suffer from high-impact velocities between the incoming fixed-wing and the recovery net, contributing to airframe and pay-load damage. The primary benefit of retrieving the fixed-wing in mid-air was that the multi-rotor aircraft carrying the net could be manoeuvred to match the speed of the incoming fixed-wing, thus preventing sudden deceleration and potential damage to the fixed-wing and on-board sensors at the time of impact.

A net system was designed to be incorporated into the same F550 used to deploy the fixed-wing, so that the F550 could both launch and recover the fixed-wing in one flight. To achieve this, it was necessary to store the net in the furled position. Once the fixed-wing was successfully launched, the net would be unfurled remotely in preparation for the recovery. Once caught, the whole net attachment, carrying the entangled fixed-wing, would be lowered to the ground by the F550 pilot and jettisoned, therefore allowing the F550 to land elsewhere unencumbered by the net and fixed-wing. To achieve this two-stage net release (unfurl and jettison entirely), two payload release brackets were designed, 3D printed, and mounted to the front arms of the F550. A servo was used to manoeuvre a pin along the bracket, causing the net to unfurl and release.

An inexpensive badminton net was selected due to its low weight, its ideal width for mounting beneath the F550, and extended length for creating a large target area for the fixed-wing to fly into. The fixed-wing was to be piloted into the net using a live first-person view (FPV) video feed. Ground testing revealed the net was difficult to see on the video feed; therefore, a bright 12 V LED strip was added to one side of the net, to improve its visibility. A simple plastic stake was added to the bottom of the net to provide the necessary weight to cause the net to unfurl and also to provide some lateral support to keep the net straight. Finally, the fixed-wing was equipped with a large hook to ensure that it caught the net (Figure 18.3).

FIGURE 18.3   Recovery net in action, successfully retrieving the fixed-wing UAV in mid-air, clear of any obstacles on the ground.

RESULTS

LAUNCH TESTS

Electromagnet Mechanism

Flight testing was first attempted using the electromagnet release mechanism. Unfortunately, it was quickly discovered that the electromagnet was not strong enough to hold the fixed-wing securely beneath the F550. Five take-offs were attempted and subsequently aborted as the fixed-wing released prematurely once the F550 reached take-off thrust. The fixed-wing could be seen being severely buffeted beneath the F550 as thrust was increased, before the point at which it detached.

Clamp Mechanism

Following the failure of the electromagnet launch mechanism, the F550 was retrofitted with the clamping launch mechanism. It was hoped that by securing the fixed-wing with the clamp's dual points of contact (nose and tail), as opposed to the single point of contact of the electromagnet release mechanism, the wing would be secure against the intense prop-wash of the F550.

The F550 successfully airlifted the fixed-wing to its deployment altitude, at which point the landing gear was retracted and the fixed-wing was released from the clamping mechanism. As per the release procedure, the fixed-wing was allowed a brief dive to gain airspeed as the motor was engaged, the pilot then successfully recovered the fixed-wing from the dive and initiated a climb towards level flight.

Due to the valley-like nature of the site used for flight testing, the F550 (with fixed-wing attached) was initially sheltered from the wind. However, once the F550 ascended above the protective valley sides, it was immediately affected by the ~10 mph winds, inducing oscillation of the F550 and the attached fixed-wing. Skilful piloting of the F550 was enough to combat the oscillations until the fixed-wing could be released, at which point the F550 regained its usual stable flight characteristics. The apparent instability of the combined aircraft in windy conditions represented a significant challenge to the system's design. Footage of the successful deployment test can be viewed here: https://goo.gl/fa4Ppn

Following successful deployment of the fixed-wing aircraft, subsequent flight-testing days were used to evaluate the net recovery process.

Recovery Tests

Early airborne recovery attempts were unsuccessful. It was found that the fixed-wing pilot could not see the F550 and suspended net until too late in the approach, largely due to high amounts of

FIGURE 18.4 GoPro cameras provide two unique viewing angles for analysing the dynamics of the fixed-wing impacting the suspended net.

static and interference in the FPV feed of the Fat Shark video goggles. Following several missed approaches, the Fat Shark goggles were removed and the fixed-wing was piloted manually using a traditional line-of-site approach. Unfortunately, the result was more missed approaches due to the difficulty of gauging the correct approach trajectory of the fixed-wing towards the net whilst standing a safe distance from the two UAVs.

To improve the stability of the fixed-wing during its approach, the Vector Autopilot's stabilisation feature was engaged and an immediate improvement in the fixed-wing's performance in the high winds was noted. The subsequent approach, after reverting back to using the Fat Shark video goggles, saw the fixed-wing successfully contact the suspended net and become entangled by its nose hook. The F550 pilot then jettisoned the net and fixed-wing to the ground. Footage of the successful recovery attempt can be viewed here: https://goo.gl/rTxHpK

Prior to beginning the recovery-testing phase, it was unknown how the F550 would respond to the fixed-wing hitting the suspended net. The intention was that the F550 would match the incoming fixed-wing's speed in order to reduce the impact velocity as it hit the net. However, the strong winds during the test caused a billowing effect on the net, effectively replicating the same effect as moving the F550 forward, thus, the F550 was left stationary. Surprisingly, the F550 scarcely registered the impact of the fixed-wing hitting the net.

Despite successful mid-air recovery of the fixed-wing during the test, it was evident that a component change was required to improve the quality of the live video feed being used by the fixed-wing pilot to target the net. Therefore, the Fat Shark video goggles were substituted for an alternative pair of video goggles.

For the second day of tests, a GoPro camera was added to both the fixed-wing and the F550 to provide additional viewing angles of the resultant net recoveries. Despite stronger and gustier winds (15 mph), two successful net recoveries were recorded. The higher success rate was attributable to the new FPV goggles providing an improved video feed for the fixed-wing pilot to use for navigating the fixed-wing into the net. The experience gained by the fixed-wing pilot on the first recovery day also played a part in the increased success rate. The GoPro cameras provided two unique viewing angles for analysing the dynamics of the fixed-wing impacting the suspended net (Figure 18.4).

DISCUSSION

MAGNETIC RELEASE

The electromagnet's inability to hold the fixed-wing beneath the F550 was a surprising and unfortunate development. The tendency of the fixed-wing to disconnect from the F550 as it reached take-off thrust indicated that the issue was related to the strengthening downward prop-wash, likely originating from the two lateral F550 motors. Therefore, it has been surmised that uneven distribution of the downward prop-wash across the surface of the wing, coupled with the sensitivity of the electromagnet to non-uniform load directions (i.e. shearing effects), likely caused the repeated failures of the electromagnetic release mechanism.

The issue could possibly be addressed by replacing the 10 kg electromagnet with a stronger variant. Consumer-grade 12 V DC electromagnets are available with up to 100 kg of rated holding strength. However, the increased strength comes at the expense of greater weight and current draw, both sensitive factors when operating UAVs. As the 10 kg electromagnet comfortably held a mass of 8 kg in laboratory testing, it should have been more than sufficient to hold the ~800 g fixed-wing in place. An alternative solution was to attempt to alleviate the shearing effect induced by the prop-wash. A ball and socket joint was designed and 3D printed for use with the 10 kg electromagnet release mechanism. It was hypothesised that by replacing the rigid connection of the electromagnet to steel plate contact with one that was free to move in response to the prop-wash, the shearing effect on the electromagnet would be sufficiently reduced to allow it to maintain its hold on the fixed-wing. This modification has yet to be tested.

A further solution for reducing the prop-wash effect was to utilise a larger multi-rotor, with increased propeller spacing so that the prop-wash was not focused directly onto the surface of the fixed-wing. Hence, a large quadcopter (in an X-4 configuration), with increased spacing between each of the motors, was designed and built. The so-called Vulcan X4 benefited not only from increased propeller spacing, mitigating the effect of downward prop-wash on the fixed-wing, but also from increased landing gear height, resulting in the fixed-wing being mounted further away from the propellers, further reducing the prop-wash. The electromagnet and clamp release mechanisms were re-designed to fit the larger Vulcan X4. It should be noted that the increased landing gear height of the Vulcan X4 may permit the launch and recovery of traditional V- and T-tail fixed-wing aircraft, as well as flying wings, which was initially impossible due to the low profile of the F550. The increased height of the Vulcan X4 system may also permit the fixed-wing UAV to be launched with its motor set to take-off thrust, which is expected to greatly improve the dynamics of the launch, avoiding the nose-down dive that is currently required to gain airspeed due to the fixed-wing being launched with its motor off. These modifications have yet to be tested.

CLAMP RELEASE

Unlike the electromagnet launch mechanism, the clamp mechanism functioned as intended, resulting in a successful airlift and deployment of the fixed-wing. Some minor issues were encountered, the most notable being the instability that the combined F550 and fixed-wing exhibited once they had ascended above the protective valley sides and were affected by the wind. The bulk of the instability/oscillations appeared to take place while the two aircraft were misaligned with the wind direction (i.e. in a cross-wind scenario). As such, it is likely that the wind-induced oscillations could have been minimised by aligning the nose of the two aircraft into the wind as early as possible after take-off. Releasing the fixed-wing into the wind would also improve the overall dynamics of the release and ensuing climb. Furthermore, the F550's instability could also be improved by configuring a gain control knob on the radio control transmitter to adjust the F550's gain settings mid-flight.

Overall, the F550 airborne deployment of the fixed-wing was a success, demonstrating a viable proof of concept for multi-rotor–assisted fixed-wing take-offs. Future use of the new, larger Vulcan X4 multi-rotor for fixed-wing deployment will likely improve the overall stability of the combined fixed-wing and multi-rotor aircraft in the wind.

A significant advantage associated with the demonstrated launch and recovery system is the non-reliance of the fixed-wing UAV on its own motor and airfoil during take-off. Traditional runway-launched fixed-wing UAVs rely upon their own motors and airfoils to generate the lift required to get airborne. As such, their airfoils and motor/propeller setups (particularly propeller pitch) are designed for maximum take-off and climb performance, resulting in airframes that are poorly optimised for efficient, cruise flight (Austin, 2010). In this study, the fixed-wing UAV's take-off is accomplished by the multi-rotor airlifter. Consequently, the fixed-wing UAV's airfoil and motor/propeller can be optimised specifically for cruise flight, extending the UAV's range and endurance capabilities.

NET RECOVERY

Traditional ground-based net recovery methods are problematic due to the severe deceleration as the fixed-wing impacts the static recovery net, which can cause damage to airframes and payloads. In this study, the recovery net was suspended (i.e. anchored only at the top), allowing the net to swing in a pendulum-like motion as the incoming fixed-wing impacted, therefore absorbing the energy of the impact and reducing the magnitude of the fixed-wing's deceleration. Additionally, the F550 and net could be manoeuvred to match the speed of the incoming fixed-wing, further reducing the fixed-wing net deceleration during impact. Consequently, the potential for airframe and payload damage has been significantly reduced compared to traditional static net recovery methods.

During the preliminary recovery attempt, the F550 remained almost completely stationary as the fixed-wing impacted the suspended net. The F550's impressive stability during the wing-net impact was largely attributable to the well-regulated approach speed of the incoming fixed-wing. When the approach speed of the fixed-wing was higher, this resulted in a noticeable wobble by the F550 as the fixed-wing impacted the net. Despite this, the F550 recovered from the impacts quickly and the fixed-wing was successfully jettisoned, demonstrating the F550's ability to handle a variety of fixed-wing to net impact speeds.

A significant issue encountered during the recovery-testing phase was the instability of the suspended net in the wind. During the second testing day, the 15 mph winds blew the net to an angle of approximately 45° from its standard vertical orientation. Therefore, it is suggested that the lightweight plastic spar used to weigh the base of the net should be replaced by a heavier alternative, to increase the net's resistance to the wind. There is, however, an accompanying trade-off set against reduced F550 flight times as the weight of the recovery net is increased. Further work is therefore required to determine optimal net weight whilst maintaining reasonable F550 flight times.

An alternative solution is to replace the '2D', flat-profile badminton net with a '3D' trawling net, commonly used in fishing applications. Trawling nets are designed to be dragged forward through a body of water. A pair of 'otterboards' (a.k.a. trawlboards), act as hydrofoils, using the forward motion of the net through the water to keep the mouth of the net spread open (Patterson and Watts, 1986). It is hypothesised that a lightweight trawling net could be re-designed to perform the same way in air. The otterboards could be replaced with a pair of airfoils, which would serve to open the mouth of the net as the F550 flew forward to match the speed of the incoming fixed-wing UAV. Unlike the badminton recovery net, which suffered in the high winds, a trawling-style net would benefit from the presence of winds, which would act to increase the spreading effect of the airfoils, keeping the mouth of the net open. Additionally, a trawling-style net would provide a wider target area for the fixed-wing pilot to aim for. The application of a trawling-style net is therefore an interesting area for future research.

The high flying speed of the fixed-wing, coupled with the initially poor FPV video quality, complicated early recovery attempts, resulting in countless failed approaches and go-arounds. Replacement of the under-performing Fat Shark video goggles significantly improved the fixed-wing pilot's view of the recovery net. However, further improvements to the FPV video link would increase the effectiveness of the net recovery process considerably. Hence, future iterations of the system would benefit significantly from a high-definition (HD) FPV link, rather than the standard-definition (SD) 5.8 GHz FPV link used in this study. The cost of off-the-shelf HD FPV systems, such as DJI's Lightbridge system, remains prohibitively high. Fortunately, recent developments within the hobbyist community have resulted in low-cost, DIY HD FPV systems. A popular example is the 'EZ-WifiBroadcast' system, capable of broadcasting low-latency 1,080 p FPV at ranges exceeding

2 km (GitHub, 2016). Such systems are readily achievable using two Raspberry Pi Zero computers, two Wi-Fi dongles, and a camera module, bringing the total cost in at under £100. Thus, future iterations of the research should consider replacing industry-standard SD 5.8 GHz FPV equipment with one of the many emergent low-cost HD alternatives.

Substituting the chosen flying wing for a slower, more controllable fixed-wing platform would further improve the effectiveness of the net recovery process by reducing approach speeds, allowing both the F550 and fixed-wing pilot more time to perfect the approach trajectory and impact speed. The Parrot Disco UAV has been identified as a suitable replacement for the flying-wing utilised in this study. The Disco is a slow-flying delta-wing UAV which comes pre-configured with heading and altitude hold as standard (due to the integrated Ardupilot FC), which would be extremely useful when lining up an approach to the recovery net. The Disco also incorporates an anti-stall algorithm (based on a pitot airspeed sensor) to allow the aircraft to be slowed down without the risk of a stall. The 720 p HD FPV downlink would eliminate the issues with SD 5.8 GHz FPV, improving the fixed-wing pilot's view of the recovery net. Additional benefits include a downward-facing gimbal-stabilised 1,080 p camera (ideal for mapping missions) and impressive 45-minute flight times in its stock configuration. Furthermore, costing just £300 (almost the same as the Vector Autopilot used in this study), the Parrot Disco is an inexpensive, yet highly capable, fixed-wing platform.

Because the entire recovery procedure was flown manually, there was considerable room for pilot error. Therefore, in order to improve the net recovery success rate and eliminate the potential for pilot error, it is suggested that future work on the system presented in this chapter focuses on the implementation of a computer vision algorithm for fully autonomous recoveries. Kim et al. (2013) successfully implemented a novel computer-vision algorithm for autonomous recovery of fixed-wing UAVs using a static, ground-based net. As such, it is probable that such a computer vision algorithm could be applied to the airborne recovery system presented in this study, once the complications associated with the non-static (dynamic) nature of the recovery net have been overcome.

Similarly, automation of the launch process would eliminate the potential for pilot error, such as the stall induced by the fixed-wing pilot in this study. Therefore, future research should investigate the possibility of implementing a flight algorithm to autonomously pilot the fixed-wing UAV out of its initial dive and into a circular loiter pattern. Such an algorithm would likely require replacement of the closed-source Vector Autopilot with an open-source equivalent, such as a Pixhawk.

COST

With the cost of many military and commercial fixed-wing recovery systems running into millions of pounds, it is important to emphasise the cost-effectiveness of the system presented in this study. The total cost of all components and aircraft (excluding the radio transmitters and batteries) was under £700. The inexpensive nature of the fixed-wing launch and recovery system presented in this study makes it ideally suited to research projects operating on limited budgets. Additionally, due to the modular nature of the manufactured release and recovery mechanisms, the airlift multi-rotor (in this case the F550) could be quickly and easily re-purposed during such projects, to complete other research tasks when not required for deploying and recovering a fixed-wing UAV.

CONCLUSIONS

This chapter has outlined a proof of concept for a multi-rotor–assisted fixed-wing UAV launch and recovery system. A fixed-wing UAV was successfully deployed in mid-air from an F550 multi-rotor, using the articulation of the F550's retractable landing gear and a simple 3D printed clamp mechanism. The clamp mechanism is lightweight, draws no power from the F550, and is easily adaptable to a range of multi-rotor platforms. An electromagnet-based release mechanism was investigated and subsequently dismissed, as the strong downward prop-wash of the F550 overcame the holding strength of the electromagnet, resulting in premature release of the fixed-wing UAV.

Following deployment, the fixed-wing UAV was successfully recovered using an airborne net suspended from the F550. The net could be manoeuvred to match the speed of the incoming fixed-wing, therefore providing a significant improvement on traditional ground based net recoveries, where high-impact speeds between the UAV and net cause airframe and payload damage.

As the launch of the fixed-wing UAV is accomplished using the multi-rotor airlift vehicle, the fixed-wing airframe can be optimised specifically for cruise flight, permitting increased range and endurance capabilities through improvements in airfoil design and motor/propeller configuration.

The total cost of the launch and recovery system, including both UAVs, is under £1,000, making it ideally suited to research applications. The fact that the fixed-wing UAV is launched and recovered at altitude removes the need for wide, open areas for conventional take-offs and landings. As such, the system has expanded the operational capabilities of fixed-wing UAVs to even the most complex environments, previously only suitable for multi-rotor operations.

Consequently, it is hoped that the system demonstrated in this chapter can act as a foundation for the increased use of fixed-wing UAVs across the UAV and remote sensing communities. At a time where environmental monitoring studies are playing an increasingly vital role in assessing the health of our planet, it is tools such as fixed-wing UAVs that will assist researchers to gain a greater understanding of the Earth's systems.

Future work should focus on automation of the launch and recovery process, to eliminate the potential for pilot error, and to improve the recovery success rate. Possible integration of V-tail and T-tail UAVs should also be investigated to allow the system to be applied to a wider range of fixed-wing UAV platforms.

REFERENCES

Anderson, K. and Gaston, K. (2013). Lightweight unmanned aerial vehicles will revolutionize spatial ecology. *Frontiers in Ecology and the Environment*, 11(3), pp. 138–146.

Austin, R. (2010). *Unmanned Aircraft Systems: UAVs Design, Development and Deployment*. Chichester: Wiley.

Bettella, A., Moretto, F., Geremia, E., Bellomo, N., Pavarin, D. and Petronio, D. (2013). Development and flight testing of a hybrid rocket booster for UAV Assisted Take Off. *49th AIAA/ASME/SAE/ASEE Joint Propulsion Conference*. San Jose, CA: AIAA, 2013–4140.

Bhatia, D. and Aljadiri, R. (2017). Electromagnetic UAV launch system. *2017 2nd IEEE International Conference on Intelligent Transportation Engineering (ICITE)*, Singapore: IEEE, pp. 280–283.

Cartwright, K. (2009). Feasibility of parachute recovery systems for small UAVs. *The UNSW Canberra at ADFA Journal of Undergraduate Engineering Research*, 1(2), p. 8.

Colomina, I. and Molina, P. (2014). Unmanned aerial systems for photogrammetry and remote sensing: A review. *ISPRS Journal of Photogrammetry and Remote Sensing*, 92, pp. 79–97.

DARPAtv. (2017). *Tern SideArm Capture System Tests & Concept Video*. [video] Available at: https://www.youtube.com/watch?v=fIX3_QCmaUM [Accessed 15 Aug. 2018].

Deploy, D. (2017). *Commercial Drone Industry Trends March 2017*. [online] Available at: https://dronedeploy-www.cdn.prismic.io/dronedeploy-www%2F7cfc75ba-3916-405b-b604-af2f538ad0bd_10m_acre_report_2017_.pdf [Accessed 28 Aug. 2018].

Erdelj, M. and Natalizio, E. (2016). UAV-assisted disaster management: Applications and open issues. *2016 International Conference on Computing, Networking and Communications (ICNC)*. Kauai, HI: IEEE. https://hal.archives-ouvertes.fr/hal-01305371

GitHub. (2016). *EZ-WifiBroadcast*. [online] Available at: https://github.com/bortek/EZ-WifiBroadcast/wiki [Accessed 28 Aug. 2018].

InSitu. (2015). *SkyHook*. [online] Available at: https://www.insitu.com/images/uploads/pdfs/Skyhook_Universal_ProductCard_PR041615.pdf [Accessed 15 Aug. 2018].

Ju, C. and Son, H. (2018). Multiple UAV systems for agricultural applications: Control, implementation, and evaluation. *Electronics*, 7(9), p. 162.

Khan, S., Aragão, L. and Iriarte, J. (2017). A UAV–LIDAR system to map Amazonian rainforest and its ancient landscape transformations. *International Journal of Remote Sensing*, 38(8–10), pp. 2313–2330.

Kim, H., Kim, M., Lim, H., Park, C., Yoon, S., Lee, D., Choi, H., Oh, G., Park, J. and Kim, Y. (2013). Fully autonomous vision-based net-recovery landing system for a fixed-wing UAV. *IEEE/ASME Transactions on Mechatronics*, 18(4), pp. 1320–1333.

Koh, L. and Wich, S. (2012). Dawn of drone ecology: Low-cost autonomous aerial vehicles for conservation. *Tropical Conservation Science*, 5(2), pp. 121–132.

Lin, J., Li, S., Zuo, H. and Zhang, B. (2012). Experimental observation and assessment of ice conditions with a fixed-wing unmanned aerial vehicle over Yellow River, China. *Journal of Applied Remote Sensing*, 6(1).

McDonnell, W. (2006). Launch and recovery system for unmanned aerial vehicles. 7,097,137 B2.

Nex, F. and Remondino, F. (2013). UAV for 3D mapping applications: a review. *Applied Geomatics*, 6(1), pp 1–15.

Novaković, Z. and Medar, N. (2013). Analysis of a UAV bungee cord launching device. *Scientific Technical Review*, 63(3), pp. 41–47.

Novaković, Z., Medar, N. and Mitrović, L. (2014). Increasing launch capability of a UAV Bungee Catapult. *Scientific Technical Review*, 64(4), pp. 17–26.

Novaković, Z., Vasić, Z., Ilić, I., Medar, N. and Stevanović, D. (2016). Integration of tactical—medium range UAV and catapult launch system. *Scientific Technical Review*, 66(4), pp. 22–28.

Patterson, R. and Watts, K. (1986). The otter board as a low-aspect-ratio wing at high angles of attack; an experimental study. *Fisheries Research*, 4(2), pp. 111–130.

Schimel, D., Asner, G. and Moorcroft, P. (2013). Observing changing ecological diversity in the Anthropocene. *Frontiers in Ecology and the Environment*, 11(3), pp. 129–137.

Silvagni, M., Tonoli, A., Zenerino, E. and Chiaberge, M. (2016). Multipurpose UAV for search and rescue operations in mountain avalanche events. *Geomatics, Natural Hazards and Risk*, 8(1), pp. 18–33.

19 Epilogue

David R. Green
University of Aberdeen

Billy J. Gregory
DroneLite

Alex R. Karachok
UCEMM – University of Aberdeen

CONTENTS

THE DRONE EVOLUTION

In a very short space of time, Unmanned Aerial Vehicles (UAVs) or drones have progressed from small toys to the current generation of very sophisticated survey platforms. There are now a wide range of both small and large aerial platforms, small and large sensors, low-cost and high-cost solutions available, many of which you can buy online or even in your local supermarket (Figure 19.1).

As this technology has evolved, the trend in recent years has gradually been moving towards the growing recreational and commercial uses of the smaller off-the-shelf Ready-to-Fly (RTF) platforms and compact sensors that are now much more sophisticated, easier to fly, and can be utilised for high-resolution aerial image and video acquisition.

However, it was only a few years ago that we had the DJI S900 and S1000, the DJI Inspire 1 and 2, and the ever-popular Phantom series as some of the most popular drones available that led the way. Whilst some of the larger drones are still available e.g. the DJI M100 and M200 series, we now also have the DJI Spark, DJI Mini, DJI Air/Air 2, and the very successful DJI Mavic/Mavic 2 Pro platform carrying either a 20-megapixel miniaturised or zoom camera the size of a small box easily providing more practicality and capability than the larger drones of the recent past; compare the newer generation of small cameras on many small platforms e.g. the Blackmagic Micro Cinema camera (https://www.blackmagicdesign.com/uk/products/blackmagicmicrocinemacamera) to a digital single-lens reflex (DSLR) camera of only a few years ago!

The miniaturisation and increasing quality of the hardware has very quickly revolutionised small platform aerial photography in a very short period of time. Even now the technology is still evolving very quickly, and no sooner has a new drone or sensor come to market then it is out of date

FIGURE 19.1   A supermarket display of affordable drones (Photograph: David R. Green).

within a very short period of time, so rapidly is this technology changing. Alongside the hardware developments have been number of other closely associated or enabling technologies including LiPo batteries permitting longer flight times of both the multi-rotor and fixed-wing aircraft. These Smart batteries now serve to provide a lot of very useful in-flight information to the controller and drone operator, something that the earlier batteries could not do. They are also lighter and include improved connections and charging, as well as battery warmers for operation in cold conditions; improved Global Positioning System (GPS) technology, including both real-time kinematic (RTK) and post-processed kinematic (PPK) GPS solutions, is also gradually becoming available to assist in providing more accurate spatial locations of the aerial imagery; a wider-range of first-person view (FPV) options including mobile phones, separate screens, built-in Smart controllers, goggles, and glasses have all become available helping the drone pilot to see exactly what the camera/sensor is seeing, to assist in manual navigation, and to facilitate single operator flying without the need for an accompanying spotter; mobile phone and tablet apps (for both Apple and Android operating systems) provide comprehensive screen-based control of the platforms and sensors and to facilitate the setup of autonomous flight (either in the office on a PC or on the fly in the field). These apps include options to acquire imagery at selected waypoints, stereo-imagery over defined areas, and imagery from a specific point of view or flight path e.g. Follow Me mode; perhaps more importantly though, most of the newer drones are now available in an RTF out-of-the-box condition with quicker and easier ways to undertake the necessary calibrations needed (e.g. the compass), built-in flight information and operator tests, a full display, and commentary on the functioning of the UAV; other developments waterproof motors and all-weather flight capability for surveying; object avoidance and vision sensors that allow for operational flight in complicated environments, and to enable more accurate positioning and landing sequences. These are just a few developments amongst many other innovations that have become available in the last five years, all of which have made UAVs safer and easier to operate safely.

NEW UAV APPLICATIONS

As can be seen from the selected chapters in this volume, there are many different applications of this technology, most of them using relatively standard approaches, apps, and the software often associated with the SfM (SFM) (Structure from Motion) type applications. To some extent, this is where the current popularity of many applications lies. Indeed the most common uses of drones often cited in the literature (e.g. journals and popular magazines) largely involve the capture of aerial imagery and video footage, platform, and building inspections, and more recently a growing range of applications involving mapping and surveying; such examples include inspections of infrastructure and the surveying of construction sites. Another big growth area recently has been that of Precision Agriculture (PA) and Precision Viticulture (PV) for the monitoring of agricultural and horticultural crops to locate and identify crop disease.

But the range of UAV applications is now also changing rapidly as new technology, improved drone control and operation, and updated legislation are gradually being put in place to allow us to realise the recognised potential for a much wider range of applications that do not necessarily focus just on image capture, interpretation, analysis, and visualisation.

Some recently announced developments include the use of drone platforms for collecting and delivering a range of different payloads. This includes examples that use a drone platform to pick up and deliver parcels or even to deliver cups of coffee! Other slightly more sophisticated examples already being trialled include the use of drones to deliver medical supplies to remote areas where there is lack of road and medical infrastructure and the dropping of survival equipment in mountain rescue and marine incidents. Recently, one researcher has even identified the use of drones to aid in the collection of fruit from the field, not only to collect the picked fruit but also to aid in the picking of fruit! Unfortunately, there are also similar but less legal examples being reported of drones being used to deliver drugs to prisons! Although these are just a few examples, they are nevertheless very illustrative of the vast potential that such platforms have and the current sophistication of the technology allowing us to achieve this. The potential of drones has recently taken a new direction with the COVID-19 pandemic. Many unique applications have already been proposed within a short period of time, some building on existing demonstrations and ideas, to those exploring new possibilities. Two streams of application have been considered: (1) delivering essential goods and services, and (2) battling the spread of coronavirus. Some examples already under development and testing include: the delivery of parcels, medical supplies, and information e.g. broadcasts; personal, health and environmental monitoring, enforcing social distancing, mapping, and spraying to disinfect.

AI AND INFORMATION EXTRACTION

Another rapidly developing area of technology with relevance to drone-related applications is artificial intelligence (AI). According to one recent article, AI and drones are a 'match made in heaven', which perhaps goes some way towards explaining why more and more novel innovations are now emerging. Smart drones are already being used in numerous applications including defence, agriculture, natural disaster relief, security, construction, and firefighting, amongst many others. Already there are Smart batteries and Smart controllers to enable safer operation and flying.

One of the key areas of development in the future will also be to develop our ability to extract information from the UAV imagery acquired. Already we are able to capture imagery from RGB, multispectral, hyperspectral, and thermal cameras, and beyond our ability to visually analyse and interpret the imagery, we can now process the imagery using digital image processing software into information such as colour composites and NDVI (Normalised Difference Vegetation Index) images. With the aid of autonomous flight apps, we are now able to fly stereo-photographic coverage, and can mosaic the images, and with softcopy photogrammetry software can generate orthophotos, Digital Terrain Models (DTMs), and Digital Surface Models (DSMs). In addition, imagery of objects and buildings can also be turned into 3D models.

More recent developments include the growing use of AI to extract information from the aerial photography for input to decision-making software. Already a number of apps have been developed to facilitate this. One example is PA applications. A perfect example of this sort of development is an app called Skippy Scout from DroneAg Limited in the UK. This application uses low-altitude aerial imagery from RTF off-the-shelf drones to generate information on crop growth stage, green area index, weeds, and disease enabling the farmer to fly, capture, process, and utilise crop information for decision-making. Other examples, such as those provided by Parrot, make use of similar technology for their fixed-wing and multi-rotor platforms and multispectral sensors to provide information on agricultural and horticultural crops processed in the cloud.

In the near future, drone data and imagery will be just one of many sources of spatial data available, and we will see growing integration of this data into multiple applications which are far removed for the current uses of such data.

SWARMING

For the most part, drones – as we know them now – have been flown as single data acquisition platforms. Most collect and store their data on-board or on a mobile device as an alternative. This data is then stored, uploaded to a desktop for viewing and sharing, or placed in the cloud for processing; particularly when a large amount of computer power is required to process large numbers of images.

However, the operation of multiple drones called swarms is also rapidly becoming possible. Together the multiple drones can operate as an individual unit communicating between each other and also the cloud.

This potential can only be realised with a combination of technologies such as AI, the Internet of Things (IoT), fog computing, and blockchain all fast becoming a reality. Fog computing e.g. allows drones to be flown autonomously, transmitting only relevant data for analysis in real time. IoT-enabled sensors on-board a drone feed data to the fog node on-board for processing and analysis with only exceptions or alerts being transmitted to the cloud. Thus, drones can be now used in applications where time is of the essence, and real-time insights are critical. For example, in the event of a flood, first responders could send a drone into a flood zone to look for stranded survivors. Each drone would be pre-programmed to fly over their designated patch of the flooded area, reporting back sightings of stranded people or animals in real time. The data can then be 'stitched' together and analysed at a central location so emergency response teams can create the optimal evacuation plan. Another example is in the event of a wildfire where firefighting agencies could send a swarm of drones into the area. Communicating with each other, the drones can share notes as they survey the area and make intelligent decisions on where to drop fire retardant for the greatest impact, based on a multitude of conditions and datasets captured by their IoT-enabled sensors. These conditions could include wind direction and speed, temperature, or the percentage of fire contained, providing a three-dimensional understanding of the topography and where firefighters are currently located. Sharing all of this information across 50 or 100 drones, the swarm can examine the problem from a big picture perspective, making more intelligent and comprehensive decisions in real time (https://www.information-age.com/ai-powered-drone-swarms-123479202/).

DRONE OPERATION SUPPORT

Aside from standard off-the-shelf solutions available, the emergence of many new sophisticated applications for UAVs has led to the creation of companies that now provide expertise and support in the form of specialist products for these developing applications. One such example is the provision of the SkyHopper system, that includes 'communication data links (SkyHopper PRO, SkyHopper One), video processing and analytics (SkyHopper PRO V), content consumption systems (SkyHopper VU), and controlling systems (SkyHopper ControlAir)' (https://www.robotics-businessreview.com/unmanned/commercial-drone-operations-rise/). This sort of expertise offers commercial businesses and operators with customised solutions where the potential role of UAVs extends beyond the simple solutions such as those required in mining, security, infrastructure inspection e.g. wind turbines, and delivery, and there are challenges such as radio communication planning perspectives; payload, control and auto pilot connectivity with the ground control units; as well as implementing more advanced features https://www.roboticsbusinessreview.com/unmanned/commercial-drone-operations-rise/).

OVERCOMING CURRENT CONSTRAINTS

As exciting as this technology is, there are also still a number of issues constraining the realisation of the full potential. Unfortunately, whilst the technology has developed very quickly, there have been a number of problems that have arisen that have put a bit of a dent in the enthusiasm that was initially associated with drones. These are not insurmountable, but as the technology continues to evolve very quickly, there is an increasing need to consider solutions and implementations to continue to enable this potential to be realised.

PUBLIC CONCERN

Public concern about drones has grown in recent years in light of a number of incidents that have taken place globally, largely precipitated by people flying too close to an object of interest, proximity to airports, and some reported UAV fly-aways. Whilst these have been few and far between, and nothing too serious, the media coverage on TV, in magazine, and newspapers, has drawn unwanted attention to drones, mostly negative. On a more positive note, it has also resulted in a general tightening up of the rules and regulations in pretty well every country around the globe (https://www.youtube.com/watch?v=Zh9XzFhGI8g).

As more drones or UAVs take to the skies – of all different makes, types, and sizes – with different sensors and controllers – growing awareness about these aerial platforms for remote sensing and many other applications is rapidly signalling the need for more education, built-in controls, new legislation, and ultimately UAV traffic management (UTM) systems for some applications to continue allow the potential of UAVs to be realised. Already many such developments are underway and are gradually being put into place and finding their way into the average drone platform.

Unfortunately, there have been numerous accounts in the media about various incidents and near misses between drones and civilian/military aircraft, examples of UAVs flying in places that they are not supposed to be flown in and plenty of YouTube footage – flights and tutorials - where photos and imagery reveal a rather casual and often risky approach to the flying of drones – both commercially and for recreational purposes. Witness the recent Gatwick and Heathrow incidents in the UK widely reported in the media and the subsequent news that has continued to circulate for some time, even accompanied by TV programmes. Whether these latter incidents were real or not, what they have served to highlight is a growing concern by the public at large that such technology could lead to a serious accident and the need to put in place mechanisms that ensure that drone operation is safe, legal, and responsible.

Notwithstanding such news items – perhaps quite typical in the early stages of the emergence of any new and unfamiliar technology that has developed very rapidly with little realisation that it would become so popular – there is clearly a growing need now to ensure that as the potential of this technology continues to evolve, some of the potential that could be realised really needs to include giving some more thought to the developments taking place: for example, the fact that more people are able to afford and fly the technology (RTF drones); the limited control over who is able to get hold of and fly drones; the need to test operators; the need for insurance; a lack of infrastructure; and the limited means to be able to enforce the use of the technology (policing).

However, already there have been marked changes to the technology available with the growing provision of e.g. online tutorials; inbuilt controller tutorials and tests, training, and development of a UTM system; better drones with more functionality; and more safety built-in such as object avoidance. All of these have helped to counter criticisms, by putting in place the necessary framework – self-learning and awareness approaches to educating drone flyers – both novice and experienced – with reminders and information designed to cover the ground needed to refresh your memory. Coupled with the updated regulations and other source of information now out there, such approaches provide the means to gently ease potential flyers into the *hot seat* more easily but all set within a safe flying environment.

There is the problem of course of how to deal with the two camps of drone flyers – the recreational and the commercial. Various different ways of doing this have so far been to try and educate people buying drones to behave responsibly and to draw their attention to some of the information e.g. documents, websites, training courses available, and by simply using signage on displays next to the shelves in a shop to raise awareness and educate potential drone operators. This is one approach that is not too invasive and one that can in practice be quite effective. It is also a positive way to engage with people without curtailing the pleasure of experiencing the technology as a recreational pilot.

Other ways that are proving useful make use of apps provided with the drones that e.g. test the drone pilot, and draw attention to regulations and tutorials. These too are potentially quite effective as they are always *in your face* when setting out to fly the drone, providing a continual reminder and a resource. Having software controls to e.g. prevent you from taking off in areas that are risky are also useful developments. The registration of drones – (e.g. by the Federal Aviation Administration (FAA) in the USA) has also been implemented – in part to focus people on the safe and responsible flying of drones. This of course involves a small cost and may be something that not everybody wants to or is willing to subscribe to even if it is beneficial in the longer term. So, despite the frequently and seemingly continuing negative perception of drones in the public domain, measures are already being put in place to help minimise the risks.

Of course, all of these developments, passing a flight test, and having all the safety equipment on-board and in place does not prevent people from flying badly nor does it prevent accidents from happening, fly-aways, flying in places you should not be in, and so on. But it does help to reduce the likelihood of many of the things mentioned above from happening and helps to reduce the chance of an incident from occurring, damage being done, or injuries being inflicted.

Acceptance by the public will continue to rely on raising awareness, educating people – as and when appropriate – to spell out what this drone technology is all about, characteristics, the value to society, appropriate contacts, regulations etc. and how to fly the equipment safely, legally, and responsibly. This is something that should increasingly also be embedded in our education system and will be even more important now as the technology continues to evolve very quickly and more applications continue to appear.

UAVs clearly have a lot of potential and lots to offer through the many different applications. Gradually more people are becoming aware of what UAVs can do and the risks they pose if not used safely, legally, and responsibly. What we still need to do, however, is to ensure that this fantastic potential is not lost as a result of misinformation and ignorance. By putting in place as many safety measures, precautions, and ensuring the public knows what they can be used for, then there is a good chance that we will not lose sight of this incredibly exciting technology and a growing an important industry.

AIRSPACE FOR DRONES

For drones to be utilised in some of the ways suggested earlier in this chapter, there will need to be some considerable thought given to how these small airborne platforms will – mostly operating in autonomous mode – be able to fly safely in built-up environments. Steps towards this inevitable development – if the potential for drone use is to be realised – have already been initiated in a number of ways – that vary from one country to another – including recreational operators operating within certain guidelines, requiring commercial operators to take and pass a test, offering a certain amount of flexibility by offering waivers to allow drone pilots to fly in special circumstances, such as flying at night, directly over people, flying multiple aircraft, flying above 400 feet, and beyond visual line of sight (BVLOS), or flying near airports or controlled airspace.

Given the intensity of use of the airspace already, the growing use of drones suggests a need for an extra layer of airspace. Many countries are now already looking into the creation of a new regulatory framework for the integration of drones into the national airspace. In the USA e.g. the FAA has recently named ten companies and programs to participate in the UAS Integration Pilot Program. To facilitate operations such as delivery, and indeed large area coverage for monitoring and surveying, the current restrictions placed on distance of operation will necessarily have to be relaxed if BVLOS operation is to be realised.

Telecommunications

Other problems facing the drone industry include the limitations imposed by having to fly within the visual line of sight (LOS); the relatively limited bandwidth currently available for image/video transmission; inaccurate tracking because the existing positioning is based on global navigation satellite system (GNSS) that may not be reliable due to potential spoofing and jamming; and limited operation time due to battery constraints. Drone operation needs wireless connectivity for communication between drones and ground control systems, among drones, and between drones and air traffic management systems.

The Internet of Drones is an emerging and underexplored field. Research by Yang et al. (n.d.) and Lin et al. (n.d.) reveals that future mobile networks could be used to extend future drone applications. 4G and 5G mobile networks are already helping to overcome some of the current obstacles through remote and real-time control, HD image/video transmission, efficient drone identification and regulation, and high-precision positioning. 5G networks gradually rolling out have the potential to provide efficient and effective mobile connectivity for larger-scale drone deployments than at present. Some of the new technologies that 5G networks will introduce include 3D coverage enhancement, high data rate, customised end-to-end QoS guarantee, and efficient identification and monitoring based on big data analysis. The new mobile networks will help to extend the communication range for real-time transmission of remote drone control information. High-speed, low-latency communications provided by mobile networks will also enable HD image/video transmission. As the number of drones in use and the possibility of fleets of drones being operatred mobile networks will be able to help to identify, track, and control them These new and advanced mobile networks will help to power drones beyond ground devices, and will provide wide-area, quality,and secure connectivity for drone operations. They will also be able to assist with drone identification, authorisation, and geo-fencing. As concluded by Yang et al. (n.d.) mobile technologies are continually evolving and will facilitate the interoperability required by the so-called drone ecosystem (a community of drones in conjunction with all the components of their environment, interacting as a system.) (see also: https://enterprise.dji.com/news/detail/dji-enterprise-empowers-commercial-drone-industry). Already 4G and 5G mobile networks can help to overcome some of these obstacles by facilitating capabilities such as remote and real-time control, HD image/video transmission, efficient drone identification and regulation, and high-precision positioning. The improved capabilities of 5G networks have the potential to provide efficient and effective mobile connectivity for large-scale drone deployments with more diverse uses. In particular, 5G networks will introduce new technologies to provide 3D coverage enhancement, high data rate, customised end-to-end QoS guarantee, and efficient identification and monitoring based on big data analysis. These mobile networks will be able to extend the communication range for the real-time transmission of remote drone control information compared to using proprietary communication technologies that are not deployed nationwide. In addition, high-speed, low-latency communications provided by mobile networks enable HD image/video transmission and thus lead to better user experience. Mobile networks are well positioned to identify, track, and control the growing fleet of drones. The drone ecosystem should take advantage of the mobile technologies for drone operations. Mobile networks stand ready nationwide to power flying drones beyond ground devices; provide wide area, quality, and secure connectivity for drone operations; can assist with drone identification, authorisation, and geo-fencing; mobile technologies are based on standards and are evolving, facilitating interoperability and vibrant global growth for the evolution of drone ecosystem.

THE FUTURE

A few years ago, when this exciting drone technology first began to appear, very few of us would ever have thought that it would have developed to the stage it has today.

The current form that UAVs have assumed is in essence the integration of a number of technologies. But what are the future developments? Where will this technology go now? Based on what we have witnessed in the past, it is really very hard to predict the future!

In part, the evolution of all this technology has been a direct result of the developments that have allowed miniaturisation to provide small platforms and sensors. The spread of autonomous control for these small platforms has also come alongside the development of similar technology for self-driving road vehicles and radar-based sensors that can detect objects – both moving and stationary – helping us to drive more safely – and reducing the risk of e.g. accidents. Not only have these technologies found their way into the air but also into the water (both fresh and marine applications). Developed a little later than their airborne equivalents, we are now seeing a number of small surface and underwater vessels, also benefiting from miniaturisation – not only of cameras but also multibeam and sonar sensors for bathymetric and hydrographic survey. Already small affordable low-cost technology is offering underwater photography and video. Slightly more expensive and larger platforms are providing hydrographic survey opportunities that minimise the risk associated with placing people and larger manned vessels in shallow water environments. Control of these vessels is also undertaken remotely.

Whilst most of the applications covered in this book have focused on the use of airborne and waterborne platforms and sensors to capture photography, video, and other image-based imagery, there are now many other uses for this technology that do not involve remote sensing. Indeed, as recently as May 2019, the FAA has announced the licensing of the first UAV Airline.

It is widely considered that the usability, safety, security, and ease of integration will ultimately define the success of commercial drone space, and the regulatory environment must clearly change to support these innovations. At present, the drone regulations have not kept pace with the available technology. We are now seeing the rise of the Aerial Enterprise. It is not just about buying and using a drone but how you connect drones and the data they can collect to other existing systems and business applications. Drones are changing the way we do things.

As an indication of the future promise of the drone, the German research firm Drone Industry Insights describes how the drone industry has already developed and will continue developing by sector. The commercial drone application space, which generated €16 billion (about $18 billion) in 2018, is projected to grow to nearly €38 billion ($42.5 billion) by 2024. But as of writing it is the COVID-19 pandemic of 2020 that has - rather unexpectedly - opened up a whole new range of drone applications, that are not only very promising but also seem set finally to be the making of this new technology making it far more widely accepted by the public with the potential they offer where remote operation really is the key and vital to their role.

REFERENCES

Xingqin Lin, Henrik Rydén, Sakib Bin Redhwan. (n.d.) *A Telecom Perspective on the Internet of Drones: From LTE-Advanced to 5G*. Ericsson. Cornell Univesrity.

Guang Yang, Yan Li, Hang Cui, Min Xu, Dan Wu. (n.d.) *A Telecom Perspective on the Internet of Drones: From LTE-Advanced to 5G*. China Mobile Research Institute. Cornell Univesrity.

Index

A

Absorption, 26, 49, 199
Accident, 16, 19, 59, 89, 198, 233, 321–322, 324
Accretion, 98, 101, 115, 124, 130–132
Accuracy, 7, 10–11, 28, 38, 41–43, 48, 50, 62–64, 66–74,
    78, 84, 88, 92–101, 103, 105–109, 112–113, 116,
    123, 126–127, 132, 142–144, 146, 152, 154–155,
    171, 175, 177, 179, 183–193, 196, 204, 207,
    219, 222, 224–226, 228, 236, 241, 244, 246,
    248, 251–254, 259, 267–268, 282–286, 288,
    300–302, 307
Acoustic, 40, 73, 198
Acquisition, 1–2, 4, 6–9, 11, 14, 18, 20, 22, 32, 66, 73,
    78, 80, 82, 88–89, 93, 104, 107, 111–113, 115,
    117, 119–121, 123, 125, 127, 132, 137, 143–148,
    150–151, 154–156, 161–162, 174–177, 179, 184,
    193, 199–200, 207, 213, 217, 219, 222, 230–233,
    237, 243, 254, 257, 278, 281–283, 286, 288,
    317, 320
Adaptable, 11, 206, 269, 304, 314
Adaptation, 8, 91, 131, 134, 145, 158, 202, 205, 254
Aerobatic, 15
Aerodynamics, 24, 201
Aeronautics, 33, 209, 257
Aerospace, 26, 33, 214, 233
Affordable, 3, 9–11, 144–147, 150, 153, 155, 169–171, 233,
    254, 262, 272, 278, 318, 324
Agricultural, 3, 16, 33, 55, 74, 91–92, 112, 170, 191, 193,
    229, 246, 284, 315, 319–320
Agriculture, 8, 99, 144, 157, 159, 161, 171, 193, 233, 243,
    304, 319
AI, 4, 317, 319–320
Ailerons, 18–19
Airbag, 205
Airborne, 3–4, 6–7, 9–10, 13–15, 17, 19–22, 24, 27,
    30–33, 49–51, 70–71, 75, 78, 83, 88, 91, 97,
    105, 107, 112, 128, 130, 136, 145, 157–158,
    161–163, 174, 192, 194–196, 198–199, 201,
    207, 214–215, 220, 235–236, 247, 254,
    271–272, 274–275, 278, 309–310, 312,
    314–315, 322, 324
Aircraft, 1–4, 6, 10–25, 27, 29–34, 37, 51, 53, 56, 73–74,
    79–80, 86–87, 89–90, 106, 108–109, 128,
    151, 162, 186–187, 195, 198, 200–202, 205,
    207–209, 214–215, 219, 223, 228–229, 233, 258,
    262–263, 267–269, 283, 303–310, 312, 314–315,
    318, 321–322
Airfoil, 17, 312–313, 315
Airframe, 20, 98, 201, 307–309, 312–313, 315
Airlift, 11, 304, 308, 312, 314–315
Airplane, 90, 92, 103
Airships, 90
Airspace, 37, 63, 107, 205, 208, 211, 213, 241, 254, 317,
    322–323
Airspeed, 306, 310, 312, 314
Algal, 3, 52, 78, 103–104
Algebraic, 25

Algorithm, 42, 49, 65, 67–68, 70–71, 88, 92–95, 100–101,
    108, 113, 117–118, 138, 142–143, 148, 154, 169,
    174, 179, 183, 187–188, 190–191, 211, 244–245,
    248, 283, 314
Alignment, 49, 62–64, 93, 175
Alluvial, 57
Alpine, 230
Altimeter, 20, 206
Altitude, 11, 15, 20–22, 25, 28, 30, 32–33, 37–39, 60–61,
    79, 90, 93–95, 98–99, 107, 112, 121, 131, 148,
    151, 154, 164, 174–175, 183, 186, 193, 200–201,
    206, 208–211, 231–232, 234, 237, 244, 246,
    248–249, 259, 263, 268, 301, 304, 306, 310,
    314–315, 319
Aluminium, 30, 213, 277
Analogue, 44
Analyse, 11, 101, 118, 170, 183, 240, 247, 253, 288, 319
Analysed, 4, 14, 28, 79, 85, 103, 138, 162, 187, 249, 254,
    288, 320
Analyses, 66, 79, 98, 100, 105, 116, 119, 179, 184, 243, 263
Analysis, 4, 6, 8, 10, 14, 32–33, 37, 41–42, 51–53, 59, 61,
    64–68, 70, 74, 82–86, 88, 91–93, 96–99, 103,
    108–109, 112–114, 120–122, 125, 127, 130, 132,
    138–139, 154, 175, 179, 183–185, 193–195, 199,
    209, 213, 229–230, 232–233, 236–237, 240,
    243, 246, 248–249, 255, 257, 259, 268–269,
    278, 281, 284, 288, 290, 293, 302, 316,
    319–320, 323
Analytical, 227–228
Analytics, 320
Anchored, 313
Android, 114, 318
Angle, 38, 42, 47, 52, 61, 63, 95–96, 125, 143, 145, 229,
    231, 247, 252, 283, 311, 313, 316
Animal, 16, 104, 155, 176, 262, 267, 320
Anodised, 277
Anomalies, 10, 240–241, 244, 247, 252
Anomaly, 247
Anoxic, 43
Antenna, 27, 221, 272
Anthropocene, 304, 316
Anthropogenic, 5, 55, 96, 103
Aperture, 27, 38–39
App, 8, 30, 145–147, 150, 152, 155–156, 161, 170, 207, 249,
    275–276, 318–319, 322
Apparatus, 119
Aquaculture, 79
Aquatic, 1, 4–5, 35–53, 56, 73, 75, 88, 270
Arable, 230
Archaeological, 9, 33, 217–237, 246, 258
Archaeology, 1, 9, 32–33, 217–223, 225–227, 229–231,
    233–237
Architect, 10
Architectural, 226, 229, 259
Architecture, 112, 143, 158, 235
Archives, 33, 50, 86, 98, 107–108, 196, 214, 235–236,
    259–260, 302, 315
Areal, 24, 81